Fresh Water Pollution Dynamics and Remediation

Humaira Qadri • Rouf Ahmad Bhat
Mohammad Aneesul Mehmood
Gowhar Hamid Dar
Editors

Fresh Water Pollution Dynamics and Remediation

Springer

Editors
Humaira Qadri
Department of Environmental Sciences,
School of Sciences
Sri Pratap College Campus, Cluster
University of Srinagar
Srinagar, Jammu and Kashmir, India

Rouf Ahmad Bhat
Department of Environmental Sciences,
School of Sciences
Sri Pratap College Campus, Cluster
University of Srinagar
Srinagar, Jammu and Kashmir, India

Mohammad Aneesul Mehmood
Department of Environmental Sciences,
School of Sciences
Sri Pratap College Campus, Cluster
University of Srinagar
Srinagar, Jammu and Kashmir, India

Gowhar Hamid Dar
Department of Environmental Sciences,
School of Sciences
Sri Pratap College Campus, Cluster
University of Srinagar
Srinagar, Jammu and Kashmir, India

ISBN 978-981-13-8276-5 ISBN 978-981-13-8277-2 (eBook)
https://doi.org/10.1007/978-981-13-8277-2

This Springer imprint is published by the registered company Springer Nature Singapore Pte Ltd.
The registered company address is: 152 Beach Road, #21-01/04 Gateway East, Singapore 189721, Singapore

To our parents

Preface

Water is the essence of life on earth and totally dominates the chemical composition of all organisms. The ubiquity of water in biota, as the fulcrum of biochemical metabolism, rests on its unique physical and chemical properties. Water ecosystems, specifically freshwater ecosystems, are some of the most important resources that nature has bestowed the planet with. Freshwater ecosystems such as lakes, ponds, rivers, streams, and wetlands are estimated to cover 15% of the world's continental surface area. These inland bodies of water are bastions of biodiversity, hosting about 10% of the world's animals and one-third of all vertebrate species. Freshwater ecosystems are highly valued for their recreational, aesthetic, and scenic qualities, and the water they contain is one of the most treasured of our natural resources. Because of proximity of most of the lacustrine habitats to human settlements, these are prone to anthropogenic pressures which lead to significant decline in their aesthetic, recreational, and aquatic ecosystem functions. Since freshwater is a finite resource, it is easily impacted by complex land and water relationships and human inputs of nutrients, particularly nitrogen and phosphorus, often leading to cultural eutrophication. Preserving the quality and availability of freshwater resources is becoming one of the most pressing environmental challenges on the international horizon. To ensure the preservation of these freshwater ecosystems, there is a need to understand the ecology of the system, pollution problems, their impacts, restoration techniques, and the conservation measures. In this backdrop, the present book, *Freshwater Pollution Dynamics and Remediation*, is in print.

The introductory chapters of the book focus on the present state of the art of the freshwater ecosystems, the pollution status, and the problems associated therewith followed by a thorough discussion about the specific issues pertaining to pesticide pollution, municipal solid waste problems, and climate change impacts.

The book provides an understanding of the changes in the physicochemical characteristics of the water and sediments along with a detailed discussion on the shift in the biological communities including plants and microbes due to pollution. The impact of deteriorating quality of the freshwater ecosystem on the animal and human health is also discussed in detail.

With the increase in the understanding regarding the ecologically unsound techniques that were previously employed for restoration and management of freshwaters, more attention is being paid to ecologically sound and economically viable restoration techniques, prominent among which is bioremediation. This book provides a comprehensive account of the techniques based on updated research on bioremediation, phyto-remediation, and nano-bioremediation along with the role of biomarkers as a remediation tool.

This book can be used as a reference by researchers, scientists, and educators who are involved in the freshwater pollution, remediation, and management studies for gaining an in-depth knowledge regarding the understanding of the freshwater ecosystems, the pollution sources, their impacts, and the ecologically sound economical techniques for remediation and restoration of the system in light of detailed case studies.

The book editors with an expertise in diverse research fields in freshwater ecosystems have congregated the most inclusive research accounts on the freshwater pollution and remediation and thus developed a repository of diverse knowledge on the subject.

Suggestions on the subject are always welcome.

Srinagar, India Humaira Qadri
Srinagar, India Rouf Ahmad Bhat
Srinagar, India Mohammad Aneesul Mehmood
Srinagar, India Gowhar Hamid Dar

Contents

Editors and Contributors

About the Editors

Humaira Qadri is heading the Department of Environment and Water Management and is teaching in the Postgraduate Department of Environmental Science in Sri Pratap College Campus, Cluster University Srinagar, J&K, India. She has her specialisation in Limnology, Nutrient Dynamics and Phytoremediation. She has published scores of papers in international journals and has more than ten books with national and international publishers. She is also the reviewer of various international journals and is the principal investigator of some major and minor research projects on phytoremediation. She is guiding a number of research students for PhD programme and has supervised more than 50 master's dissertations. A gold medallist at her master's level, she has bagged a number of awards and certificates of merit. With a number of national scientific events to her credit, she also has been on the scientific board of various international conferences. She is an active participant in national and international scientific events and has organised a number of national conferences on science. Moreover, she holds life memberships of various international organisations.

Rouf Ahmad Bhat has his specialisation in Limnology, Toxicology, Phytochemistry and Phytoremediation. He has been teaching graduate and postgraduate students for the past 2 years. He is an author of more than 50 scientific articles and 15 book chapters and has published 9 books with international publishers. He has presented and participated in numerous state, national and international conferences, seminars, workshops and symposium and has worked as an associate environmental expert in the World Bank-funded flood recovery project and also environmental support staff in Asian Development Bank-funded development projects. He has received many awards, appreciations and recognitions for his services to the science of water testing and air and noise analysis. He has served as editorial board member and reviewer of reputed international journals. Currently, he is working as assistant professor and is still writing and experimenting with diverse capacities of plants for use in aquatic pollution.

Mohammad Aneesul Mehmood has his specialisation in Limnology and Environmental Toxicology. He completed his doctorate from the Division of Environmental Science, SKUAST- K, with a meritorious certificate from the university. He received Dr. Mumtaz Ahmad Khan Gold Medal and Shri Bhushan Memorial Gold Medal for being a topper during his master's programme. He has qualified various national competitive tests in the discipline of Environmental Science. He was also awarded with INSPIRE Merit Fellowship (JRF & SRF) by the Department of Science and Technology, GoI, during his doctoral programme. He has been teaching graduate and postgraduate students for the past 2 years in the Department of Environmental Science, School of Sciences, Sri Pratap College Campus, Cluster University Srinagar, J&K, India. He has been supervising many students for their MSc projects. Furthermore, he has a number of publications in international journals of repute and a number of books with international publishers.

Gowhar Hamid Dar is currently working as an assistant professor in Environmental Science, Sri Pratap College, Cluster University Srinagar, Department of Higher Education, J&K, India. He has a PhD in Environmental Science with specialisation in Environmental Microbiology (Fish Microbiology, Fish Pathology, Industrial Microbiology, Taxonomy and Limnology). He has been teaching graduate and postgraduate students for the past many years. He has more than 40 papers in international journals of repute and a number of books with international publishers. He is guiding a number of students for master's thesis. He has been working on the Isolation, Identification and Characterization of microbes, their pathogenic behaviour and impact of pollution on development of diseases in fish fauna for the past several years. He has received many awards and appreciations for his services towards the science and development. Besides, he is also the acting member of various research and academic committees.

Contributors

Md. Niamat Ali Department of Environmental Science, University of Kashmir, Srinagar, Jammu and Kashmir, India

Cytogenetics and Molecular Biology Research Laboratory, Centre of Research for Development, University of Kashmir, Srinagar, Jammu and Kashmir, India

Mohd. Muzamil Bhat Faculty- Environmental Science, School of Sciences, Islamic University of Science and Technology, Pulwama, Jammu and Kashmir, India

Rouf Ahmad Bhat Department of Environmental Sciences, School of Sciences, Sri Pratap College Campus, Cluster University of Srinagar, Srinagar, Jammu and Kashmir, India

Division of Environmental Sciences, Sher-e-Kashmir University of Agricultural Sciences and Technology of Kashmir, Srinagar, India

Sartaj Ahmad Bhat River Basin Research Center, Gifu University, Gifu, Japan

Umar Nazir Bhat Department of Ecology and Environmental Sciences, Pondicherry Central University, Puducherry, India

Mohammad Z. Chishti School of Life Sciences, Central University of Kashmir, Srinagar, India

Guangyu Cui Graduate School of Engineering, Gifu University, Gifu, Japan

Ashaq Ahmad Dar Department of Ecology and Environmental Science, Pondicherry University, Puducherry, India

Gowhar Hamid Dar Department of Environmental Sciences, School of Sciences, Sri Pratap College Campus, Cluster University of Srinagar, Srinagar, Jammu and Kashmir, India

Rubiya Dar Centre of Research for Development, University of Kashmir, Srinagar, India

Shakeel Ahmad Dar Division of Environmental Sciences, Sher-e-Kashmir University of Agricultural Sciences and Technology of Kashmir, Srinagar, India

Moonisa Aslam Dervash Division of Environmental Science, Sher-e-Kashmir University of Agricultural Sciences and Technology, Shalimar, India

Muneeb A. Faiq Medical Biotechnology Laboratory, Dr. B. Ambedkar Center for Biomedical Research, University of Delhi, New Delhi, India

Laboratory for Molecular Reproduction and Genetics, Department of Anatomy, All India Institute of Medical Sciences, New Delhi, India

Neuroimaging and Visual Science Laboratory, Department of Ophthalmology, Langone Health Center, New York University School of Medicine, New York, NY, USA

Bashir Ahmad Ganai Department of Environmental Science/Centre of Research for Development, University of Kashmir, Srinagar, India

Jaskaran Kaur Gill Department of Botanical and Environmental Sciences, Guru Nanak Dev University, Amritsar, India

Mehvish Hameed College of Agricultural Engineering, Division of Soil and Water Engineering, Sher-e-Kashmir University of Agricultural Sciences and Technology, Shalimar, India

Omar bin Hameed College of Natural Sciences, Arba Minch University, Arba Minch, Ethiopia

Birjees Hassan Department of Environmental Science, University of Kashmir, Srinagar, Jammu and Kashmir, India

Maheen Javaid Department of Environmental Science, University of Kashmir, Srinagar, Jammu and Kashmir, India

Sabiha Javeed Department of Environmental Science, University of Kashmir, Srinagar, Jammu and Kashmir, India

Azra N. Kamili Centre of Research for Development, University of Kashmir, Srinagar, India

Jatinder Kaur Katnoria Department of Botanical and Environmental Sciences, Guru Nanak Dev University, Amritsar, India

Inderpreet Kaur Department of Chemistry, Centre for advance Studies, Guru Nanak Dev University, Amritsar, India

Jagdeep Kaur Department of Botanical and Environmental Sciences, Guru Nanak Dev University, Amritsar, India

Anisa Basheer Khan Department of Ecology and Environmental Sciences, Pondicherry Central University, Puducherry, India

Nissar Ahmad Khan Department of Botany, Punjabi University, Patiala, Punjab, India

T R Kumaraswamy Department of Environmental Science, Bangalore University, Bengaluru, Karnataka, India

Fusheng Li River Basin Research Center, Gifu University, Gifu, Japan

Rafiq Lone Department of Botany, SBBS University, Jalandhar, Punjab, India

Javid Manzoor Department of Environmental Science, Jiwaji University Gwalior, Gwalior, India

Mohammad Aneesul Mehmood Department of Environmental Sciences, School of Sciences, Sri Pratap College Campus, Cluster University of Srinagar, Srinagar, Jammu and Kashmir, India

Nighat Mushtaq Division of Vegetable Science, Sher-e-Kashmir University of Agricultural Sciences and Technology, Shalimar, India

Kumara Naika Department of Environmental Science, Bangalore University, Bengaluru, Karnataka, India

Humaira Qadri Department of Environmental Sciences, School of Sciences, Sri Pratap College Campus, Cluster University of Srinagar, Srinagar, Jammu and Kashmir, India

Rizwana Qadri Department of Laboratory Medicine, All India Institute of Medical Sciences, New Delhi, India

Genetics Unit, Department of Pediatrics, All India Institute of Medical Sciences, New Delhi, India

U. N. Rai Plant Ecology and Environmental Science Division, CSIR-NBRI, Lucknow, India

Nowsheeba Rashid Amity Institute of Food Technology, Amity University, Noida, Uttar Pradesh, India

Saraswati Saini Department of Botanical and Environmental Sciences, Guru Nanak Dev University, Amritsar, India

Hridoy Ranjan Saikia Department of Botanical and Environmental Sciences, Guru Nanak Dev University, Amritsar, India

Sadiqa Shafiq Institute of Home Science, University of Kashmir, Srinagar, Jammu and Kashmir, India

Abdul Barey Shah Department of Environmental Science, Babasaheb Bhimrao Ambedkar (Central) University, Lucknow, India

Razia Shuab Department of Botany, University of Kashmir, Srinagar, India

R. N. Shukla School of Environmental Biology, Awadesh Pratap Singh University, Rewa, Madhya Pradesh, India

Dig Vijay Singh School of Environmental Science, DES, Babasaheb Bhimrao Ambedkar University, Lucknow, India

Navdeep Singh Department of Botanical and Environmental Sciences, Guru Nanak Dev University, Amritsar, India

Rana Pratap Singh Department of Environmental Science, Babasaheb Bhimrao Ambedkar (Central) University, Lucknow, India

Bhat Mohd Skinder Department of Environmental Science/Centre of Research for Development, University of Kashmir, Srinagar, India

Irfan-ur-Rauf Tak Department of Zoology, SP College, Cluster University Srinagar, Srinagar, India

Baba Uqab Department of Environmental Science/Centre of Research for Development, University of Kashmir, Srinagar, India

Khursheed Ahmad Wani Department of Environmental Science, Govt Degree College, Bijbehara, Jammu and Kashmir, India

Ali Mohd Yatoo Department of Environmental Science, University of Kashmir, Srinagar, Jammu and Kashmir, India

M. Yunus Mohammad Ali Jauhar University, Rampur, Uttar Pradesh, India

The Concerns for Global Sustainability of Freshwater Ecosystems

Humaira Qadri and Rouf Ahmad Bhat

Abstract Water is a fundamental compound for survival of life on earth. Its unique physico-chemical properties are a hinge for biochemical metabolism in any form of biota. Environmental pollution and issues has degraded the quality of freshwater and depletes its resources from finite to limited quantity. Preservation of freshwater sources is nowadays the most pressing environmental concern due to ever increasing anthropogenic pressures. For the global sustainability of freshwater ecosystems, it becomes imperative to adopt ecologically sound restoration and management practices for the sustainability of life on earth.

Keywords Pollution · Freshwater biota · Restoration · Sustainability · Heavy metal · Pesticides · Wastewater

1 Introduction

Water is a fundamental compound for survival of life on earth. Its unique physico-chemical properties are a hinge for biochemical metabolism in any form of biota. The nature and properties of water have intrigue scientists as antiquity (Sharp 2001). This is because water is anomalous in many of its physical and chemical properties. Some of exceptional properties of water are literally crucial for biota, while others have insightful special effects on the features and characteristics of living organisms (Pandey et al. 2017). Due to its unique properties, water is called the foundation of life and undoubtedly the most important requirement for economic social development. Most of the earth's surface is covered with water (70–72%), but, the accessible freshwater resources for direct human requirement is <1%. This is the water

H. Qadri (✉) · R. A. Bhat
Department of Environmental Sciences, School of Sciences, Sri Pratap College Campus, Cluster University of Srinagar, Srinagar, Jammu and Kashmir, India

H. Qadri et al. (eds.), *Fresh Water Pollution Dynamics and Remediation*,
https://doi.org/10.1007/978-981-13-8277-2_1

Table 1 Earth's total water availability

Type of water	Quantity (%)
Saline	97.3
Fresh	2.7

Table 2 Freshwater resources supply

Freshwater resource	Quantity (%)
Polar ice-caps	81.1
Aquifers	15.3
Surface water	2.3
Air and soil	1.3

found in lakes, reservoirs, rivers and streams, and underground aquifers that are shallow enough to be tapped at an affordable cost (Mishra and Dubey 2015). The atmosphere potentially week doesn't hold a large percentage of Earth's freshwater at anytime, and large quantities of water are continually cycling through the atmospheric reservoir on very rapid timescales (Boberg 2005). Out of total water available on earth only 2.8% is available for human consumption. The other 97.2% is in the oceans and it is too salty to use it for most purposes (Table 1). Most of the Earth's fresh water is frozen in polar ice caps, icebergs and glaciers as is shown in Table 2 (Oana et al. 2010). The global freshwater distribution is elaborated in Fig. 1 (Shiklomanov and Rodda 2003; UNWWDP 2015).

2 Water Pollution

Freshwater resources on the earth diminish fast. The resources are concise but the degradation in the form of pollutants is continue non-stop process and the situation of dealings is alarming (Tripathi and Pandey 2009). Pollution is created in water by industrial as well as commercial waste, agricultural practices, anthropogenic activities and most notably, modes of transportation (Owa 2014). Some of the major factors which are responsible for causing water pollution are growing population, rapid industrialization, and urbanisation, use of science and technology and modern agriculture practices (Haseena et al. 2017). The variety and number of pollutants increases yearly as new compounds are synthesised (Abel 1996). Every day, millions of tons of sewage, industrial and agricultural waste are discharged into the world's water while the amount of wastewater produced yearly is estimated to be about 1.5×10^3 Km^3, 06 times more water than exists in all the rivers of the world (UN WWAP 2003). Pollutants such as herbicides, pesticides, fertilizers and hazardous chemicals from agriculture and industry are getting discharged into our freshwater resources (Oana et al. 2010; Smical et al. 2010a, b) leading to deleterious changes.

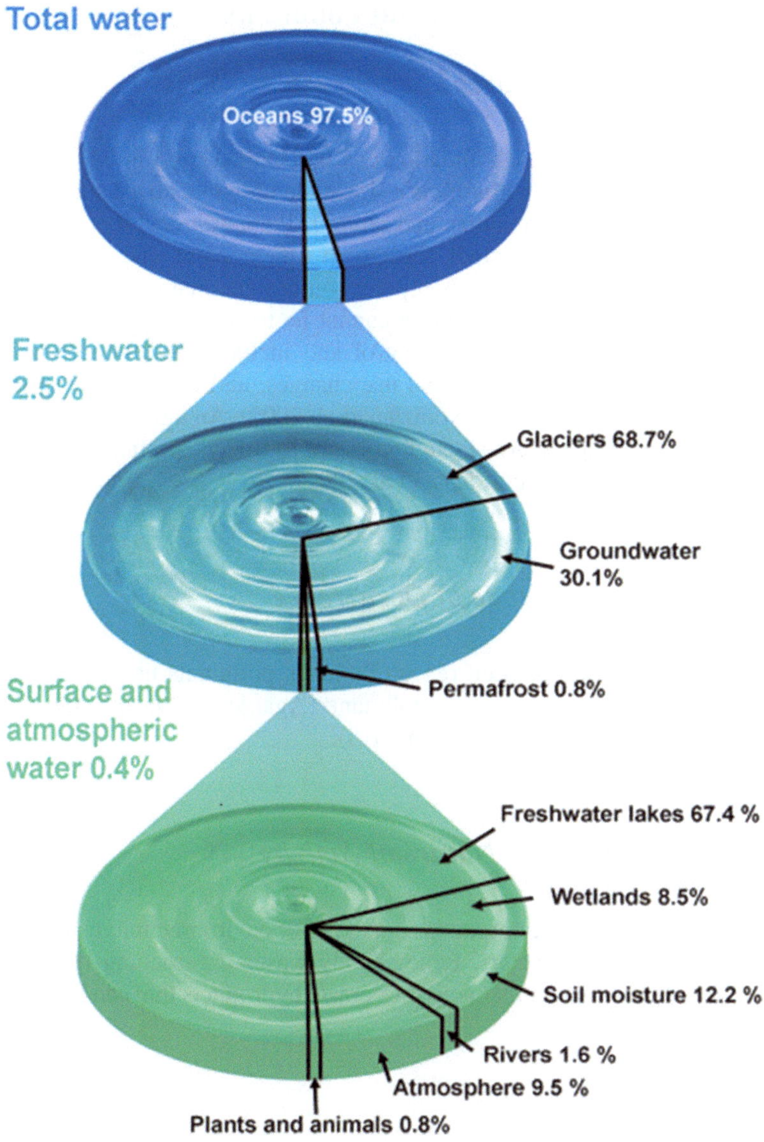

Fig. 1 Global freshwater distribution

3 Classification of Pollution and Pollutants

3.1 On the Basis of Source

Depending on the source, contamination can be characterized in two types, point and non point. Point source pollution is a single identified localized source which is relatively easy to identify, quantify and control. Point sources of water pollution include discharge from municipal sewage treatment plant and industrial plant (Oroian and Viman 2010). While, non point sources are characterized by multiple discharge points. This type of pollution cannot be traced to a single point of discharge, is difficult to monitor and control and includes pollution from diffuse sources, such as human land use, land use changes, and runoff from agricultural areas draining into a water body (Norazian et al. 2009). An agriculture activity is a major non point pollution source including use of fertilizers, pesticides and application of livestock manure (Wan Ruslan et al. 2002).

3.2 On the Basis of Mode of Occurrence

According to their mode of occurrence (Table 3), pollutants have been classified into physical, chemical and biological pollutants (Agarwal 2009a, b; Schwarzenbach et al. 2010; Akinbo and Tawari-Fufeyin 2014).

3.3 On the Basis of the Nature of Activity

Virtually all human activities produce some kind of environmental disturbance that contaminate surrounding waters. Eating (body wastes), gardening (pesticide and sediment runoff) and many other activities create byproducts that can find their way

Table 3 Classification of pollutants

Occurrence	Nature	Examples
Physical	Temperature turbidity	Waste heat from industry
	Colour	Dyes and pigments
	Suspended and floating matter	Soil particles, rubber and leather, wooden twigs, MSW carcasses
Chemical	Inorganic	N, P, Cl, F, etc.
		Detergents plastic,
	Organic	Pesticides
Biological	Pathogenic	Microorganism and worms
	Nuisance organisms	Algae

into the water cycle. For convenience, we can assign the large majority of sources of water pollution to three broad categories of waste (McKinney and Schoch 2003).

3.3.1 Industrial Wastes

Wastes from industry serve as major sources for all water pollutants (McKinney and Schoch 2003). In many developing countries, more than 70% of industrial waste is dumped untreated into waters, polluting the otherwise usable water supply (UNEP-UN-HABITAT 2010). Manufacturing industries like chemical, oil refining, steel etc. contribute many of the most highly toxic pollutants, including a variety of organic chemicals and heavy metals which are highly reactive and toxic. Other industries have less potential impact but are still considered highly problematic when it comes to pollution. These industries include the textile, leather tanning, paint, plastics, pharmaceutical, paper and pulp industries (Raja and Venkatesan 2010). Industrial waste chemicals can only be treated by using special waste treatment plants as they cannot be treated by sewage treatment plants (EPA 2011). Heavy metals which constitute a major portion of industrial waste are the most widespread pollutants of great environmental concern as they are non-degradable, toxic and persistent with serious ecological ramification on aquatic ecology (Jumbe and Nandini 2009).

Power generating industries are the major contributors of heat and radioactivity (Gambhir et al. 2012). Nearly all power plants, whatever the fuel, are major sources of thermal (heat) pollution. Radioactivity from nuclear power plants can pollute waters in a variety of ways, including discharge of mildly radioactive waste water and ground water pollution by buried radioactive waste (McKinney and Schoch 2003). Radioactivity may be found in ground waters as well as surface waters. In ground waters it may be due to radioactive material present in underground rocks, while in surface waters it may have been passed on with effluents from uranium mining and enrichment plants (Rao 2001).

3.3.2 Agricultural Wastes

These are generated by the cultivation of crops and rearing of animals. Globally, agriculture is the leading source of sediment pollution which includes plowing and other activities that remove plant cover and disturb the soil (Gambhir et al. 2012). Sedimentation due to runoff from agricultural areas affects water quality (Tundu et al. 2018). It decreases the capacity of freshwater bodies and also decreases the penetration of light into water due to which under water flora is disturbed. So the fishes and other fauna feeding on that flora are also disturbed and whole food chain is affected. Sediment particles also attach to fish gills and causes fish death (Swinkels et al. 2014). Pollutants like pesticides, nutrients and dangerous chemicals are transported and accumulated due to sedimentation (Dudal 1981; Letchinger 2000).

Agriculture is also a major contributor of organic chemicals, especially pesticides (McKinney and Schoch 2003). Pesticides are widely used in modern agriculture in most countries throughout the world and in a large range of environments. But environmental monitoring increasingly indicates that trace amounts of pesticides are present in surface and underground water bodies, far from the sites of pesticide application (Voltz et al. 2007; Gilliom 2007). Globally, 4.6 million tons of chemical pesticides are annually sprayed into the environment (Zhang et al. 2011) and are leached into the water bodies (Environmental Fate of Pesticides 2015). Pesticides cause widespread pollution of various fresh water bodies like rivers, lakes and estuaries (Agrawal 2009a, b).

Agricultural run-off causes eutrophication with phosphate being the main contributor promoting cyanobacteria and algal growth which ultimately reduces dissolved oxygen in water (Werner 2002). Harmful toxins which accumulate in food chain are produced by cyanobacterial blooms (Schmidt et al. 2013). The use of nitrogen fertilizers can be a problem in areas where agriculture is becoming increasingly intensified. These fertilizers increase the concentration of nitrates in groundwater, leading to high nitrate levels in underground drinking water sources, which can cause methemoglobinemia, the life threatening "blue baby" syndrome, in very young children, which is a significant problem in parts of rural Eastern Europe (Yassi et al. 2001). Nitrogen rich fertilizer compounds cause dissolved oxygen deficiencies in rivers, lakes and coastal zones which have devastating effects on oceanic fauna. Nitrogen fertilizers have high water solubility and increased runoff and leaching rate which results in ground water pollution (Rosen and Horgan 2005; NOFA 2004; Singh et al. 2006).

Agricultural practices lead to landscape changes resulting in many effects on aquatic system. These might include effects on water chemistry (Haygarth et al. 2005; Agouridis et al. 2005; James et al. 2005; Mehaffey et al. 2005; Olson et al. 2005) with consequent eutrophication and food web modification (Pretty et al. 2003; Moss et al. 2004), biocide leaching (Hanazato 2001; Corsolini et al. 2002; Van den Brink et al. 2002; Cold and Forbes 2004; Traas et al. 2004; Christensen et al. 2005) and suspended loads from soil erosion (Brodie and Mitchell 2005).

3.3.3 Domestic Wastes

These are the wastes that are produced by households. Most domestic waste is from sewage or septic tank leakage that ends up in natural waters. These come not only from human waste, but also from fertilizers used extensively in household lawns and gardens (McKinney and Schoch 2003). Today, many people dump their garbage into streams, lakes, rivers, and seas, thus making water bodies the final resting place of cans, bottles, plastics, and other household products (Groundwater Quality 2003). Most of today's cleaning products are synthetic detergents and come from the petrochemical industry. Most detergents and washing powders contain phosphates, which are used to soften the water among other things. These and other chemicals contained in washing powders affect the health of all forms of life in the water. The water discharged from untreated or inadequately treated sewage which goes into rivers, lakes, wells etc. causes serious infectious diseases like typhoid, cholera, dysentery and other skin diseases (Paranjape 2013).

Depending on the pollution type as well as the primary source of pollution a variety of effects occur in the freshwater bodies affecting various physical, chemical as well as biological parameters as shown in Table 4 (Revenga and Mock 2000; Taylor and Smith 1997; Shiklomanov 1997; UNWWD 2015).

Table 4 Freshwater pollution sources, effects and constituents of concern

Nature of contamination	Source	Impacts	Concerns
Organic matter	Effluents from various industries and built-up (sewage)	Decreases O_2 levels due to high rate of decomposition, affecting aquatic life	Biological Oxygen Demand (BOD), Dissolved Organic Carbon (DOC), Dissolved Oxygen (DO)
Pathogens (Microbes)	Sewage and livestock	Spreads diseases via contaminated drinking water supplies leading to diarrhoeal disease and increased childhood mortality, intestinal infections	Fecal coliform (Coliform), *Escherichia coli*, *Shigella* and *Salmonella*
Nutrients	Runoff from agricultural activities and industrial discharge	Growth of algae (eutrophication) which then decomposes, robbing water of oxygen and harming aquatic life	Total N and P
Salinization	Leached from alkaline soils, over irrigation of saltwater, over pumping coastal aquifers	Salt build-up in soils which kills crops or reduces yields. Renders freshwater supplies undrinkable	EC, p H and sodium toxicity.
Heavy metals	Variety of sources viz., industries, mining, agricultural activities and vehicles	Biomagnification, nervous, bone and blood disorders	Hg, Pb, Cd, Cr, Hg and As
Toxic organic compounds	MSW, industries, automobiles	A series of toxic effects in aquatic fauna and humans	PAHs, PCBs, pesticides (lindane, DDT, PCP, Aldrin, Dieldrin, Endrin, Isodrin

4 Water Pollution and Health

Variety of pathogenic and undesirable disease spreading microorganisms is present in domestic and hospital sewage. In adequate treatment of domestic and hospital sewage may cause outbreak of chronic diseases. The disease spreading agents involved as given in Table 5, include bacteria, viruses and protozoa. Most of them are widely spread in the entire world. Coliform species is used for the detection of water pollution. Disease causing bacterial species includes *Cryptosporidium parvum, Burkholderia pseudomallei, Giardia lamblia,* (EPA 2003; USGS Reston 2014; Homas 2000).

Numerous water borne diseases are thriving in humans (Halder and Islam 2005). Intense precipitation associated with floods is linked to severe weather and creating diverse diseases in different regions (Ahmad et al. 2014). More than 10% of the world population reliable on food that is cultivated in contaminated waters (Corcoran et al. 2010). Many waterborne infectious diseases are linked with faecal pollution of water sources (Nel and Markotter 2009). Health risk connected with contaminated water includes wide variety of diseases *viz.*, respiratory, diarrheal diseases, cancer, neurological disorder and cardiovascular disease (Krishnan and Indu 2006; Ullah et al. 2014). Mortality rate due to cancer is higher in rural areas than urban areas because urban inhabitants use treated water for drinking while rural people don't have facility of treated water and use unprocessed water (Jabeen et al. 2011). Polluted water has huge adverse impacts to the women exposed to toxic pollutants during pregnancy; it leads to the increased rate of low birth weight as a result foetal health is affected (Currie et al. 2013).

Degraded water quality damages the crop yield and contaminates food (Khan and Ghouri 2011). Pollutants have negative impact on food chain (Halder and Islam 2005), heavy metals, mainly Fe affect the respiratory tract of fishes. Fe clog in to gills of fishes hence increases mass motility of fishes. Furthermore, when these contaminated fishes are eaten by human leads to the major health issue (Ahmed et al. 2013). There are many other disorders of metal related contamination in water *viz.*, hair loss, renal failure (Salem et al. 2000) and neural disorder (Chowdhury et al. 2015).

Table 5 Sources of water related diseases

Source or infecting agent	Diseases	Cited from
Ingesting disease spreading pathogens (bacteria, viruses or parasites)	Cholera, shigellosis, typhoid diarrhea and hepatitis	Nel and Markotter (2009), Obasohan et al. (2010), Gambhir et al. (2012), Pandey (2006) and Ullah et al. (2014)
Water and personal hygiene	Skin ulcers, conjunctivitis and trachoma	
Penetration of human skin by infective forms (aquatic hosts)	Schistosomiasis, clonorchiasis and paragonimiasis	Pandey (2006), Obasohan et al. (2010) and Gambhir et al. (2012)
Insect bite	Malaria, dengue and yellow fever	Pandey (2006), Obasohan et al. (2010) and Gambhir et al. (2012)

5 Restoration of Water Bodies

The anthropogenic activities have fastened the deprivation of freshwater ecosystems which has enhanced the need to protect and reinstate these ecosystems. To overcome the global water pollution nuisance, it is necessary to build up an effective approach that fits at technological, sustainable, economical and within policies limit to meet desired treatment standard. The remediation of different pollutants affecting the ecology of a water body involves the following principles (Ramachandra et al. 2002; Rekha et al. 2005; Paz-Alberto and Sigua 2013; Mani and Kumar 2014; Hwang et al. 2014; Hong et al. 2018) which need to be followed to achieve the goal of restoration of the freshwater bodies:

- Preserve and protect aquatic resources for conserving biodiversity.
- Develop adequate and measurable restoration goals given the natural potential of the area, and socio-economically, given the available resources and the extent of community support for the project.
- Focus on feasibility taking into account scientific, financial, social and other considerations.
- Address ongoing causes of degradation and eliminate or remediate ongoing stresses wherever possible.
- Use ecologically sound techniques based on bioremediation involving the use of microbes and plants for pollution remediation.
- Use natural fixes and bioengineering techniques, where possible involving a method of construction combining live plants with dead plants or inorganic materials, to produce living, functioning systems to prevent erosion, control sediment and other pollutants, and provide habitat.
- Restore ecological integrity and natural function which can bring back beneficial functions.
- Restore native species and eliminate the non-native species invasives which outcompete the natives
- Anticipate future changes which may help integrate any foreseeable ecological and societal changes into restoration design.
- Design for self-sustainability which involves favouring ecological integrity, as an ecosystem in good condition is more likely to have the ability to adapt to changes.
- Adopt adaptive management through continuous monitoring and incorporating changes wherever necessary.

6 Conclusion

The deterioration of freshwater bodies around the world and the increasing potable water crisis brings in the need to understand the causes underlying the drastic changes in the physico-chemical and biological features of these water bodies. A sound ecological understanding of the processes that operate in these systems becomes necessary for the revival and management of these ecosystems. There arises a need to pay more attention towards controlling the sources that contribute pollutants to these ecosystems. Newer ecologically sound technologies like bio-remediation, phyto-remediation nano-bioremediation and biomarkers coupled with the traditional practices are required for restoring these fragile ecosystems. For successful management of these water bodies, participatory management and a sustainable approach needs to be practiced so that these freshwater bodies can be preserved and conserved sustainably.

References

Abel, P. D. (1996). *Water pollution biology* (2nd ed., p. 29). London: Taylor and Francis.

Agarwal, S. K. (2009a). *Water pollution* (p. 54). New Delhi: APH Publishing Corporation.

Agarwal, S. K. (2009b). *Water pollution* (p. 33). New Delhi: APH Publishing Corporation.

Agouridis, C. T., Workman, S. R., Warner, R. C., & Jennings, G. D. (2005). Livestock grazing management impacts on stream water quality: A review. *Journal of the American Water Resources Association, 41*, 591–606.

Ahmad, S. M., Yusafzai, F., Bari, T., et al. (2014). Assessment of heavy metals in surface water of River Panjkora Dir Lower, KPK Pakistan. *Journal of Biological and Environmental Sciences, 5*, 144–152.

Ahmed, T., Scholz, F., Al-Faraj, W., et al. (2013). Water-related impacts of climate change on agriculture and subsequently on public health: A review for generalists with particular reference to Pakistan. *International Journal of Environmental Research and Public Health, 13*, 1–16.

Akinbo, M., & Tawari-Fufeyin, P. (2014). Physical, chemical and biological parameters of water from medical waste dumpsites in South-Western Niger Delta, Nigeria. *Asian Journal of Water, Environment and Pollution, 11*(04), 83–88.

Boberg, J. (2005). *Freshwater availability, liquid assets: How demographic changes and water management policies affect freshwater resources* (pp. 15–28). Santa Monica: RAND Corporation.

Brodie, J. E., & Mitchell, A. W. (2005). Nutrients in Australian tropical rivers: Changes with agricultural development and implications for receiving environments. *Marine and Freshwater Research, 56*, 279–302.

Chowdhury, S., Annabelle, K., & Klaus, F. Z. (2015). Arsenic contamination of drinking water and mental health, ZEF Discussion Papers on Development Policy, No. 222, University of Bonn, Center for Development Research (ZEF), Bonn.

Christensen, B. T., Lauridesen, T. L., Ravn, H. W., & Bayley, M. (2005). A comparison of feeding efficiency and swimming ability of Daphnia magna exposed to cypermethrin. *Aquatic Toxicology, 73*, 210–220.

Cold, A., & Forbes, V. E. (2004). Consequences of a short pulse of pesticide exposure for survival and reproduction of *Gammarus pulex*. *Aquatic Toxicology, 67*, 287–299.

Corcoran, E., Nellemann, C., Baker, E., et al. (2010). *Sick water? The central role of wastewater management in sustainable development. A Rapid Response Assessment*. Arendal: United Nations Environment Programme.

Corsolini, S., Romeo, T., Ademolla, N., Greco, S., & Focardi, S. (2002). POPs in key species of marine Antarctic ecosystem. *Microchemical Journal, 73*, 187–193.

Currie, J., Joshua, G. Z., & Katherine, M. (2013). Something in the water: Contaminated drinking water and infant health. *Canadian Journal of Economics, 46*(3), 791–810.

Dudal, R. (1981). An evaluation of conservation needs. In R. P. C. Morgan (Ed.), *Soil conservation, problems and prospects*. Chichester: Wiley.

Environmental Fate of Pesticides. (2015). *Pesticide Wise*. Victoria: British Columbia Ministry of Agriculture.

EPA. (2003). U.S. Environmental Protection Agency Nonpoint Source Control Branch (4503T) 1200 Pennsylvania Avenue, NW Washington, DC 20460.

EPA. (2011). U. S. Environmental Protection Agency, Introduction to the National Pre-treatment Program. https://www3.epa.gov/npdes/pubs/pretreatment_program_intro_2011.pdf

Gambhir, R. S., Kapoor, V., Nirola, A., Sohi, R., & Vikram, B. (2012). Pollution: Impact of pollutants and new promising techniques in purification process. *Journal of Human Ecology, 37*(2), 103–109.

Gilliom, R. J. (2007). Pesticides in U.S. streams and groundwater. *Environmental Science & Technology, 41*, 3407–3413.

Groundwater Quality. (2003). http://groundwater.ucdavis.edu/files/136273.pdf

Halder, J. N., & Islam, M. N. (2005). Water pollution and its impact on the human health. *Journal of Environment and Human, 2*(1), 36–46.

Hanazato, T. (2001). Pesticide effects on freshwater zooplankton: An ecological perspective. *Environmental Pollution, 112*, 1–10.

Haseena, M., Malik, F. M., Javed, A., Arshad, S., Asif, N., Zulfiqar, S., & Hanif, J. (2017). Water pollution and human health. *Environmental Risk Assessment and Remediation, 01*(03), 223–230.

Haygarth, P. M., Wood, F. L., Heathwaite, A. L., & Butler, P. J. (2005). Phosphorus dynamics observed through increasing scales in a nested headwater-to-river channel study. *Science of the Total Environment, 344*, 83–106.

Homas, R. S. (2000). *Microbes and urban watersheds: Concentrations, sources, & pathways*. Ellicott City: Centre for Watershed Protection.

Hong, C.-Y., Chang, H., & Chung, E.-S. (2018). Resident perceptions of urban stream restoration and water quality in South Korea. *River Research and Applications, 34*(5), 481–492.

Hwang, S. J., Lee, S. W., & Yoo, B. (2014). Ecological conservation and the restoration of freshwater environments in Korea. *Paddy and Water Environment, 12*(1), 1–5.

Jabeen, S. Q., Mehmood, S., & Tariq, B. (2011). Health impact caused by poor water and sanitation in district Abbottabad. *Journal of Ayub Medical College, Abbottabad, 23*(1), 47–50.

James, C. S., Fisher, J., Russell, V., Collings, S., & Moss, B. (2005). Nitrate availability and hydrophyte species richness in shallow lakes. *Freshwater Biology, 50*, 1049–1063.

Jumbe, A. S., & Nandini, N. (2009). Heavy metals analysis and sediment quality values in urban lakes. *American Journal of Environmental Sciences, 5*, 678–687.

Khan, M. A., & Ghouri, A. M. (2011). Environmental pollution: Its effects on life and its remedies. *Journal of Arts, Science and Commerce, 2*(2), 276–285.

Krishnan, S., & Indu, R. (2006). *Groundwater contamination in India: Discussing physical processes, health and socio behavioral dimensions*. IWMI-Tata, Water Policy Research Programmes, Anand, India.

Letchinger, M. (2000). *Pollution and water quality, neighbourhood water quality assessment*. Project oceanography.

Mani, D., & Kumar, C. (2014). Biotechnological advances in bioremediation of heavy metals contaminated ecosystems: An overview with special reference to phytoremediation. *International journal of Environmental Science and Technology, 11*, 843–872.

McKinney, M. L., & Schoch, R. M. (2003). *Environmental science: Systems and solutions*. London: Jones & Bartlett.

Mehaffey, M. H., Nash, M. S., Wade, T. G., Ebert, D. W., Jones, K. B., & Rager, A. (2005). Linking land cover and water quality in New York City's water supply watersheds. *Environmental Monitoring and Assessment, 107*, 29–44.

Mishra, R. K., & Dubey, S. C. (2015). Fresh water availability and its global challenge. *International Journal of Engineering Science Invention Research & Development, 2*(4), 2349–6185.

Moss, B., et al. (2004). Continental-scale patterns of nutrient and fish effects on shallow lakes: Synthesis of a pan-European mesocosm experiment. *Freshwater Biology, 49*, 1633–1649.

Nel, L. H., & Markotter, W. (2009). New and emerging waterborne infectious diseases. *Encyclopedia of Life Support System, 1*, 1–10.

NOFA Interstate Council. (2004) *Natural farmer. Ecologically sound nitrogen management*. Mark Schonbeck.

Norazian, M. N., Kamarudin, M. N., Azhar, S. S., Ragunathan, K., Nihla, A., & Khairuddin, M. I. (2009). Water pollution. In Introduction to environmental engineering. Pauh Putra Campus, Arau: Universiti Malaysia Perlis (UniMAP) Publication.

Oana, V., Oroian, V. I., & Fleşeriu, A. (2010). Types of water pollution: Point source and nonpoint source aquaculture, aquarium, Conservation & Legislation. *International Journal of the Bioflux Society, 3*(5), 393–395.

Obasohan, E. E., Agbonlahor, D. E., & Obano, E. E. (2010). Water pollution: A review of microbial quality and health concerns of water, sediment and fish in the aquatic system. *African Journal of Biotechnology, 9*(4), 423–427.

Olson, B. M., Miller, J. J., Rodvang, S. J., & Yanke, J. L. (2005). Soil and groundwater quality under a cattle feedlot in southern Alberta. *Water Quality Research Journal of Canada, 40*, 131–144.

Oroian, I., & Viman, O. (2010). Damaging effects of overall water pollution. *Environment, 3*, 203–207.

Owa, F. D. (2014). Water pollution: Sources, effects, control and management. *Mediterranean Journal of Social Sciences*. Rome: MCSER Publishing.

Pandey, S. (2006). Water pollution and health. *Kathmandu University Medical Journal (KUMJ), 4*(1), 128–134.

Pandey, S., Singh, A., & Mishra, S. (2017). Dynamism of structure and properties of water molecule. *International Journal of Advance Research and Innovative Ideas in Education, 3*(1), 2395–2436.

Paranjape, V. N. (2013). *Environmental law* (p. 117). Allahabad: Central Law Agency.

Paz-Alberto, A. M., & Sigua, G. C. (2013). Phytoremediation: A green technology to remove environmental pollutants. *American Journal of Climate Change, 2*, 71–86.

Pretty, J. N., Mason, C. F., Nedwell, D. B., & Hine, R. E. (2003). Environmental costs of freshwater eutrophication in England and Wales. *Environmental Science & Technology, 37*, 201–208.

Raja, G., & Venkatesan, P. (2010). Assessment of groundwater pollution and its impact in and around Punnam Area of Karur District, Tamil Nadu, India. *E-Journal of Chemistry, 7*(2), 473–478.

Ramachandra, T. V., Rajashekariah, K., & Ahalya, N. (2002). *Status conservation and management of wetlands*. Bangalore: Allied Publishers Pvt Ltd.

Rao, K. R. (2001). Radioactive waste: The problem and its management. *Current Science, 81*(12), 1534–1546.

Rekha, P., Raj, D. S., Aparna, C., Bindu, V. H., & Anjaneyulu, Y. (2005). Bioremediation of contaminated lake sediments and evaluation of maturity indices as indicators of compost stability. *International Journal of Environmental Research and Public Health, 2*(2), 251–262.

Revenga, C., & Mock, G. (2000). *Pilot analysis of global ecosystems: Freshwater systems* (pp. 1–78). Washington, DC: World Resource Institute.

Rosen, C. J. & Horgan, B. P. (2005). Preventing pollution problems from Lawn and Garden fertilizers. St. Paul: University of Minnesota Extension Service. Retrieved from the University of Minnesota Digital Conservancy, http://hdl.handle.net/11299/93999.

Salem, H. M., Eweida, E. A., & Farag, A. (2000). Heavy metals in drinking water and their environmental impact on human health. *ICEHM, 2000*, 542–556.

Schmidt, J. R., Shaskus, M., Estenik, J. F., Oesch, C., Khidekel, R., et al. (2013). Variations in the microcystin content of different fish species collected from a eutrophic lake. *Toxins (Basel), 5*, 992–1009.

Schwarzenbach, R. P., Egli, T., Hofstetter, T. B., von Gunten, U., & Wehrli, B. (2010). Global water pollution and human health. *Annual Review of Environment and Resources, 35*, 109–136.

Sharp, K. A. (2001). *Water: Structure and properties. Encyclopaedia of life sciences*. John Wiley & Sons, Ltd. www.els.net.

Shiklomanov, I. A. (1997). *Comprehensive assessment of the freshwater resources of the world* (p. 88). Stockholm: World Meteorological Organization.
Shiklomanov, I. A., & Rodda, J. C. (2003). *World water resources at the beginning of the twenty-first century*. Cambridge: Cambridge University Press.
Singh, B., Singh, Y., & Sekhon, G. S. (2006). Fertilizer-N use and nitrate pollution of groundwater in developing countries. *Journal of Contaminant Hydrology, 20*, 167–184.
Smical, I., Mihaly-Cozmuta, L., & Costin, D. (2010a). Research concerning the influence of several factors on Pb^{2+}, Cu^{2+} and Zn^{2+} ions adsorption by natural zeolite tuff from Maramureş county, Northern Romania. *AES Bioflux, 2*(2), 171–180.
Smical, I., Mihaly-Cozmuta, L., & Costin, D. (2010b). Use of natural zeolites from Maramures county (Romania) in removal of Cu2+, Pb2+, Zn2+ ions from industrial wastewaters. *AES Bioflux, 2*(2), 181–188.
Swinkels, L. H., Van de Ven, M. W. P. M., Stassen, M. J. M., Van der Velde, G., Lenders, H. J. R., & Smolders, A. J. P. (2014). Suspended sediment causes annual acute fish mortality in the Pilcomayo River (Bolivia). *Hydrological Processes, 28*, 8–15.
Taylor, R., & Smith, I. (1997). *State of New Zealand's environment*. Wellington: The Ministry for the Environment.
The United Nations World Water Development (UNWWD) Report. (2015). Section 2: Changing Natural Systems, Chapter 4, Part 1. Global Hydrology and Water Resources, p. 121.
Traas, T. P., Janse, J. H., Van den Brink, P. J., Brock, T. C. M., & Aldenberg, T. (2004). A freshwater food web model for the combined effects of nutrients and insecticide stress and subsequent recovery. *Environmental Toxicology and Chemistry, 23*, 521–529.
Tripathi, A. K., & Pandey, S. N. (2009). *Water pollution* (p. 3). New Delhi: APH Publishing Corporation.
Tundu, C., Tumbare, M. J., & Onema, J. M. K. (2018). Sedimentation and its impacts/effects on river system and reservoir water quality: Case study of Mazowe Catchment, Zimbabwe. *Proceedings of the International Association of Hydrological Sciences, 377*, 57–66.
Ullah, S., Javed, M. W., Shafique, M., et al. (2014). An integrated approach for quality assessment of drinking water using GIS: A case study of Lower Dir. *Journal of Himalayan Earth Sciences, 47*(2), 163–174.
UN WWAP. (2003). *United Nations world water assessment programme. The world water development report 1: Water for people, water for life*. Paris: UNESCO.
UNEP-UN-HABITAT. (2010). http://unhabitat.org/un-habitat-annual-report-2010/
USGS Reston VA. (2014, February 18). *A primer on water quality*. U.S. Department of the Interior, U.S. Geological Survey.
Van den Brink, P. J., Roelsma, J., Van Nes, E. H., Scheffer, M., & Brock, T. C. M. (2002). PERPEST model, a case-based reasoning approach to predict ecological risks of pesticides. *Environmental Toxicology and Chemistry, 21*, 2500–2506.
Voltz, M., Louchart, X., Andrieux, P., & Lennartz, B. (2007). Process of water contamination by pesticides at catchment scale in Mediterranean areas. *Geophysical Research Abstracts, 37*(2). Retrieved 5 March 2014 from http://krepublishers.com/02-Journals/JHE/JHE.
Wan Ruslan, I., Khairul, R. A. B., Zullyadini, A. R., & Norhidayat, K. (2002). *Storage of sediment and nutrients in littoral zone of a shallow tropical reservoir: A case of Timah Tasoh reservoir, Perlis, Malaysia* (p. 276). Wallingford: IAHS Publications.
Webster.com. (2010). Definition from Webster Dictionary 08–13. Retrieved August 26, 2010.
Werner, W. (2002). Fertilizers; Environmental aspects. In *Ullmann's encyclopaedia of industrial chemistry*. Weinheim: Wiley-VCH.
Yassi, A. L., Kjellstrom, T., DeKok, T., & Guidotti, T. (2001). *Basic environmental health*. New York: Oxford University Press.
Zhang, S., Qiu, C. B., Zhou, Y., Jin, Z. P., & Yang, H. (2011). Bioaccumulation and degradation of pesticide fluroxypyr are associated with toxic tolerance in green alga *Chlamydomonas reinhardtii*. *Ecotoxicology, 20*, 337–347.

Freshwater Pollution: Effects on Aquatic Life and Human Health

Rizwana Qadri and Muneeb A. Faiq

Abstract An all-inclusive academic anecdote on water or freshwater and its role in the maintenance and wellbeing of biological systems is a momentous task which goes beyond mere rattling of statistics. Therefore, a discussion about any aspect of water right from basic chemistry to biochemistry and then to biological realm has to have certain restrictions in terms of their scope and area of focus. Phenomena from basic combustion to all life processes are mediated through production and/or consumption of molecules of water. There are dedicated biochemical processes in the biological systems that are important to regulate life at the fundamental level and these processes have a lot to do with molecular water and water as a solvent. There are even certain channels in the most fundamental parts of cellular life-the cell membranes; these channels are called aquaporins which are dedicated to the flow of water across the bio-membrane system to maintain life. With this view, it becomes easy to understand that the most abundant and ardently fundamental biochemical on earth is water. And it goes without saying that more that 70% of the earth's surface is water and – curiously- the composition of any living organism is 70% water (or more); and this holds true for humans also. Based on this elementary verity, it becomes a mathematical reality that around 70–80% of the diseases must be waterborne. This gives birth to an enormously important field of medicine dealing with the investigations related to water in health and disease. Rapid advancements of unregulated technology and unchecked lifestyle changes have led to an intense

R. Qadri
Department of Laboratory Medicine, All India Institute of Medical Sciences, New Delhi, India

Genetics Unit, Department of Pediatrics, All India Institute of Medical Sciences, New Delhi, India

M. A. Faiq (✉)
Medical Biotechnology Laboratory, Dr. B. Ambedkar Center for Biomedical Research, University of Delhi, New Delhi, India

Laboratory for Molecular Reproduction and Genetics, Department of Anatomy, All India Institute of Medical Sciences, New Delhi, India

Neuroimaging and Visual Science Laboratory, Department of Ophthalmology, Langone Health Center, New York University School of Medicine, New York, NY, USA

H. Qadri et al. (eds.), *Fresh Water Pollution Dynamics and Remediation*,
https://doi.org/10.1007/978-981-13-8277-2_2

upsurge in water pollution and- as a consequence- water borne diseases. A direct impact of industrialization, radioactive discharge, pesticide use etc. on water is generally encompassed within the umbrella term called water pollution. This phenomenon has ill effects on all biological systems including humans. In this chapter we shall restrict our discussion to ill effects of water pollution on human health and wellbeing with occasional narratives on freshwater aquatic life. Water pollution, to start with, is synonymous with typhoid, cholera, encephalitis, hepatitis, giardiasis, cholera, poliomyelitis, shigellosis, diarrhea, death etc. Diseases like lead toxicity, are also included in this group though they are not pathogen borne. There is, therefore, a dire need to have regulatory interventions to bring down the severity, incidence and prevalence of diseases secondary to water pollution which includes agricultural and domestic waste also.

Keywords Human health · Water pollution · Water borne diseases

1 Introduction

Water pollution may be understood as the phenomenon of addition and ingress of substances and materials into water/water bodies which are not the natural part of the water cycle on earth. Such an addition leads to a compromise in the quality of water (Alrumman et al. 2016) and poses overbearing threat to the ecosystem and balance of the natural forces which are otherwise important in keeping the biosphere working in the well-modulated form. Hitches in the environment and human wellbeing are, for this reason, inevitable (Briggs 2003). Water is presumably the most important natural resource second only to air. Food may attain the third position but for production of food, water is indispensable. Water is required for majority of the biochemical transformations that take place inside the bodies of living being from bacteria to the most advanced organisms. In addition to that, most of the life processes happen in a four dimensional world in which the three dimensions of space are being occupied by water and the fourth dimension of time also interacts with water in certain important ways. Indications can, therefore, be seen towards the very process of life happening in water as a medium even in the terrestrial forms. Water is important all through the origin, birth and maintenance of life as a continuous and self-describing algorithm. Or in biological terms water is important for generation and development of living organisms which by simple extrapolation means that water is essential for all our life processes, our living processes, our domestic needs, and our development (Bibi et al. 2016). Before industrial revolution most of the fresh water used to be safe for drinking but post industrialization and urbanization, fresh water is not synonymous with drink water though it is now the term used to distinguish it from the saline water of the oceans and the seas.

From medical point of view, safe drinking water is indispensable for human health world over. There are various reasons for this notion and the best premise comes from the fact that all life happens in water. Water is the most important medium for life to happen and hence most of the organisms (many of which are pathogens) while others are relatively harmless or even useful. Extrapolating on this context puts forth what the World Health Organization (WHO) has stated – 80% of the human diseases are water borne. The simultaneous contrast becomes conspicuous when we come face to face with the verity that majority of countries world over do not meet the criteria laid down by WHO for safe drinking water (Khan et al. 2013). This is unfortunate for those countries which have ample fresh water available naturally yet the human activities are so ruthless that the naturally occurring fresh water is not fit for consumption. Around 3% of the deaths in any age group occur just due to unhygienic water (Pawari and Gawande 2015) that is being supplied for human consumption which may be thought as an immediate result of thoughtless human activities and lack of effective regulations.

There are numerous sources of water pollution, but the ones that can be thought to be representative and major players in this process are industrial wastes, unregulated domestic wastes, marine dumping of the wastes produced by human activity, air pollution, deforestation and consequent soil erosion, use of pesticides and harmful chemicals in agriculture, untreated wastes etc. Heavy metals and their related by-products produced by active industries and disposed off without proper processing and treatment accumulate finally in water bodies including oceans and fresh water bodies (lakes, rivers and ponds) thereby exposing humans and other animals to hazardous substances. Exposure of the livestock to the contaminated water ultimately leads to deposition of these hazardous chemicals in human bodies because humans ultimately consume the livestock and their products. These substances are toxic and cause of a variety of malfunctions in human body and precipitate ill health. The major ailments caused in this way are immune compromise (which exposes to a gamut of many other pathogen based diseases and pose life threatening conditions), inflammation, reproductive malfunction including infertility, respiratory and gastrointestinal disorders (a major concern being hepatotoxicity), cancer and even death. In addition to these maladies many infectious diseases like cholera and typhoid (Juneja and Chauhdary 2013) and even certain cancers are a result of water pollution. The list does not stop here, renal diseases, dermatitis, dementia, diarrhea etc. are also thought to be primarily spread through infected/contaminated water (Khan and Ghouri 2011).

Humans live as an integral part of their environment which includes the flora and the fauna as well as the abiotic realm. Since water is ambient and a necessary ingredient of the environment that harbors life, it follows that water pollution will directly affect the plants as well as the animals. Substances that pollute water (generally called water pollutants) affect flora and fauna in many important ways and also get accumulated in them. Water pollution also kills sea weeds, mollusks, water birds, fishes, and many other marine and fresh water organisms that have a potential to serve as a part of human nutrition. Many chemicals accumulate in the egg shell of these birds and cause the bird population to diminish. Also, when these plants and animals are consumed for nutritional purposes, these chemicals accumulate in human bodies in fat and other tissues causing toxicity, inflammation, immunosuppression, allergies and many other ailments (Owa 2013).

2 Causes of Water Pollution

As already mentioned, there are many sources of water pollution but some of the representative basis are discusses as follows:

2.1 *Domestic Sewage*

It is also called domestic wastewater and sometimes municipal wastewater. It is produced by domestic and community activities of people in the form of waste water. Domestic sewage is generally described by the amount of its production, dynamics of its flow, physical characteristics, chemical constituents, toxin concentration and pathogenic load. Examples of domestic sewage include greywater (from sinks, tubs, showers etc.), blackwater (The water that comes from flushed toilets, and cleansing activities of human waste) and other chemical substances like soaps and detergents.

2.2 *Industrialization*

Industries generally take cold water and return back hot adding a significant momentum to global warming and climate changes. These effects in turn have harmful effects on the ecosystems which affects human health. Industrial pollution brings forth its effects through industrial wastes such as heavy metals, harmful substances/chemicals, industrial by-products, organic pollutants etc.

2.3 *Pesticides and Fertilizers*

Pesticides, which are mostly sulfur based substances, are rampant in modern agricultural practices. They are potentially toxic, hence their use in killing pests. Their toxicity is not specific, therefore, they pose a huge threat in terms of chronic diseases, congenital deformities, and many other development issues in humans. Given the amount of pesticides used, the situation is alarming from the health point of view. Fertilizers are mostly nitrogen based substances which serve as nutritional supplements for agricultural purposes. While a proper, regulated and careful use of fertilizers many not be as harmful as pesticide use, nevertheless the excessive use (which is a regular practice nowadays) is hazardous. Both pesticides and fertilizers run off into water and pollute the waterbodies thereby precipitating a serious hazard. Rains and over-irrigation wash off the fields and lawns and the resultant solutions drain into the waterbodies, thereby polluting them and exposing the humans to detrimental effects. Excess fertilizer content in lakes and other water bodies stimulate the growth of aquatic plants and algae which hamper the normal flow of water and cause stagnant waters which then breed infectious pathogens.

2.4 *Plastics/Polythene Bags*

Plastic bags and other polythene items accumulate in our environment and seldom decompose and hence the amount is ever increasing. It clogs the flow of water leading to pollution and pathogen growth. Plastic bags are one of the greatest menaces of the modern society. It is a huge threat to our water bodies and is immensely detrimental to human health and wellbeing. It is also one of the most easily removable menace we have seen in our times. All it takes is a personal decision of not using plastic bags. That does most of the job. And it is also important to note that plastic bags are easily replaceable with more environment friendly options which are as convenient.

Other cause of water pollution include, but are not limited to urbanization, poor management system, weak policy making, lack of education, selfish public behavior and poor attention of government and appellate bodies.

By any estimates, domestic sewage is one of the most common cause of water pollution accounting for more than 75% of the contribution to water pollution. Sugar industries, textile mills, metallurgic industries and pesticide use are sources to mention (Kamble 2014). Due to these wastes most of the rivers world over are polluted, infected and contain a lot of pathogens with less than ever population of useful organisms.

Momentous amount of domestic sewage gets added to and accumulates into the rivers and lakes. Since most of the domestic sewage is untreated and unregulated, it is bound to contain high amounts of toxins, litters, garbage, plastic, pathogens, and other substances which are sufficient to pollute any large and clean water body. When this pollution is augmented with the waste and pollutants coming out of industries, the pollution level becomes extreme. These industries lead to production of many detrimental substances which not only pollutes the water bodies like lakes, ponds and rivers but also seeps into the underground water, thereby, rendering it unsafe for any use. Such a seepage also leads to the accumulation of these pollutants in plants that we consume. This phenomenon in turn leads to many hazardous effects the list of which is quite exhaustive. One important component of the industrial waste is heavy metals which are highly toxic (Ho et al. 2012). Though the water pollution caused by industries is about one fourth of the total sources of water pollution but they are arguably the most detrimental sources of water pollution (Desai and Smt Vanitaben 2014).

It is presumably the increase in population density rather than the population increase and also the change in lifestyles and needs that is harmful rather than a simple increase in population. Increase in the number of people per unit area leads to increase in the generation of all kinds of pollutants among which solid waste is the main issue to mention (Jabeen et al. 2011). All these waste material whether solid or liquid, are drained into the water bodies and get dumped there. In terms of the human excretion and the mixing of human excretory substances into water, a huge number of pathogens spread in the society which leads to many diseases and some of these diseases have spread to endemic proportions. Due to this, we witness many epidemics and deaths many a times. Infants are particularly susceptible to such epidemic disease which include cholera, typhoid, malaria and other deadly diseases. Increase in population leads to lower supply of various vaccines and other useful and safeguarding substances to the people, which in turn, exposes them to the threat water pollution upsurges with. Though increase in population should lead to increase in man power and consequent growth in resources, but governments have not been able to harness the potential completely and hence increase in population leads to diminishing of the resources and the availability of these resources to common people. Increase in population density is an obvious cause of urbanization which causes ill hygienic conditions, dumping of more garbage and other waste products per unit area, production of more unhygienic water contaminated with excreta, etc. All these factors lead to a high risk of urban population to suffer from diseases caused by water pollution.

Pesticides are used to kill pathogens and other organisms which cause threat to the production levels and efficiency of our agricultural practices. Though the term pesticides literally means substances that kill pests, but they are in no way specific to pest. These substances are chemically toxins that can be toxic to the organisms we recognize as pests as well as harmful to those we don't identify as pests. Excessive and unregulated use of pesticides is a great hazard to the ecology and environment (Khurana and Sen 2008). It is nearly more than half of the total pesticides that remain in the soil and yet the rest of these pesticide based toxins move into the water thereby polluting it. When fertilizers drain off from the soil and go to water, addition of nutrients lead to hypertrophication (when a water body is nutrient enriched, excessive growth of plants and algae ensues) leading to increase in plant density and death and decomposition of animals giving rise to many diseases. Whether pesticides or excessive fertilizers, they are potentially toxic and their accumulation in soil or water means the accumulation in plants that grow over that soil or water. The vegetables and other products that grow out of this environment contain these chemicals (Ebenstein 2008) and are a serious threat to human health. The products that are pharmaceutical in nature are also not benign in this context (http://research.gsd.harvard.edu/hapi/).

3 Human Health and Water Pollution: The Crosstalk

The effects of pollution or any type of the disturbance of any sort in the homeostasis of the environment has health effects and such effects cannot be over-estimated. Water pollution has direct as well as indirect effects on health of all organisms but we will restrict our focus primarily to human health in this discussion. Water pollution is the etiology of a majority of human ailments which range from relatively benign to the most dreadful diseases and even epidemics seen throughout the human history. Any organism that is capable of causing a disease is called a pathogen. Pathogens are of many types and are responsible for causing and spreading diseases in humans and other organisms. Pathogens can be classified on many basis like the species, mode of transmission etc. Certain types are pathogens are found throughout the world and global programs have been initiated to combat the diseases caused by them yet other types are restricted to certain specific areas, geographies and climatic conditions. It is important to note that a significant percentage of water borne diseases are spread by human to human mode of transmission (Halder and Islam 2015) which makes it all the more imperative to tackle the diseases right from the cause -which is water pollution. Global warming and other environmental changes have caused vicissitudes in the climate. Hence, heavy rainfalls and floods are rampant. These floods and eroding rainfalls caused by human activities lead to a variety of diseases (Ahmad et al. 2014) in every country though the proportion is higher in the developing countries. More than 10% of the human population is living on food that is cultivated on contaminated water (Corcoran et al. 2010). Consumption of contaminated food by such a significant proportion of human population leads to a growing chain/cascade towards increasing risk of diseases in those individuals also who do not eat even that food. This happens due to human to human transmission of the disease. It is further accelerated by the fact that waterborne diseases enter the fresh water through fecal matter (Nel and Markotter 2009) and hence contaminate it and then get transmitted to people who may not be primarily eating the contaminated food. The list of diseases that are caused by or related to water pollution is exhaustive but a representative list includes, while not being limited to, diarrhea, cholera, malaria, typhoid, respiratory illnesses, various cancers, skin conditions, genetic issues, congenital developmental problems, neurological disorders, gastrointestinal diseases and hepatic diseases (Ullah et al. 2014). Pollutants containing nitrogen have been associated etiologically to blue baby syndrome (Krishnan and Indu 2006; Ahada and Suthar 2018), Blue baby syndrome is a combination of congenital heart conditions that lead to cyanosis (bluish skin discoloration due to inadequate circulation/oxygenation of the blood and hence body hypoxia) in infants. Even the Zika virus based microcephaly has also been attributed to pesticide use. Our group recently explained the mechanism of Zika virus mediated developmental brain defects (Kumar et al. 2016; Faiq et al. 2018). Regarding the proliferative disorders, a high mortality rate has already been established in literature to be secondary to cancers arising from water pollution. This condition is relatively prevalent in rural societies due to supply to water which is not properly treated with cleaning processes and protocols. In addition to rural environment, poverty may also be seen as one of the important factors for such conditions as economically deprived societies/people don't have much access to clean water and live is unhygienic conditions. One of the main issues of grave concern are women of reproductive age who may suffer infertility due to contaminated water. Also there are high chances of mutations and genetic disorder in the developing fetuses due to exposure to contaminated water. Many children are, therefore, born with a lot of medically alarming conditions

(Currie et al. 2013). Contaminated and infected water has many damaging effects on the crop and livestock they come in contact with which in turn poses a hazard to human life. Such a scenario leads to shift of the equilibrium in the food chain in addition to many health problems in aquatic life. One of the representative functions in the fresh water fishes is the threat posed by heavy metals like iron which accumulates in the gills and leads to respiratory distress. This leads to increased death rates and decomposition of the carcasses in the water. That in turn becomes the ripe platform for many pathogens to develop and spread. Exposure to such water or eating the ailing fishes leads to many diseases in humans (Salem et al. 2000; Chowdhury et al. 2015; Ahmed et al. 2013)

Though there are many ways to classify the water borne diseases, but the best and most widely accepted classification is the one based on the pathogen type. Hence water borne diseases can grossly be classified into bacterial diseases, viral diseases and parasitic diseases. We will briefly discuss all these one by one.

3.1 Bacterial Diseases

Contaminated and untreated water for drinking is the source of numerous diseases. One of these diseases is diarrhea (which is more of a symptom than a disease itself). The organism mainly responsible for the spread of diarrhea through contaminated and untreated water is *Campylobacter jejuni* which is responsible for almost 15% of the diarrhea cases in the world. *C. Jejuni* is a bacterial pathogen which causes many disease including enteritis characterized by abdominal pain, diarrhea (from loose to bloody stools) and fever. This pathogen belongs to the genus *Campylobacter*. And responds to antibiotics (azithromycin, ciprofloxacin, erythromycin, and norfloxacin). *Campylobacter jejuni* can also cause arthritis. Another representative disease of water pollution is cholera caused by a bacterial agent called *Vibrio cholerae*. This is a somewhat comma shaped gram negative bacteria found generally in brackish waters. Contaminated water containing *V. cholerae* can cause diarrhea and vomiting within hours of ingestion. It secretes a substance called cholera toxin (an oligomeric proteinaceous complex composed of six protein subunits mediating its effects through adenylate cyclase activity) which then leads to frequent diarrhea. This organism is responsible for a significant number of deaths worldwide. Secondary to diarrhea, *V cholera* can also cause, dry mucous membranes, hypotension, slowed and diminished radial pulse, tachycardia, renal failure, seizures, coma and, in many cases, death. Frequent washing of hands with soap, drinking safe water, avoiding open defecation, cooking food thoroughly are some of the measures recommended for preventing cholera. A vaccine is also recommended for those who are supposed to travel to cholera prone areas.

Shigella is another group of gram negative, non-spore forming, rod shaped bacterial pathogens that cause shigellosis (a disease characterized by diarrhea, fever, and stomach cramps). The symptoms appear within a day or two after infection. Bacteria of genus *Shigella* are closely related to *E. coli*. This class of pathogens

(*Shigella*) is considered to be one of the leading causes of diarrhea worldwide with bacterial etiology. It is responsible for 80–160 million cases of diarrhea and around half a million deaths worldwide annually. Well defined recommendation of maintaining proper hygiene for prevention of shigellosis have been documented. There are many antibiotics to which the infection responds (https://www.nps.gov).

3.2 Viral Diseases

Viral disease that are spread by contaminated water primarily target the liver. The basic reason for that is the viruses mostly present in the contaminated and polluted waters are hepatotrophic in nature and, therefore, target the liver causing hepatic inflammation and hepatotoxicity. Hepatic inflammation and toxicity is covered under the umbrella term hepatitis. Hepatitis is a variable disease which sometimes presents with no symptoms while other times may present with a gamut of signs and clinical features. Sometime hepatitis produces acute symptoms and sometime relatively chronic condition is precipitated. One of the most common symptoms of hepatitis are yellow discoloration of skin and sclera, loss of appetite, vomiting, abdominal pain, diarrhea, tenderness on right side of abdomen, nausea, headaches etc. Two main types of hepatitis on the basis of its time course are chronic and acute hepatitis. In certain cases acute hepatitis may resolve on its own while in other cases it progresses into the chronic hepatitis. At times acute hepatitis can lead to acute liver failure. Persisted hepatitis leads to scarring of the liver tissue causing cirrhosis (which is untreatable and causes death), liver failure. In many instances chronic hepatitis can also cause liver cancer generally referred to as hepatocellular carcinoma. The best way to prevent hepatitis is to maintain proper hygienic conditions, avoid drinking untreated and contaminated water and also getting a vaccine. Since there are many types of hepatitis (depending on the type of virus on which the etiology is based) there are different vaccines for different hepatitis types (Ballester and Sunyer 2000).

Infective encephalitis is an inflammatory condition of the brain based on viral etiology in which a certain type of mosquito (termed Culex) acts as a vector. Encephalitis presents with symptoms akin to flu including fever, headache. Yet sometimes there are no symptoms. Encephalitis leads to confusion, nausea, seizures and problems in motor and sensory processes. Encephalitis is not generally life threatening but death can occur. The Culex mosquito lays its eggs (contaminated with the virus) on stagnant water surfaces. Consumption of the contaminated water causes encephalitis. It is important to mention that there is no vaccine available for viral encephalitis and hence the best measure to control is to stop mosquito breeding and maintaining recommended hygienic conditions.

One of the most prevalent and highly regulated campaigns run by the WHO is against polio. It is caused by the virus Poliomyelitis. Pulse polio immunization program has been regularly running in many countries throughout the world and it has helped to bring down the incidence of the disease drastically. Polio can lead to nausea, fever, vomiting and in certain cases paralysis (of one or more limbs). It also arrests muscle development.

In addition to the above mentioned viral diseases there are many other water borne viral diseases that need attention but are beyond the scope of this chapter. For gastroenteritis (presenting with fever, nausea and headache), the causative agents can be many including, but not limited to, rotaviruses and adenoviruses.

3.3 Parasitic Diseases

Parasitic disease are a group of ailments caused by certain organisms called parasites. Parasites are a heterogeneous group of organisms which infest/infect other living organisms and masquerade the host from nutrients and cause diseases also. Diseases caused by parasites are referred to as parasitosis. Cryptosporidiosis in one of the representative examples of a parasitic disease. It is caused by the parasite *Cryptosporidium pavum*. It is a protozoan intracellular parasite often leading to opportunistic infections due to precipitation of severe immunocompromised in their hosts. The organism lives inside the cells in the intestinal lumen and is spread through the oral fecal route. It easily contaminates water. Crypto (as the disease and the parasite is generally referred to as) can be asymptomatic or may present with symptoms akin to diarrhea and gastrointestinal upset. Another parasite called *Entamoeba histolytica* also causes disease with similar symptoms. The diseases may be referred to as amoebiasis or Galloping amoebia. This organism mainly affects the lining of the stomach. Since its life cycle is accompanied by a cyst formation stage, the cysts easily contaminate food and water and, therefore, can infect the host through contaminated water. Once infection shows up, symptoms like fever and diarrhea are generated within hours or days. Another disease to mention here is giardiasis caused by the parasite *Giardia lamblia* also known as *Giardia intestinalis.* It is a flagellated parasite that lives and reproduces in the small intestine. By means of an adhesive disc, it attaches to the epithelial wall of the intestinal cells. Giardiasis commonly spreads by ingestion of untreated sewage. There are around 20,000 cases of giardiasis reported in the United States annually. Its spread is preventable like many other water borne diseases.

4 Conclusions

Water is indispensable to life. All life happens in the medium of water which includes the terrestrial forms also. So the proportion of all living organisms is more than 70% water. Hence, water is important and yet may be a major source of infections. This problems has seen an upsurge in modern times because of industrialization, urbanization and population density growth. Additional problems have led to compromise in quality of water. These problems include use of polythene bags, plastic usage, change in lifestyle, non-treatment of sewage, poor regulations and government policies, poverty and poorly educated populations. It appears that almost all waterborne diseases are preventable (though vaccines are available for many of them). Effective government policies, maintenance of good hygiene and personal discipline are the major factors that will prevent water pollution and spread of water borne diseases.

References

Ahada, C. P. S., & Suthar, S. (2018). Groundwater nitrate contamination and associated human health risk assessment in southern districts of Punjab, India. *Environmental Science and Pollution Research International, 25*(25), 25336–25347.

Ahmad, S. M., Yusafzai, F., Bari, T., et al. (2014). Assessment of heavy metals in surface water of River Panjkora Dir Lower, KPK Pakistan. *Journal of Biological and Environmental Sciences, 5*, 144–152.

Ahmed, T., Scholz, F., Al-Faraj, W., et al. (2013). Water-related impacts of climate change on agriculture and subsequently on public health: A review for generalists with particular reference to Pakistan. *International Journal of Environmental Research and Public Health, 13*, 1–16.

Alrumman, S. A., El-kott, A. F., & Kehsk, M. A. (2016). Water pollution: Source and treatment. *American Journal of Environmental Engineering, 6*(3), 88–98.

Ballester, F., & Sunyer, J. (2000). Challenges to public health in the new millennium. *Journal of Epidemiology and Community Health, 54*, 2–5.

Bibi, S., Khan, R. L., Nazir, R., et al. (2016). Heavy metals in drinking water of Lakki Marwat District, KPK, Pakistan. *World Applied Sciences Journal, 34*(1), 15–19.

Briggs, D. (2003). Environmental pollution and the global burden of disease. *British Medical Bulletin, 68*, 1–24.

Chowdhury, S., Annabelle, K., & Klaus, F. Z. (2015). *Arsenic contamination of drinking water and mental health* (pp. 1–28). London: Centre for Economic Policy Research.

Corcoran, E., Nellemann, C., Baker, E., et al. (2010). *Sick water? The central role of wastewater management in sustainable development. A Rapid Response Assessment*. Arendal: United Nations Environment Programme.

Currie, J., Joshua, G. Z., Katherine, M., et al. (2013). Something in the water: Contaminated drinking water and infant health. *Canadian Journal of Economics, 46*(3), 791–810.

Desai, N., & Smt Vanitaben. (2014). A study on the water pollution based on the environmental problem. *Indian Journal of Research, 3*(12), 95–96.

Ebenstein, A. Y. (2008). *Water pollution and digestive cancer in China*. Institutions and governance programs, pp. 1–45.

Faiq, M. A., Kumar, A., Singh, H. N., Pareek, V., & Commentary, K. P. (2018). A possible mechanism of Zika virus associated microcephaly: Imperative role of retinoic acid response element (RARE) consensus sequence repeats in the viral genome. *Frontiers in Microbiology*. https://doi.org/10.3389/fmicb.2018.00190. eCollection 2018.

Halder, J. N., & Islam, M. N. (2015). Water pollution and its impact on the human health. *Journal of Environment and Human, 2*(1), 36–46.

Ho, Y. C., Show, K. Y., Guo, X. X., et al. (2012). Industrial discharge and their affects to the environment. In *Industrial waste* (pp. 1–32). London: InTech.

http://research.gsd.harvard.edu/hapi/

http://www.in.gov/isdh/22963.htm

https://www.nps.gov

Jabeen, S. Q., Mehmood, S., Tariq, B., et al. (2011). Health impact caused by poor water and sanitation in district Abbottabad. *Journal of Ayub Medical College, Abbottabad, 23*(1), 47–50.

Juneja, T., & Chauhdary, A. (2013). Assessment of water quality and its effect on the health of residents of Jhunjhunu district, Rajasthan: A cross sectional study. *Journal of Public Health and Epidemiology, 5*(4), 186–191.

Kamble, S. M. (2014). Water pollution and public health issues in Kolhapur city in Maharashtra. *International Journal of Scientific and Research Publications, 4*(1), 1–6.

Khan, M. A., & Ghouri, A. M. (2011). Environmental pollution: Its effects on life and its remedies. *Journal of Arts, Science and Commerce, 2*(2), 276–285.

Khan, N., Hussain, S. T., Saboor, A., et al. (2013). Physiochemical investigation of the drinking water sources from Mardan, Khyber Pakhtunkhwa, Pakistan. *International Journal of Physical Sciences, 8*(33), 1661–1671.

Khurana, I., & Sen, R. (2008). *Drinking water quality in rural India: Issues and approaches-Water Aid.* India water Portal.

Krishnan, S., & Indu, R. (2006). *Groundwater contamination in India: Discussing physical processes, health and sociobehavioral dimensions.* IWMI-Tata, Water Policy Research Programmes, Anand, India.

Kumar, A., Singh, H. N., Pareek, V., Raza, K., Dantham, S., Kumar, P., Mochan, S., & Faiq, M. A. (2016). A possible mechanism of Zika virus associated microcephaly: Imperative role of retinoic acid response element (RARE) consensus sequence repeats in the viral genome. *Frontiers in Human Neuroscience, 9*(995), 403. https://doi.org/10.3389/fnhum.2016.00403. eCollection 2016.

Lu, Y., Song, S., Wang, R., et al. (2015). Impacts of soil and water pollution on food safety and health risks in China. *Environment International, 77*, 5–15.

Nel, L. H., & Markotter, W. (2009). New and emerging waterborne infectious diseases. *Encyclopedia of Life Support System, 1*, 1–10.

Owa, F. D. (2013). Water pollution: Sources, effects, control and management. *Mediterranean Journal of Social Sciences, 4*(8), 65–68.

Pawari, M. J., & Gawande, S. (2015). Ground water pollution & its consequence. *International Journal of Engineering Research and General Science, 3*(4), 773–776.

Salem, H. M., Eweida, E. A., & Farag, A. (2000). Heavy metals in drinking water and their environmental impact on human health. *ICEHM, 2000*, 542–556.

Ullah, S., Javed, M. W., Shafique, M., et al. (2014). An integrated approach for quality assessment of drinking water using GIS: A case study of Lower Dir. *Journal of Himalayan Earth Sciences, 47*(2), 163–174.

Freshwater Contamination: Sources and Hazards to Aquatic Biota

Nighat Mushtaq, Dig Vijay Singh, Rouf Ahmad Bhat, Moonisa Aslam Dervash, and Omar bin Hameed

Abstract Water is an essential compound for supporting the biota on earth. There are various sources of water to support the life in freshwater ecosystems. Freshwater ecosystems play magical role as they provide services to support life process in living creatures. But, growing population, urbanization, industrialization put drastic pressure on the freshwater ecosystems, with the result altered the quality scenario of freshwaters by adding huge quantities of contamination. Water contaminations not only degrade the quality of freshwater, but simultaneously pose harmful risks to the whole environment. The chemical substances in freshwater ecosystem can't be neutralized easily due to their complex structure and have great potential to remain intact in any kind of environments. These substances nowadays are continuously added into the freshwater ecosystem on daily basis by way of discharging untreated domestic, industrial and agricultural wastewater. Most of these substances get accumulated in the bottom sediments and very minute concentration in the form of organic and inorganic constituents remain either in suspended form or solution in liquid medium of freshwater ecosystem. Presences of these kinds of pollutants in freshwater ecosystem have long-term impacts on aquatic and associated biota. Therefore, need of an hour is to monitor the quality of freshwater ecosystem on regular basis and focus should be given to the treatment of effluents prior to its discharge into the freshwater ecosystem.

Keywords Aquatic ecosystem · Freshwater · Eutrophication · Toxic substances · Health hazard · Effluent

---

N. Mushtaq
Division of Vegetable Science, Sher-e-Kashmir University of Agricultural Sciences and Technology, Jammu, Jammu and Kashmir, India

D. V. Singh · R. A. Bhat (✉) · M. A. Dervash
Division of Environmental Science, Sher-e-Kashmir University of Agricultural Sciences and Technology, Jammu, Jammu and Kashmir, India

O. b. Hameed
College of Natural Sciences, Arba Minch University, Arba Minch, Ethiopia

H. Qadri et al. (eds.), *Fresh Water Pollution Dynamics and Remediation*,
https://doi.org/10.1007/978-981-13-8277-2_3

1 Introduction

Water is considered as an essential natural resource and covers almost entire surface of earth and contributing 70% in the form of fresh waters and oceans and the availability varies being only 3% by fresh water bodies (Dikio 2010; Adesuyi et al. 2015). Worldwide, lakes and rivers in the form of fresh water embrace 10^5 Km^3 that counts 10^{-4} less than total water on earth (Jackson et al. 2001). Water bodies are important as they contribute major source of water for sustaining life on earth and its scarcity is likely to affect agricultural as well as domestic life. Rapid population growth during recent years has led to increased use of freshwaters by 10% from 2000 to 2010 (Vorosmarty et al. 2005; Rijsberman 2004) major portion of fresh waters goes to irrigation of fields (40%) globally with 20–40 L is used for consumption per day. Freshwater flora and fauna is prime source of valuable items and services for human use (Halder and Islam 2015), contributes 42% of total fish (Lundberg et al. 2008), 25% of fresh water molluscus Rijsberman and Moden (2001), some percentage of phytoplanktons and crabs Water bodies serve as an essential input for generating power and for disposal of sewage and industrial effluents (Eunice et al. 2017) but at the same time their regular disposal has generated environmental concerns as it has led to pollution problems and has become threat to aquatic flora and fauna and is considered unfit for human use (Ekubo and Abowei 2011; Dulo 2008). Water is contaminated mainly by chemical, physical, biotic factors coming from various sources (Richardson et al. 2007). There are many defining causes of water being getting polluted like discharge of field (Bhat et al. 2012) and industrial debris to waters, leakage of oil from oil tankers, use of excessive fertilizers and pesticides for crop protection, sewage sludge and much more (Aboyeji 2013; Bhat et al. 2017). Major pollution of water bodies is due to overcrowding in urban areas activities like polythene bags directly released to rivers and streams (Bhat et al. 2014, 2018), excessive use of chemicals, dirt, dust and debris (Master et al. 1998; Carpenter et al. 1998; Kalff 2002; Moss 2008). Water bodies have capacity to assimilate pollutants to the extent that may be hazardous to the life of biota in the form of number, variety or structure, the term known as assimilative capacity (Adekunle 2009; Adekunle and Eniola 2008).there exists a positive correlation between pollutants and health of organisms (Otukunefor and Obiukwu 2005). When excess pollutants are being discharged they compete for dissolved oxygen because there is conversion of organic substances to inorganic ones requiring oxygen for transformation creating a condition of biological oxygen demand, loss of biodiversity and eutrophication (Beeby 1993). Among aquatic biota, fishes are being severely affected as metal ions released from industries cause respiratory breakdown in them (Arimoro 2009), eg. Excessive release of iron cause iron clog in the gills of fishes and their consumption by humans affects human health (Ahmed et al. 2013). Pathogens are exuded from untreated sewage (Helmer and Hespanhol 1997) and many harmful substances from nuclear plants (Master et al. 1998). This has led to outbreak of numerous harmful diseases (APHA 1996) and has led to shortage of potable water therby affecting humam health (Eunice et al. 2017). Water pollution has given rise to health issues like discharge of faecal matter to water resources has led to infectious fecal-oral route disorders (Nel and Markotter 2009), cancer, lung diseases, abdominal disorders, heart problems, low birth rates, etc. (Ullah et al. 2014). Recent survey has concluded that water pollution is affecting at a greater rate in rural areas than in urban regions due to lack of infrastructures for treating water before its use (Jabeen et al. 2011; Currie et al. 2013). More over poor quality water used for irrigating agricultural fields has affected our economy by producing poor quality produce (Khan and Ghouri 2011). Renewable fresh water is an indispensable resource for life and deserves special attention because it is very impaired and seriously threatened by human activities (Togue et al. 2017). In fact, population growth accompanied by rapid urbanization is major factor for causing disturbances (Fig. 3) in natural environments (Kinney 2002; Postel and Richter 2012).

2 Sources of Water Pollution

2.1 Sewage

Sewage consists of all waste material coming out from homes, industries, schools, farms, cities and towns and is mixture of degradable and non degradable substances with major portion consisting of human excreta (Cumberlidge et al. 2009; Chowdhury et al. 2015). Wastewaters are an assortment of likely organic and inorganic constituents associated with minute quantity of cultural substances (human excreta) and sewage from all point and non-point sources (Sulaiman et al. 2016). Large portion of industrial waste waters comes from big cites and in combination of domestic waste waters known as municipal waste waters (de Mora and Harrison 2013). Biodegradable wastes come from agriculture farm lands, pastures, cowsheds and dairy farms (Vikranthpridhvi and Musalaiah 2015). Almost all domestic sewage from cities and towns is discharged into rivers and streams sometimes untreated sewage are discharged into freshwater ecosystems (Banks et al. 1997; Barker and Stuckey 1999; Baron and Poff 2004; Burton and Pitt 2001) and cause freshwater pollution (Haseena et al. 2017) as shown in Fig. 1. Freshwater pollution are responsible for enhancing bacterial load in water bodies, which are directly and indirectly and trigger for serious health hazards (Desai and Vanitaben 2014). Most of water body pollution (70%) is due to domestic wastes (Sankhla et al. 2018) as it contains toxicants, solid and liquid wastes, polythene bags, (Haseena and Malik 2017) (Fig. 2).

2.2 Agricultural Waste

Widespread use of fertilizers and pesticides for increasing production of agricultural crops and improved livestock practices for doubling farmers income have adverse effect on quality of water used for drinking and irrigation purposes

(Mali et al. 2015) via flushing into lakes and rivers. This has also led to ground water contamination through seepage (Anonymous 2002). Pesticides pose major threat to aquatic biota because of their rapid fat solubility action and their accumulation in non target organisms (Dutta et al. 2009), thereby having harmful effect on their lives (Dapena-Mora et al. 2007; Das et al. 2010). Animal, human wastes and chemical pesticides used in agriculture or released directly into ponds constitute the major sources of pollutants of the water (Anonymous 2002).

2.3 *Petroleum Slicks*

Transportation of petroleum tanks may sometimes cause leakage of pipes and tapes and thus are ultimately discharged into water bodies which result in a condition known as anoxia as the living organisms in water bodies and they fail to take up dissolved oxygen due to the presence of oil film formed on the surface (Arora and Kakkar 2017).

2.4 *Nuclear and Thermal Pollution*

Thermal power plants contribute major sources power generation in India, but are also an important cause of environmental pollution, because it uses coal as fuel as it is the major fossil fuel resource in India for producing electricity (EEB 2000; Elhatip and Gullu 2005). Most of the coal based thermal power plants produce as

Fig. 1 Discharge of untreated domestic sewage into river ecosystem

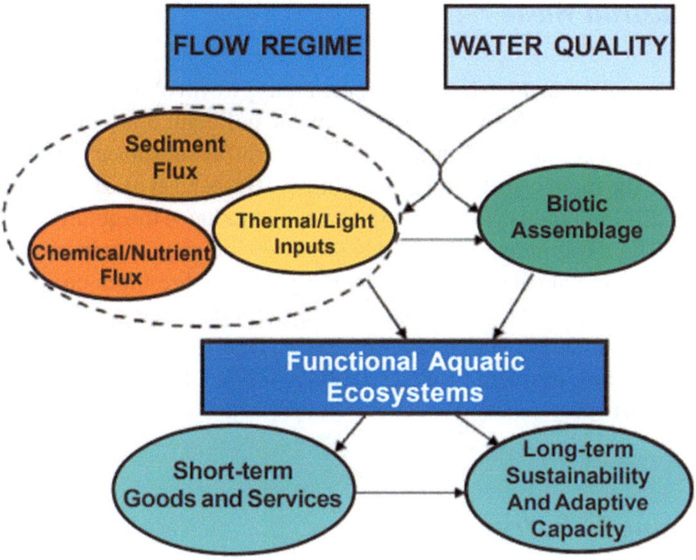

Fig. 2 Major forces that influence freshwater ecosystems. (Baron et al. 2003)

much as 70% of electricity in India (Dhadse and Bhagia 2008). Fly ash produced in such power plants is threatening aquatic life as well (Dart and Stretton 1980; Nair 1985) as it decreases dissolved content of oxygen in water and responsible for loss of ecological balance (Rao and Ravindhranath 2013; Arora and Kakkar 2017).

2.5 *Pesticides and Fertilizers*

The term pesticides include wide range of substances like insecticides, fungicides, growth regulators etc. and are now widely used for plant protection and their excessive usage has created havoc to the life forms on earth. These are directly or indirectly discharged to waters, polluting it and render it unsafe for use (Sankhla et al. 2018; Khurana and Sen 2008). Leached out chemicals after drained to water bodies lead to eutrophication (Sankhla et al. 2018) or get accumulated in fruits or vegetables that are consumed as such (Ebenstein 2008; Kamble 2014). Though pesticide usage cannot be completely controlled but can be managed efficiently to overcome its harmful effects (Yonglong et al. 2015; Khurana and Sen 2008). Pesticide exposure to the drinking water can be checked as given by NAS (National Academy of Sciences 1993) to prevent risks on biotic life.

2.6 Industrial Waste

With rise in population, there has been tremendous growth of industries and factories and produce large quantities of smoke as major pollutant and contributes 25% of overall pollution (Desai and Vanitaben 2014). Harmful substances discharged from industries are the main causes of water pollution (Haseena et al. 2017). Toxic metals released by different industries enter into water and reduced the quality of freshwater (Ho et al. 2012). The industrial mishaps have now become a matter of concern as these are rich in metal substances *viz.,* mercury, leads, chromium etc., that are very harmful to biota and sometimes make water unfit for drinking purposes (Owa 2013; Arora and Kakkar 2017; Haseena et al. 2017). Due to these reasons most of the population is facing problem of water shortage (EPA 2002; Farenzena et al. 2005; Garcia et al. 2014) .

2.7 Mining Industry

Large quantities of wastes are also produced by mining industry during the process of extraction and manufacturing (Das and Choudhury 2013). During the process of extraction waste substances pollute the water used in the process making it acidic which dissolves the metals from residue deposits (Musingafi and Tom 2014). The left out material results in both surface as well as ground water pollution leading to destruction of soil profile and soil quality (Adler and Rascher 2007). Mining of heavy metals pollute the surrounding environment and prove hazardous to human as well as animal life (Hetrick et al. 2000; Gong et al. 2008). Thus this uncontrollable discharge is a matter of concern and requires serious attention to prevent its ill effects on environment.

2.8 Population Growth and Urbanization

With increase in population there have been many issues with pollution to environment and is having negative influence on it (Ho et al. 2012). There is generation of large quantities of wastes, both solid and liquid wastes, by growing humans (Jabeen et al. 2011) and are being directly thrown to rivers, canals and streams. Contaminated waters are sources to emerging bacterial and other microbial diseases (Desai and Vanitaben 2014). Plastic bags containing large amount of wastes are also being discharged to fresh water sources and contribute major source of fresh water pollution (Desai and Vanitaben 2014; Sankhla et al. 2018). Population explosion, carrying capacity, in adequate management of waste and wastewater are major concerns to deplete fresh water qualities and create grave health issues, in urban localities (Kamble 2014).

3 Impact of Pollution on Water Physico-Chemical Characteristics

3.1 pH

It is an important factor to describe water health (Fakayode 2005) and play a crucial role in bio-chemical life cycles in aquatic environs (Chapman 1996; Lokhande et al. 2011; Smitha et al. 2013). Variation of pH from acidic to basic and vice versa could have deadly impact on aquatic biota. Aquatic organisms are sensitive to pH changes and biological treatment requires pH control or monitoring (Lokhande et al. 2011). Availability of heavy metals in water ecosystems are also affected by alteration of water pH. Thus, pH is having primary importance in deciding the quality of wastewater effluent (Lokhande et al. 2011). The high pH is probably due to the direct disposal of refuse into the water body and also to sea water intrusion (Ogbonna 2014). Water with a pH outside the normal range may cause a nutritional imbalance or may contain a toxic ion which can adversely affect the growth and development of aquatic life (Bolawa and Gbenle 2012). It is a known fact that variations in pH affect chemical and biological processes in water and low pH increases the availability of metals and other toxins for intake by aquatic life and high pH may be due to the presence of other pollutants introduced into the water (Ogbonna 2014).

3.2 Electrical Conductivity

It's value in aquatic environs depends on the concentration furnishing ions as well as the water temperature (Uqab et al. 2017). Electrical conductivity is increasing due to the influence of sewage water coming out from hotels coming out from kitchens, bathrooms and washrooms (Bhat and Pandit 2001). Increase in levels of electrical conductivity and cations and input of sewage water may be the result of decomposition and mineralisation of organic materials (Abida and Harikrishna 2008).

3.3 Temperature

It is a well known fact that the pace at which chemical reactions occur amplify with rise in temperature and the rate of biochemical processes usually double for every 10.0 °C rise in temperature (Ogbonna 2014). The temperature of untreated discharged from domestic sources sewage ranges between 8 and 12 °C in winter to 17–20 °C in summer (Sun et al. 2015). Increased temperature increases respiration leading to increased oxygen consumption and increased decomposition of organic matter (Pierce et al. 1998). Hence, population of bacteria and phytoplankton would double in warm weather in a very short time (Chapman 1996). The temperature plays an important role in the metabolic activities of the organisms and is considered as a biologically most significant factor (Varunprasath and Daniel 2010).

3.4 Suspended and Total Dissolved Solids (TDS)

Suspended solids consist of materials originating from the surface of the catchment area, eroded from river banks or lake shores and suspended from the bed of the water body (Chapman 1996). It includes vey minute substances *viz.,* silts, clays, zooplankton, phytoplankton and dead particulate matter (Davis and Day 1998; Mahananda 2010). According to Lester and Birkett (1999), suspended solid values of less than 25 mg/l have no harmful effect on fisheries. High suspended solids might beo due to run off from many bathing Ghats, drain water discharge (Pawar and Vaidya 2012). Suspended solids generally cause damage to fish gills affecting their oxygen consumption and ultimately causing death at high concentrations (Ogbonna 2014). Total dissolved solid depends on various factors such as geological character of watershed, rainfall and amount of surface runoffs and gives an indication of the degree of dissolved substances (Smitha et al. 2013). Higher concentrations of total dissolved solids (TDS) in sewage water may be attributed to higher concentrations of carbonates, bicarbonates, chlorides, sulphates, phosphates, nitrates, nitrogen and calcium (Kannan et al. 2004).

3.5 Dissolved Oxygen

High TDS, BOD and COD content cause decrease in DO of the water system creating stress condition to the aquatic living organisms (Kambole 2003). Dissolved organic contents consume a large amount of oxygen and increase BOD level which leads to anaerobic fermentation and produces organic acids and hydrolysis of these organic acids causes the decrease in pH values (Ahmed et al. 2011). According to Cunningham and Saigo (1999), the addition of certain organic materials due to discharge of sewage into water stimulates oxygen consumption by decomposers. Dissolved Oxygen thus indicates the ability of water body to support aquatic life and volume of oxygen present in the water (Sharma and John 2009; Shivayogimath et al. 2012; Smitha et al. 2013). Low oxygen in water can kill fish and other aquatic organisms (Smitha et al. 2013). Lower values of dissolved oxygen may be due to the organic matter in sewage which leads to the rapid decrease in this oxygen availability (Gray 2004). Ahipathy and Puttaiah (2006) reported that lower values of dissolved oxygen may be due bioaccumulation, biomagnifications and active utilization in bacterial decomposition of organic matter (Thilaga et al. 2005; Lone et al. 2017).

3.6 Biological Oxygen and Chemical Oxygen Demand

Biochemical Oxygen Demand (BOD) is used as an index for determining the amount of decomposing organic materials as well as the rate of biological activities in waste because oxygen is required for respiration by microorganisms involved in the decomposition of organic materials (Nartey et al. 2012). The slightly high BOD values may be attributed to the discharge of organic waste into water bodies resulting in the uptake of DO in the oxidative breakdown of these wastes (Akuffo 1998; Milovanovic 2007). High BOD values may be attributed to the percolation of waste water loaded with biodegradable compounds (Pitchammal et al. 2009), which might be the result of untreated sewage, solid and industrial waste discharge directly into water (Milovanovic 2007). Sources of BOD in aquatic environment include leaves and woody debris, dead plants and animals, animal manure, industrial effluents, wastewater treatment plants, feedlots, and food-processing plants, failing septic systems, and urban storm water runoff (Lokhande et al. 2011). In untreated effluent, high BOD may be due to fibre residues and suspended solids (Yusuff and Sonibare 2004; Pawar and Vaidya 2012).

Chemical Oxygen Demand (COD) is the quantity of O_2 needed for the oxidation of inorganic constituents in water environs (Eunice et al. 2017). The higher level of COD in the effluent discharged indicate that it contains high oxygen demanding materials, which causes depletion of dissolved oxygen in water thereby limiting its use for other purposes (Hogan 2014; Jeswani and Mukherji 2015; Kahiluoto et al. 2015; Kozai et al. 2014), such as irrigation and recreational purposes (Eunice et al. 2017). Thus, high BOD and COD levels in water indicated that the water is highly polluted (Eunice et al. 2017) (Fig. 3).

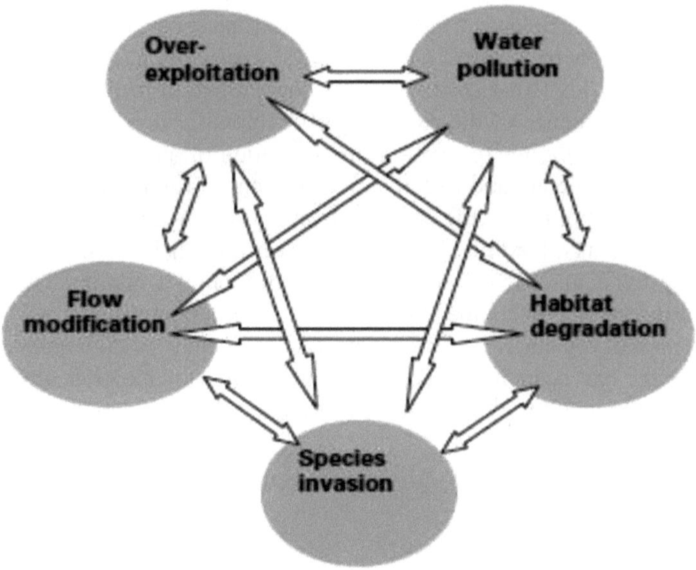

Fig. 3 Major threats to freshwater biodiversity. (Dudgeon et al. 2006)

3.7 *Chloride*

High amount of Cl^- due to discharge of untreated sewage into water reacts with Na^+ turns the water salty and also enhances dissolved solids (Little 1971; NRCC 1977; Malik et al. 2012; Smitha et al. 2013). Chloride is present in all types of water and its concentration shows the presence of sewage pollution (McConnell et al. 1993; Smitha et al. 2013). Higher concentration of chloride in water may result from the higher usage of washing agents like detergents, soaps and faecal matter (Sawhney 2008).

4 Impact of Increased Nitrogen Content on Freshwater

Nitrate in water is an important factor for water quality assessment (Jones and Burt 1993). The presence of nitrate in freshwater may be due to the result of waste being disposed off at the dumpsites or from agro- based industries (Ogbonna 2014). Excess amounts of nitrogen availability in water systems lead to eutrophication and algae blooms (Smitha et al. 2013). The natural background levels of nitrate may come from rocks, land drainage and plant and animal matter and extremely high concentration of nitrate in freshwater body is toxic (Ogbonna 2014). Invariably, nitrate is seldom abundant in natural surface water because it is incorporated into cells and chemically reduced by microbes and converted into atmospheric nitrogen (Chapman 1996). This phenomenon may account for the low concentration of nitrate in surface waters (Ogbonna 2014). Nitrate is highly oxidized form of nitrogen compound and is commonly present in surface and ground waters (Uqab et al. 2017), since it is final product of aerobic decomposition of organic nitrogenous matter (Bartram and Balance 1996). Nitrate concentration in surface and ground waters is usually low but can reach high levels as a result of leeching or runoff from the agricultural fields (Bhat et al. 2017) or contamination from human or animal wastes as a consequence of the oxidation of ammonia and similar sources (Uqab et al. 2017). Water nitrates result from the oxidation of ammoniacal nitrogen and nitrites and the mineralization of the river biomass (Togue et al. 2017). Major sources of nitrate pollution (Fig. 4) vary from agrochemicals, human and animal wastes, sewage leaks, landfills, application of wastewater for irrigation, industrial wastes (Adesuyi et al. 2015). Nitrates are generally found in nature they are the end product of the aerobic decomposition of organic nitrogenous matter as well as the decomposition of organic micro-organisms (Adesuyi et al. 2015) Contamination in drinking water has been implicated to be the causes of major health problems (Krishnan and Indu 2006; WHO 2007; Jagessar and Sooknundun 2011).

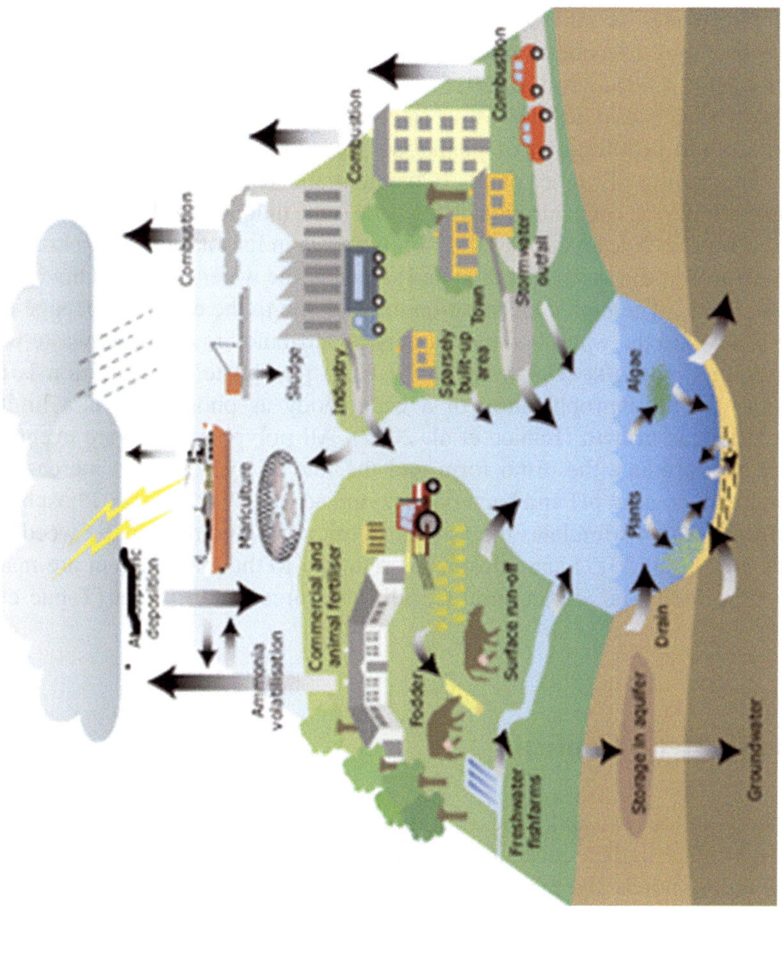

Fig. 4 Various sources of nitrogen pollution in water

5 Impact of Increased Phosphorus Content on Freshwater

Phosphorus is an essential nutrient and can exist in water in both dissolved and particulate forms (Ogbonna 2014). It gets into the water through various sources including leached or weathered soils from igneous rocks, phosphates from detergents in industrial effluents, run offs from fertilized farm lands and domestic sewage containing human excrement (Abowei et al. 2010; Bhat et al. 2017). Its compounds are used in detergent formulation as water softeners (Eunice et al. 2017). Elevated level of phosphates in surface water is one of the most serious environmental problems because of its contribution to the eutrophication process and impairment of water quality (Adesuyi et al. 2015). The elevated phosphate concentration in water have been linked to increasing rates of plant growth, changes in species composition and proliferation of planktonic and epiphytic and epibenthic algae, resulting in shading of higher plants (Chapman and Kimstach 1996). A plausible reason underlying the concentration differential is the unique behavior of phosphorus in shallow waters (Adesuyi et al. 2015) Phosphorus in its soluble state (phosphate) quickly adsorbs at the surface of mud and re-enters the water column (Onwugbuta-Enyi et al. 2008). The high concentration may be due to the effect of seepage from the dumpsites into the water bodies and may be attributed to domestic waste water and agricultural run-offs (Ogbonna 2014). High phosphate concentration is also responsible for the eutrophication of a water body as phosphorus is a limiting nutrient for algae growth (Eunice et al. 2017). All polyphosphates are eventually hydrolysed to produce the ortho form and the rate of hydrolysis is increased by temperature, decreased pH and bacterial enzyme action (WHO 2004). Phosphate, a plant nutrient which stimulates the growth of microbes, molds, aquatic weeds and algae (Eunice et al. 2017) and phosphates returned to the environment are mainly derived from industrial, agricultural (fertilizer) sources and excreta (Togue et al. 2017).

6 Heavy Metal Content

Atmospheric fallout is usually the most important source of lead in the freshwaters (Paar 1998). The higher concentration of cadmium and lead can be due to the potential tailing ponds and the consequences of mining and unregulated landfills in the catchment area (Georgieva et al. 2018). Acute toxicity of Pb in invertebrates is reported at concentration of 0.1–10 mg/L (Paar 1998). Iron is the fourth most abundant element in the earth's crust and is considered as a main factor determining the adsorption capacity (Wright and Welbourn 2002). The presence of high concentration of Fe may increase the hazard of pathogenic organisms; since most of these organisms need Fe for their growth (Centre for Ecological Sciences 2001). Copper is highly toxic to most fishes, invertebrates and aquatic plants than any other heavy metal except mercury by reducing growth and rate of reproduction in plants and animals (Lokhande et al. 2011). Aquatic plants absorb three times more Cu than plants on dry lands (Moore and Ramamoorthy 1984). Excessive Cu content can cause damage to roots, by attacking the cell membrane and destroying the normal membrane structure; inhibited root growth and formation of numerous short, brownish secondary roots (Lokhande et al. 2011). Cr compounds are used as pigments, mordents and dyes in the textiles and as a tanning agent in the leather (Lokhande et al. 2011). Acute toxicity of Cr to invertebrates is highly variable, depending upon species (Paar 1998). Cr is generally more toxic at higher temperatures and its compounds are known to cause cancer in humans. The toxic effect of Cr on plants indicate that the roots remain small and the leaves narrow, exhibit reddish brown discoloration with small necrotic blotches (Moore and Ramamoorthy 1984). Cd is contributed to the surface waters through paints, pigments, glass enamel, deterioration of the galvanized pipes etc. (Lokhande et al. 2011). It is less toxic to plants than Cu similar in toxicity to Pb and Cr and is equally toxic to invertebrates and fishes (Paar 1998). Chromium is a potential pollutant and well known for its mutagenicity (Cheng and Dixon 1998) and carcinogenicity (Wang et al. 1999) effects in humans, animals and plants. Soil profile, surface water bodies (ponds and rivers), human health, fishes and other aquatic biodiversities are at risk of serious threat due to the extensive use of chromium in tanning industries and discharge of wastewater (Mohanta et al. 2010). Cu, Zn, and Pb are closely associated with crude oil and municipal wastes disposal (Chindah et al. 2004). These metals tend to increase in less saline and highly turbid media (Selvaraj et al. 2010; Wright and Welbourn 2002). The enrichments of heavy metals in water due to the discharge of waste are affecting the properties of the water (Shah et al. 2013; Singh et al. 2011). Discharge of pollutants directly in the water is deteriorating the water quality by decreasing the dissolved oxygen content of water thus making it unfit for aquatic life (Bulut and Aksoy 2008). The various potential sources of heavy metals pollution in any environment are presented in Table 1.

Among the pollutants, toxic metals are of serious concern because they accumulate through the food chain and create environmental problems (Praveena et al. 2010; Paul and Sinha 2015). Higher concentrations of heavy metals can form harmful complex compounds, which critically effect different biological functions (Rajbanshi 2009). The presence of heavy metals in the wastewater of industry is a

Table 1 Potential industrial and agricultural sources for metals in the environment. (Fifield and Haines 2000; Yahya et al. 2018)

S. No.	Metal	Sources
1	Fe	Textiles, paints, fuels and refineries
2	Zn	Pesticides, fertilizers, glass, fuel, etc.
3	Pb	Fertilizers, acid-batteries, alloys, wear and tear of tyres and plastic, vehicles
4	Cd	Batteries and electrical; pigments and paints; alloys and solids; fuel; plastic; fertilizers
5	Ni	Batteries and electrical; pigments and paints; alloys and solids; fuel; catalysts; fertilizers
6	Cu	Vehicles, electrical gadgets, utensils, catalysts
7	Cr	Fertilizers, pigments

potential risk to aquatic ecosystem, animal, and human. High concentrations of heavy metals often pose a serious threat to biota and the environment of any ecosystem (Cheng 2003). Heavy metal pollution can be a much more serious problem because they cannot be degraded by natural processes and persist in soil and sediment from where they are released gradually into water bodies as sink. "Heavy metals" is a collective term, which applies to the group of metals and metalloids with a atomic density greater than 4 g/cm 3, or 5 times or more, greater than water (Hawkes 1997). Heavy-metal contamination is not a modern problem arising out of industrialization e it began when humans started processing ores (Renberg et al. 1994; Sharma et al. 2003).

Generally, most of the heavy metals enter the in river from different sources (Lettinga 1995; Maekawa 2003; Mehmood et al. 2019), it be can be either natural by erosion and weathering and or anthropogenic (Gupta et al. 2013; Sheykhi Moore 2016). In view of the intense human activity, natural sources of heavy metals from leaching and weathering of rocks in the environment are usually of little importance (Dixit et al. 2015). The most important anthropogenic sources of heavy metal are various industries, vehicles and domestic sewage (Paul 2017; Singh et al. 2018). The practice of discharging waste from industries and untreated domestic sewage into the aquatic ecosystem is continually going on that leads to the increase in the concentration of heavy metals in river water (Wang et al. 2011; Capangpangan et al. 2016). The industries which attribute heavy metals in river water are generally metal industries, paints, pigment, varnishes, pulp and paper, tannery, distillery, rayon, cotton textiles, rubber, thermal power plant, steel plant, galvanization of iron products and mining industries as well as unsystematic use of heavy metal-containing pesticides (Morrison et al. 2001; Mohan et al. 2006; Ogbonna et al. 2008; Parveen et al. 2017) and fertilizer in agricultural fields (Suthar et al. 2009; Sindern et al. 2016). These heavy metals have accumulative effect at the low level in drinking water and ground water (Prabha and Selvapathy 1997). Ansari et al. (1999) determined the concentrations of Al, Cd, Co, Cr, Cu, Fe, Mn, Ni, Pb, Sn and Zn in sediments of the river Ganga and according to them about 90% of the contents of Cd, Cr, and Sn; 50–75% of organic carbon, Cu and Zn; and 25% of Co, Ni and Pb in sediments are derived from the anthropogenic input in relation to the natural background values. Sarkar et al. (2007) also analyzed the level of dissolved heavy metals such as Fe, Zn, Mn, Cu, Pb, Hg at three ecologically distinct zones along the course of the river Ganga- Babughat, Diamond Harbour and Gangasagar in West Bengal and reported high values for Hg and Pb which can be attributed to the discharge from pulp and paper manufacturing units and to atmospheric input (Pretty 2002; Pulles et al. 2005; Ram et al. 2007; Rast 2009) and runoff of automobile emission (Singh et al. 2018; Rashid et al. 2019). The most important heavy metals from the point of view of water pollution are Zn, As, Cu, Pb, Cd, Hg, Ni and Cr as some of these metals (e.g. Cu, Fe, Mn, Ni, and Zn) are required as nutrients in trace amount for life processes in plants and microorganisms but become toxic at higher concentrations (Paul 2017) while other such as Pb, Cr, and Cd has no known biological function but are toxic elements (Ghannam et al. 2015). These heavy metals are not readily degradable in nature and accumulate in the animal as well as human bodies to a very high toxic amount leading to undesirable effects beyond a certain limit (Adakole and Abolude 2012: Govind and Madhuri 2014). The fatal diseases such as eyelid edema, nephritis, renal tumor, extensive lesions in the kidneys, anuria, nasal mucous membranes (Pei et al. 2012; Pawari and Gawande 2015) and pharynx congestion, increase blood pressure and cardiovascular diseases, osteoporosis, cancer, headache (Sartor et al. 1974; Salem et al. 2000; Scannell and Duffy 2007; Wosnie and Wondie 2014; Yuan et al. (2014) and malfunctions of different systems of the body caused by heavy metals (Jaishankar et al. 2014; Solenkova et al. 2014). They are also known to interfere with synthesis and metabolism of the hormones (Sharma et al. 2014).

7 Conclusion

Untreated wastewater drastically changes the physico-chemical and biological characteristics of freshwater ecosystems. Wastewater contains persistent pollutants and harmful microbes which not only decrease the quality of freshwater environment but also damages permanently aquatic and associated biota. The foremost threat due to discharge of untreated wastewater causes eutrophication in stagnant freshwater ecosystems. Eutrophication not only decreases the water holding capacity in wetlands and lakes, but also depletes the dissolved oxygen level below the service to fish fauna. Besides this untreated wastewater becomes pollutant factory for freshwater ecosystems. Presences of these kinds of pollutants in freshwater ecosystem have long-term impacts on aquatic and associated biota. Therefore, need of an hour is to monitor the quality of freshwater ecosystem on regular basis and focus should be given to the treatment of effluents prior to its discharge into the freshwater ecosystem.

References

Abida, & Harikrishna. (2008). Study on the quality of water in some streams of Cauvery River. *Journal of Chemistry Pollution Research, 5*(2), 377–384.

Abowei, J. F. N., Davies, O. A., & Eli, A. (2010). Physicochemistry, morphology and abundance of fin fish of Nkoro River, Niger Delta, Nigeria. *International Journal of Pharma and Bio Sciences, 6*(2), 1–11.

Aboyeji, O. O. (2013). Freshwater pollution in some Nigerian local communities, causes, consequences and probable solutions. *Academic Journal of Interdisciplinary Studies, 2*(13), 111–117.

Adakole, J. A., & Abolude, D. S. (2012). Pollutional status of Kubanni Lake through metal concentrations in water and sediment columns, Zaria, Nigeria. *Research Journal of Environmental and Earth Sciences, 4*, 424–427.

Adekunle, A. S. (2009). Effects of industrial effluent on quality of well water within Asa Dam Industrial Estate, Ilorin, Nigeria. *Nature and Science, 7*, 1545–0740.

Adekunle, A. S., & Eniola, I. T. (2008). Impact of industrial effluents on quality of segment of Asa river within an industrial estate in Ilorin, Nigeria. *New York Science Journal, 1*, 17–21.

Adesuyi, A. A., Nnodu, V. C., Kelechi Longinus Njoku, K. N., & Anuoluwapo, J. (2015). Nitrate and phosphate pollution in surface water of Nwaja Creek, Port Harcourt, Niger Delta, Nigeria. *International Journal of Geology, Agriculture and Environmental Sciences, 3*(5), 14–20.

Adler, R., & Rascher, J. (2007). *A strategy for the management of acid mine drainage from gold mines in Gauteng*. (Report No. CSIR/NRE/PW/ER/2007/0053/C). Pretoria: CSIR.

Ahipathy, M. V., & Puttaiah, E. T. (2006). Ecological characteristics of Vrisha Bhavathy River in Bangalore, India. *Environ Goyal Pollution Research, 49*, 1217–1222.

Ahmed, M. K., Das, M., Islam, M. M., Akter, M. S., Islam, S., & Abdullah, M. A. (2011). Physico-chemical properties of tannery and textile effluents and surface water of river Buriganga and Karnatoli, Bangladesh. *Journal World Applied Science, 12*(2), 152–159.

Ahmed, T., Scholz, F., Al-Faraj, W., et al. (2013). Water-related impacts of climate change on agriculture and subsequently on public health: A review for generalists with particular reference to Pakistan. *International Journal of Environmental Research and Public Health, 13*, 1–16.

Akuffo, S. B. (1998). Pollution control in a developing country: A case study of the situation in Ghana, 2nd Edition, Ghana Universities Press. *World Journal of Scientific Research and Reviews, 2*(2), 1–19.

American Public Health Association (Ed.). (1996). *Standard methods of examination of water and wastewater* (19th ed.). Washington, DC: American Public Health Association.

Anonymous. (2002). *Guidelines for the allocation and management of water for maintaining the ecological functions of wetlands.* Ramsar Resolution VIII.1 on allocation and management of water.

Ansari, A. A., Singh, I. B., & Tobschall, H. J. (1999). Status of anthropogenically induced metal pollution in the Kanpur-Unnao industrial region of the Ganga Plain, India. *Environmental Geology, 38*, 25–33.

Arimoro, F. O. (2009). Impact of rubber effluent discharges on the water quality and macroinvertebrate community assemblages in a forest stream in the Niger Delta. *Chemosphere, 77*, 440–449.

Arora, S., & Kakkar, P. (2017). Water pollution in India-causes & policy implications. *Two days international conference on recent innovations in engineering, science, humanities and management*, pp. 721–728.

Banks, D., Younger, P. L., Arnesen, R. T., Iversen, E. R., & Banks, S. B. (1997). Mine-water chemistry: The good, the bad and the ugly. *Environmental Geology, 32*(3), 157–174.

Barker, D. J., & Stuckey, D. C. (1999). A review of soluble microbial products (SMP) in wastewater treatment systems. *Water Research, 33*, 3063–3082.

Baron, J. S., & Poff, N. L. (2004). Sustaining healthy freshwater ecosystems. *Water Resources Update, 127*, 52–58.

Baron, J. S., Poff, N. L., Angermeier, P. L., Dahm, C. N., Gleick, P. H., Hairston, N. G., Jackson, R. B., Johnston, C. A., Richter, B. D., & Steinman, A. D. (2003). Sustaining healthy freshwater ecosystems. *Issues in Ecology, 10*, 2–16.

Bartram, J., & Balance, R. (1996). *Water quality monitoring. A practical guide to the design and implementation of fresh water quality studies and monitoring programs* (Pollution Research 383). UNEP/WHO, Chapman and Hall.

Beeby, A. (1993). Measuring the effect of pollution. In *Applying ecology*. London/New York: Chapman and Hall.

Bhat, S. A., & Pandit, A. K. (2001). Impact of effluents from Sheri- Kashmir institute of medical sciences (SKIMS), Soura on Anchar Lake. *Journal of Research and development Pollution Research, 1*, 30–37.

Bhat, R. A., Dar, G. A., Jehangir, A., Bhat, B. A., & Yousuf, A. R. (2012). Municipal solid waste generation and present scenario of waste management during Yatra season in Pahalgam: A tourist health resort of Kashmir valley. *International Journal of current Research, 4*(10), 004–009.

Bhat, R. A., Nazir, R., Ashraf, S., Ali, M., Bandh, S. A., & Kamili, A. N. (2014). Municipal solid waste generation rates and its management at Yusmarg forest ecosystem, a tourist resort in Kashmir. *Waste Management, 32*(2), 165–169.

Bhat, R. A., Shafiq-ur-Rehman, Mehmood, M. A., Dervash, M. A., Mushtaq, N., Bhat, J. I. A., & Dar, G. H. (2017). Current status of nutrient load in Dal Lake of Kashmir Himalaya. *Journal of Pharmacognosy and Phytochemistry, 6*(6), 165–169.

Bhat, R. A., Dervash, M. A., Mehmood, M. A., & Hakeem, K. R. (2018). Municipal solid waste generation and its management, a growing threat to fragile ecosystem in Kashmir Himalaya. *American Journal of Environmental Sciences, 13*(6), 388–397.

Bolawa, O. E., & Gbenle, G. O. (2012). Analysis of industrial impact on physiochemical parameters and heavy metal concentrations in waters of river Majidun, Molatori and Ibeshe around Ikorodu in Lagos, Nigeria. *Journal of Environmental Science and Water Resources, 1*(2), 34–38.

Bulut, E., & Aksoy, A. (2008). Impact of fertilizer usage on phosphorus loads to Lake Uluabat. *Desalination, 226*, 289–297.

Burton, G. A., & Pitt, R. (2001). *Strom water effects handbook, toolbox for watershed managers, scientists and engineers*. Boca Raton: Lewis.

Capangpangan, R. Y., Pagapong, N. K., Pineda, C. P., & Sanchez, P. B. (2016). Evaluation of potential ecological risk and contamination assessment of heavy metals in sediment samples using different environmental quality indicesea case study in Agusan river, Caraga Philippines. *Journal of Biodiversity and Environmental Sciences, 8*, 1–16.

Carpenter, S. R., Caraco, N. F., Correll, D. L., Howarth, R. W., Sharpley, A. N., & Smith, V. H. (1998). Nonpoint pollution of surface waters with phosphorus and nitrogen. *Ecological Applications, 8*(3), 559–568.

Centre for Ecological Sciences. (2001). *IISc environmental hand book – Documentation on monitoring and evaluating environmental impacts, compendium of environmental standards* (Vol. 3). Bangalore: Indian Institute of Science. Available: http://wgbis.ces.iisc.ernet.in/energy/HC270799/HDL/ENV/START.HTM

Chapman, D. (1996). *Water quality assessment. A guide to the use of biota, sediments and water in environmental monitoring* (2nd ed.). New York: E & FN Spon.

Chapman, D., & Kimstach, V. (1996). Selection of water quality variables. In D. Chapman (Ed.), *Water quality assessment: A guide to the use of biota, sediments and water in environmental monitoring* (2nd ed.). Cambridge: University Press.

Cheng, S. (2003). Heavy metal pollution in China: Origin, pattern and control. *Environmental Science and Pollution Research International, 10*(3), 192–198.

Cheng, L., & Dixon, K. (1998). Analysis of repair and mutagenesis of chromium induced DNA damage in yeast mammalian cells and transgenic mice. *Environmental Health Perspectives, 106*, 1027–1032.

Chindah, A. C., Braide, S. A., & Sibeudu, O. C. (2004). Distribution of hydrocarbon and heavy metal in sediments and crustacean (shrimps – *Paneaus notialia*) from Bonny/New Calabar River Estuary Niger Delta, Nigeria. *African Journal of Environmental Management, 9*, 1–17.

Chowdhury, S., Annabelle, K., & Klaus, F. Z. (2015). Arsenic contamination of drinking water and mental health. *Importance of Freshwater Biodiversity, 1*, 28.

Cumberlidge, N., Ng, P. K., Yeo, D. C., Magalhães, C., Campos, M. R., Alvarez, F., et al. (2009). Freshwater crabs and the biodiversity crisis: Importance, threats, status, and conservation challenges. *Biological Conservation, 142*(8), 1665–1673.

Cunningham, W. P., & Saigo, B. W. (1999). *Environmental science: A global concern* (5th ed.). Boston: McGraw-Hill Publishers.

Currie, J., Joshua, G. Z., Katherine, M., et al. (2013). Something in the water: Contaminated drinking water and infant health. *Canadian Journal of Economics, 46*(3), 791–810.

Dapena-Mora, A., Fernandez, I., Campos, J., Mosquera-Corral, A., Mendez, R., & Jetten, M. (2007). Evaluation of activity and inhibition effects on Anammox process by batch tests based on the nitrogen gas production. *Enzyme and Microbial Technology, 40*, 859–865.

Dart, R. K., & Stretton, R. J. (1980). *Thermal pollution, in microbiological aspects of pollution control* (pp. 185–191). Amsterdam: Elsevier.

Das, R., & Choudhury, I. (2013). Waste management in mining industry Indian. *Journal of Scientific Research, 4*(2), 139–142.

Das, M., Ahmed, M. K., Islam, M. M., Akter, M. S., Islam, M. S., & Mansur, M. A. (2010). Heavy metal concentrations in industrial effluents (tannery and textile) and water of adjacent river. *Terrestrial and Aquatic Environmental Toxicology, 5*(1), 8–13.

Davis, B. R., & Day, J. A. (1998). *Vanishing waters*. Cape Town: University of Cape Town Press/ Juta Press.

De Mora, S., & Harrison, R. M. (2013). Physicochemical speciation of inorganic compounds. In *Hazard assessment of chemicals: Current developments* (Vol. 3, p. 1).

Desai, N., & Vanitaben. (2014). A study on the water pollution based on the environmental problem. *Indian Journal of Research, 3*(12), 95–96.

Dhadse, S. K., & Bhagia, L. J. (2008). Fly ash characterization and utilization and government initiatives in India—A review. *Journal of Scientific and Industrial Research, 67*, 11–18.

Dikio, E. D. (2010). Water quality evaluation of Vaal River, Sharpeville and Bedworth lakes in the Vaal Region of South Africa. *Research Journal of Applied Sciences, Engineering and Technology, 2*(6), 574–579.

Dixit, R., Malaviya, D., Pandiyan, K., Singh, U. B., Sahu, A., Shukla, R., Singh, B. P., Rai, J. P., Sharma, P. K., Lade, H., & Paul, D. (2015). Bioremediation of heavy metals from soil and aquatic environment: An overview of principles and criteria of fundamental processes. *Sustainability, 7*, 2189–2212.

Dudgeon, D., Arthington, A. H., Gessner, M. O., Kawabata, Z. I., Knowler, D. J., Leveque, C., et al. (2006). Freshwater biodiversity: Importance, threats, status and conservation challenges. *Biological Reviews, 81*(2), 163–182.

Dulo, S. O. (2008). Determination of some physico-chemical parameters of the Nairobi River, Kenya. *Journal of Applied Sciences and Environmental Management, 12*(1), 57–56.

Dutta, B. K., Khanra, S., & Mallick, D. (2009). Leaching of elements from coal fly ash: Assessment of its potential for use in filling abandoned coal mines. *Fuel, 88*, 1314–1323.

Ebenstein, A. Y. (2008). *Water pollution and digestive cancer in China*. Institutions and governance programs, pp. 1–45.

EEB. (2000). *The environmental performance of the mining industry and the action necessary to strengthen European legislation in the wake of the Tisza-Danube pollution* (Document No. 2000/016). European Environmental Bureau.

Ekubo, A. T., & Abowei, J. F. N. (2011). Aspects of aquatic pollution in Nigeria. *Research Journal of Environmental and Earth Sciences, 3*(6), 673–693.

Elhatip, H., & Gullu, O. (2005). Influence of wastewater discharges on the water quality of Mamasin dam watershed in Aksaray, Central Anatolian part of Turkey. *Environmental Geology, 48*, 829–834.

EPA. (2002). *Ghana landfills guidelines*. Accra.

Eunice, O. E., Frank, O., Voke, U., & Godwin, A. (2017). Assessment of the impacts of refinery effluent on the physico-chemical properties of Ubeji creek, Delta State, Nigeria. *Journal of Environmental & Analytical Toxicology, 7*, 428.

Fakayode, S. O. (2005). Impact assessment of industrial effluent on water quality of the receiving Alaro river in Ibadan, Nigeria. *Ajeam-Ragee, 10*, 1–13.

Farenzena, M., Ferreira, L., Trierweiler, J. O., & Aquim, P. M. (2005). Tanneries: From waste to sustainability. *Brazilian Archives of Biology and Technology, 48*, 281–289.

Fifield, F. W., & Haines, P. J. (2000). *Environmental analytical chemistry*. Malden: Wiley-Blackwell.

Garcia, L. S. M., Martínez, C. A. T., & Diaz, A. E. (2014). Analisis of waste water treatment plant processes (WWTP) "sedimentation". *Visión Electrónica: algo más que un estado sólido, 8*, 172–185.

Georgieva, S., Gartsiyanova, K., Ivanova, V., & Vladimirova, L. (2018). *Assessment of physical-chemical characteristics of surface water from key sites of the Mesta River: State and environmental implications* (IOP conference series: Materials science and engineering, 374).

Ghannam, H. E., El Haddad, E. S., & Talab, A. S. (2015). Bioaccumulation of heavy metals in tilapia fish organs. *Journal of Biodiversity and Environmental Sciences, 7*, 88–99.

Gong, J., Liu, Y., & Sun, X. (2008). O3 and UV/O3 oxidation of organic constituents of biotreated municipal wastewater. *Water Research, 42*, 1238–1244.

Govind, P., & Madhuri, S. (2014). Heavy metals causing toxicity in animals and fishes. *Research Journal of Animal, Veterinary and Fishery Sciences, 2*, 17–23.

Gray, L. (2004). Changes in water quality and macro invertebrate communities resulting from urban storm flows in the Provo river, Utah, USA. *Hydrobiologia, 518*, 33–49.

Gupta, N., Yadav, K. K., Kumar, V., & Singh, D. (2013). Assessment of physicochemical properties of Yamuna river in Agra city. *International Journal of ChemTech Research, 5*, 528–531.

Halder, J. N., & Islam, M. N. (2015). Water pollution and its impact on the human health. *Journal of Environment and Human, 2*(1), 36–46.

Haseena, M., & Malik, M. F. (2017). Water pollution and human health. *Environmental Risk Assessment and Remediation, 1*(3), 16–19.

Haseena, M., Malik, M. F., Javed, A., Arshad, S., Asif, N., Zulfiqar, S., & Hanif, J. (2017). Water pollution and human health. *Environmental Risk Assessment and Remediation, 1*(3), 16–19.

Hawkes, J. S. (1997). What is a "heavy metal". *Journal of Chemical Education, 74*, 1374.

Helmer, R., & Hespanhol, I. (1997). *Water pollution control: A guide to the use of water quality management principles* (pp. 1–449). London: E & FN Spon.

Hetrick, J. R., Parker, R., Pisigan JR, & Thurman, N. (2000). *Progress report on estimating pesticide concentration in drinking water and assessing water treatment effects on pesticide removal and transformation*. Briefing Document for a Presentation to the FIFRA Scientific Advisory Panel (SAP)

Ho, Y. C., Show, K. Y., Guo, X. X., et al. (2012). Industrial discharge and their affects to the environment. In *Industrial waste* (pp. 1–32). China: InTech.

Hogan, C. (2014). *Water pollution*. Retrieved from http://www.eoearth.org/view/article/156920

Jabeen, S. Q., Mehmood, S., Tariq, B., et al. (2011). Health impact caused by poor water and sanitation in district Abbottabad. *Journal of Ayub Medical College, Abbottabad, 23*(1), 47–50.

Jackson, R. B., Carpenter, S. R., Dahm, C. N., McKnight, D. M., Naiman, R. J., Postel, S. L., & Running, S. W. (2001). Water in a changing world. *Ecological Applications, 11*(4), 1027–1045.

Jagessar, R. C., & Sooknundun, L. (2011). Determination of Nitrate anion in waste water from nine selected areas of coastal Guyana via a spectrophotometric method. *JJRRAS, 7*(2), 203–212.

Jaishankar, M., Tseten, T., Anbalagan, N., Mathew, B. B., & Beeregowda, K. N. (2014). Toxicity, mechanism and health effects of some heavy metals. *Interdisciplinary Toxicology, 7*, 60–72.

Jeswani, H., & Mukherji, S. (2015). Treatment of simulated biomass gasification wastewater of varying strength in a three stage rotating biological contactor. *Chemical Engineering Journal, 259*, 303–312.

Jones, P. J., & Burt, T. P. (1993). Nitrate in surface water. In T. P. Burt, A. L. Heathwaite, & S. T. Trudgill (Eds.), *Nitrate: Process, patterns and management cedes* (pp. 269–310). Chichester: Wiley.

Kahiluoto, H., Kuisma, M., Ketoja, E., Salo, T., & Heikkinen, J. (2015). Phosphorus in manure and sewage sludge more recyclable than in soluble inorganic fertilizer. *Environmental Science & Technology, 49*, 2115–2122.

Kalff, J. (2002). *Limnology: Inland water ecosystems* (Vol. 592). Upper Saddle River: Prentice Hall.

Kamble, S. M. (2014). Water pollution and public health issues in Kolhapur city in Maharashtra. *International Journal of Scientific and Research Publications, 4*(1), 1–6.

Kambole, M. S. (2003). Managing the water quality of the Kafue River. *Physics and Chemistry of the Earth, Parts A/B/C, 28*(20–27), 1105–1109.

Kannan, N., Karthikeyan, G., Vallinayagam, P., & Mil, S. N. (2004). A study on assessment of pollution load of sugar industrial effluent.*Indian Journal of Environmental Protection, 24*(11), 856–862.

Khan, M. A., & Ghouri, A. M. (2011). Environmental pollution: Its effects on life and its remedies. *Journal of Arts, Science and Commerce, 2*(2), 276–285.

Khurana, I., & Sen, R. (2008). *Drinking water quality in rural India: Issues and approaches-Water Aid.* India water Portal.

Kinney, M. L. (2002). Thèse de Doctorat. Université de Limoges, France.

Kozai, N., Suzuki, S., Aoyagi, N., Sakamoto, F., & Ohnuki, T. (2014). *Of Presentation Chemical states of 137 Cs in sewage sludge ash and radioactivity concentration reduction method.*

Krishnan, S., & Indu, R. (2006). *Groundwater contamination in India: Discussing physical processes, health and sociobehavioral dimensions.* IWMI-Tata, Water Policy Research Programmes, Anand, India

Lester, J. N., & Birkett, J. W. (1999). *Microbiology and chemistry for environmental scientists and engineers* (Pd edn.). New York: E & FN Spon.

Lettinga, G. (1995). Anaerobic digestion and wastewater treatment systems. *Antonie Van Leeuwenhoek, 67*, 3–28.

Little, A. D. (1971). *Inorganic chemical pollution of freshwater.* Washington, DC: U.S. Environmental Protection Agency.

Lokhande, R. S., Singare, P. U., & Pimple, D. S. (2011). Toxicity study of heavy metals pollutants in waste water effluent samples collected from Taloja industrial estate resources and environment of Mumbai, India. *Resources and Environment, 1*(1), 13–19.

Lone, J. F., Rasoo, A., & Unnisa, S. A. (2017). Assessment of physico-chemical parameters of water in Kashmir region with reference to Dal Lake. *Journal of Environmental & Analytical Toxicology, 7*, 435.

Lundberg, J. G., Kottelat, M., Smith, G. R., Stiassny, M. L., & Gill, A. C. (2008). So many fishes, so little time: An overview of recent ichthyological discovery in continental waters. *Annals of the Missouri Botanical Garden, 87*(1), 26–62.

Maekawa, T. (2003). *Method and apparatus for treatment of organic matter-containing wastewater.* Google Patents.

Mahananda, M. R. (2010). Physico-chemical analysis of surface water and ground water of Bargarh District, Orissa, India. *International Journal of Research and Review in Applied Sciences, 2*(3), 284–295.

Mali, S. S., Sanyal, S. K., Bhatt, B. P., & Pathak, H. (2015). Water pollution and agriculture. In H. Pathak, B. P. Bhatt, & S. K. Gupta (Eds.), *State of Indian agriculture-water* (pp. 39–47). New Delhi: National Academy of Agricultural Sciences.

Malik, G. M., Raval, V. H., Zadafiya, S. K., & Patel, A. V. (2012). Idol immersion and physico-chemical properties of South Gujarat Rivers, India. *Research Journal of Chemical Sciences, 2*(3), 21–25.

Master, L. L., Flack, S. R., & Stein, B. A. (1998). *Rivers of life.* Nature Conservancy in cooperation with natural heritage programs and Association for Biodiversity Information, pp 1–77.

McConnell, et al. (1993). *Health hazard evaluation report in pesticides in the diets of infants and children.* Washington, DC: Pesticides in the Diets of Infants and Children National Academy Press.

Mehmood, M. A., Qadri, H., Bhat, R. A., Rashid, A., Ganie, S. A., Dar, G. H., & Shafiq-ur-Rehman. (2019). Heavy metal contamination in two commercial fish species of a trans-Himalayan freshwater ecosystem. *Environmental Monitoring and Assessment, 191*, 104. https://doi.org/10.1007/s10661-019-7245-2.

Milovanovic, M. (2007). Water quality assessment and determination of pollution sources along the Axios/Vardar River, Southeastern Europe. *Desalination, 213*, 159–173.

Mohan, D., Kunwar, P. S., & Vinod, K. S. (2006). Trivalent chromium removal from waste water using low cost activated carbon derived from agricultural waste material and activated carbon fabric cloth. *Journal of Hazardous Materials, 135*, 280–295.

Mohanta, M. K., Salam, M. A., Saha, A. K., & Hasan, A. (2010). Roy A.K. Effect of tannery effluents on survival and histopathological changes in different organs of Channa punctatus. *Asian Journal of Experimental Biological Sciences, 1*, 294–302.

Moore, J. W., & Ramamoorthy, S. (1984). *Heavy metals in natural waters: Applied monitoring and impact assessment* (pp. 28–246). New York: Springer.

Morrison, G., Fatoki, O. S., Persson, L., & Ekberg, A. (2001). Assessment of the impact of point source pollution from the Keiskammahoek Sewage Treatment Plant on the Keiskamma River – pH, electrical conductivity, oxygen-demanding substance (COD) and nutrients. *Water SA, 27*, 475–480.

Moss, B. (2008). Water pollution by agriculture. *Philosophical Transactions of the Royal Society B: Biological Sciences, 363*(1491), 659–666.

Musingafi, M. C. C., & Tom, T. (2014). Fresh water sources pollution: A human related threat to fresh water security in south. *Africa Journal of Public policy and Governance, 1*(2), 72–81.

Nair, K. V. K. (1985). Impact of a nuclear power station on the hydrobiological characteristics of Katpakkam waters. In *Proc. Sea Water Qual. Demands Natl. Chem.* Methodological Laboratory, Bombay.

Nartey, V. K., Hayford, E. K., & Ametsi, S. K. (2012). Assessment of the Impact of Solid Waste Dumpsites on some surface water systems in the Accra Metropolitan Area, Ghana. *Journal of Water Resource and Protection, 4*, 605–661.

National Academy of Sciences. (1993). (Ref. #8). Pages 261: 360–361.

Nel, L. H., & Markotter, W. (2009). New and emerging waterborne infectious diseases. *Encyclopedia of Life Support System, 1*, 1–10.

NRCC. (1977). *The effects of Alkali Halides in the Canadian environment* (NRCC No. 15019). Ottawa: Associate Committee on Scientific Criteria for Environmental Quality.

Ogbonna, D. N. (2014). The impact of untreated sewage wastes discharge on the physico-chemical properties of rivers in Port Harcourt metropolis. *World Journal of Scientific Research and Reviews, 2*(2), 1–19.

Ogbonna, J. O., Lawal, F. A., Owoeye, L. D., & Udeh, M. U. (2008). Chemical characteristics and fertilizing value of primary sludge from tannery effluent treatment plant. In *6th Ann. Natl. Conf. Nigerian Inst. Sci. Technol. Univ. Ibadan, 24–27th.*

Onwugbuta-Enyi, J., Zabbey, N., & Erondu, E. S. (2008). Water quality of Bodo Creek in the lower Niger Delta basin. *Advances in Environmental Biology, 2*(3), 132–136.

Otukunefor, T. V., & Obiukwu, C. O. (2005). Impact of refinery influent on physicochemical properties of a water body on the Niger Delta. *Applied Ecology and Environmental Research, 3*, 61–72.

Owa, F. D. (2013). Water pollution: Sources, effects, control and management. *Mediterranean Journal of Social Sciences, 4*(8), 65–68.

Paar, A. (1998). *'Microwave sample preparation system' –instruction handbook* (p. 128). Graz: Anton Paar GmbH.

Parveen, S., Bharose, R., & Singh, D. (2017). Assessment of physico-chemical properties of tannery waste water and its impact on fresh water quality. *International Journal of Current Microbiology and Applied Sciences, 6*(4), 1879–1887.

Paul, D. (2017). Research on heavy metal pollution of river ganga: A review. *Annals of Agrarian Science, 15*, 278–286.

Paul, D., & Sinha, S. N. (2015). Isolation and characterization of a phosphate solubilizing heavy metal tolerant bacterium from river Ganga, West Bengal, India. *Songklanakarin Journal of Science and Technology, 37*, 651–657.

Pawar, S., & Vaidya, R. (2012). Studies on physico chemical characteristics and level of sewage pollution in Krishna River at Wai, Dist: Satara. In *Proceeding of international conference SWRDM*, pp. 129–131.

Pawari, M. J., & Gawande, S. (2015). Ground water pollution & its consequence. *International Journal of Engineering Research and General Science, 3*(4), 773–776.

Pei, Y., Li, L., Li, Z., Wu, C., & Wang, J. (2012). Partitioning behavior of wastewater proteins in some ionic liquids-based aqueous two-phase systems. *Separation Science and Technology, 47*, 277–283.

Pierce, J. J., Weiner, R. F., & Vesihind, A. P. (1998). *Environ- mental pollution and control* (4th ed.). Boston: Butterworth- Heinemann Press.

Pitchammal, V., Subramanian, G., Ramadevi, P., & Ramanathan, R. (2009). The study of water quality at Madurai, Tamilnadu, India. *Nature Environment and Pollution Technology, 8*(2), 355–358.

Postel, S., & Richter, B. (2012). *Rivers for life: Managing water for people and nature* (pp. 1–41). Washington, DC: Island Press.

Prabha, S., & Selvapathy, P. (1997). Toxic metal pollution in Indian Rivers. *Indian Journal of Environmental Protection, 17*, 641–649.

Praveena, S. M., Aris, A. Z., & Radojevic, M. (2010). Heavy metals dynamics and source in intertidal mangrove sediment of Sabah, Borneo Island. *Environment Asia, 3*, 79–83.

Pretty, J. (2002). *Agri-culture*. London: Earthscan.

Pulles, W., Banister, S., & Van Biljon, M. (2005). *The development of appropriate procedures towards and after closure of underground gold mines from a water management perspective* (Report No. 1215/1/05). Pretoria: Water Research Commission.

Rajbanshi, A. (2009). Study on heavy metal resistant bacteria in Guheswori sewage treatment plant. *Our Nature, 6*, 52–57.

Ram, C. L., Srivastava, K. N., Tripathi, C. R., Thakur, K. S., Sinha, K. A., Jha, K. A., Masto, E. R., & Mitra, S. (2007). *Environmental Geology, 51*, 1119.

Rao, Y. H., & Ravindhranath, K. (2013). Thermal and heavy metal ions pollution assessment in near by water bodies of Vijayawada thermal power station. *Asian Journal of Chemistry, 25*(3), 1547–1554.

Rashid, A., Bhat, R. A., Qadri, H., & Mehmood, M. A. (2019). Environmental and socioeconomic factors induced blood lead in children: An investigation from Kashmir, India. *Environmental Monitoring and Assessment, 191*(2), 76. https://doi.org/10.1007/s10661-019-7220-y.

Rast, W. (2009). *Lakes: Freshwater storehouses and mirrors of human activities, briefing note*. Assessment Programme Office for Global Water Assessment, Division of Water Sciences, pp 1–4.

Renberg, I., Persson, M. W., & Emteryd, O. (1994). Pre-industrial atmospheric lead contamination detected in Swedish lake sediments. *Nature, 368*, 323–326.

Richardson, S. D., Plewa, M. J., Wagner, E. D., Schoeny, R., & Demarini, D. M. (2007). Occurrence, genotoxicity, and carcinogenicity of regulated and emerging disinfection by-products in drinking water: A review and roadmap for research. *Mutation Research/Reviews in Mutation Research, 636*(1), 178–242.

Rijsberman, F. (2004). *The challenge of sanitation and water*. Copenhagen Consensus, pp. 1–39.

Rijsberman, F. R., & Molden, D. (2001). Balancing water uses: Water for food and water for nature. In *International conference on freshwater* 3(7), 1–20.

Salem, H. M., Eweida, E. A., & Farag, A. (2000). Heavy metals in drinking water and their environmental impact on human health. *ICEHM, 2000*, 542–556.

Sankhla, M. S., Kumari, M., Sharma, K., Kushwah, R. S., & Kumar, R. (2018). Water contamination through pesticide & their toxic effect on human health. *International Journal for Research in Applied Science & Engineering Technology (IJRASET), 6*(1), 967–970.

Sarkar, S. K., Saha, M., Takada, H., Bhattacharya, A., Mishra, P., & Bhattacharya, B. (2007). Water quality management in the lower stretch of the river Ganges, east coast of India: An approach through environmental education. *Journal of Cleaner Production, 15*, 1559–1567.

Sartor, J. D., Boyd, G. B., & Agardy, F. J. (1974). Water pollution aspects of street surface contaminants. *Journal (Water Pollution Control Federation), 46*(3), 458–467.

Sawhney, N. (2008). PhD thesis, University of Jammu, Jammu.

Scannell, W. P. K., & Duffy, I. K. (2007). Effects of total dissolved solids on aquatic organisms: A review of literature and recommendation for salmonid species. *American Journal of Environmental Sciences, 3*(1), 1–6.

Selvaraj, K., Partiban, G., Chen-Tung, A. C., & Jiann-Yuh, L. J. (2010). Anthropogenic effects on sediment quality offshore South-Western Taiwan: Assessing the sediment core geochemical record. *Continental Shelf Research, 30*(10–11), 1200–1210.

Shah, B. A., Mistry, C. B., & Shah, A. V. (2013). Sequestration of Cu (II) and Ni (II) from wastewater by synthesized zeolitic materials: Equilibrium, kinetics and column dynamics. *Chemical Engineering Journal, 220*, 172–184.

Sharma, G., & John, R. V. (2009). Study of physico-chemical parameters of waste water from dyeing units in Agra city. *Pollution Research, 28*, 439–442.

Sharma, M., Tobschall, H., & Singh, I. (2003). Environmental impact assessment in the Moradabad industrial area (rivers Ramganga-Ganga interfluve), Ganga Plain, India. *Environmental Geology, 43*, 957–967.

Sharma, B., Singh, S., & Siddiqi, N. J. (2014). Biomedical implications of heavy metals induced imbalances in redox systems. *BioMed Research International, 2014*, 640754.

Sheykhi, V., & Moore, F. (2016). Environmental risk assessment of heavy metals pollution in aquatic ecosystem-a case study: Sediment of Kor river, Iran. *Human and Ecological Risk Assessment, 22*, 899–910.

Shivayogimath, C. B., Kalburgi, P. B., Deshannavar, U. B., & Virupakshaiah, D. B. M. (2012). Water quality evaluation of river Ghataprabha, India. *International Journal of Environmental Science, 1*(1), 12–18.

Sindern, S., Trem€ohlen, M., Dsikowitzky, L., Gronen, L., Schwarzbauer, J., Siregar, T. H., Ariyani, F., & Irianto, H. E. (2016). Heavy metals in river and coast sediments of the Jakarta Bay region (Indonesia)-geogenic versus anthropogenic sources. *Marine Pollution Bulletin, 110*, 624–633.

Singh, R., Gautam, N., Mishra, A., & Gupta, R. (2011). Heavy metals and living systems: An overview. *Indian Journal of Pharmacology, 3*, 246–253.

Singh, D. V., Bhat, J. I. A., Bhat, R. A., Dervash, M. A., & Ganei, S. A. (2018). Vehicular stress a cause for heavy metal accumulation and change in physico-chemical characteristics of road side soils in Pahalgam. *Environmental Monitoring and Assessment, 190*, 353. https://doi.org/10.1007/s10661-018-6731-2.

Smitha, Ajay, D., & Shivashankar, P. (2013). Physico chemical analysis of the freshwater at river Kapila, Nanjangudu industrial area, Mysore. *International Research Journal of Environment Sciences, 2*(8), 59–65.

Solenkova, N. V., Newman, J. D., Berger, J. S., Thurston, G., Hochman, J. S., & Lamas, G. A. (2014). Metal pollutants and cardiovascular disease: Mechanisms and consequences of exposure. *American Heart Journal, 168*, 812–822.

Sulaiman, A., Alrumman, Attalla, F., El-kott, Sherif, M. A. S., & Keshk. (2016). Water pollution: Source & treatment (2016). *American Journal of Environmental Engineering, 6*(3), 88–98.

Sun, Y., Jin, B., Wu, W., Zuo, W., Zhang, Y., Zhang, Y., & Huang, Y. (2015). Effects of temperature and composite alumina on pyrolysis of sewage sludge. *Journal of Environmental Sciences, 30*, 1–8.

Suthar, S., Nema, A. K., Chabukdhara, M., & Gupta, S. K. (2009). Assessment of metals in water and sediments of Hindon river, India: Impact of industrial and urban discharges. *Journal of Hazardous Materials, 171*, 1088–1095.

Thilaga, A., Subhashini, Sobhana, S., & Kumar, K. L. (2005). Studies on nutrient content of the Ooty Lake with reference to pollution. *Nature Environment and Pollution Technology, 2*, 299–302.

Togue, F. K., Kuate, G. L. O., & Oben, L. M. (2017). Physico-chemical characterization of the surface water of Nkam River using the Principal Component Analysis. *JMES, 8*(6), 1910–1920.

Ullah, S., Javed, M. W., Shafique, M., et al. (2014). An integrated approach for quality assessment of drinking water using GIS: A case study of Lower Dir. *Journal of Himalayan Earth Sciences, 47*(2), 163–174.

Uqab, B., Singh, A., & Mudasir, S. (2017). Impact of sewage on physico-chemical water quality of Tawi River in Jammu city. *Environmental Risk Assessment and Remediation, 1*(2), 56–61.

Varunprasath, K., & Daniel, A. N. (2010). Comparison studies of three freshwater Rivers (Cauvery, Bhavani and Noyyal) in Tamil Nadu, India, Iran. *Journal of Energy and Environmental*, 1(4), 315–320.

Vikranthpridhvi, Y., & Musalaiah, M. (2015). A review on water and sewage water treatment process. *Journal of Pharmaceutical Chemical and Biological Sciences*, 5(1), 225–231.

Vorosmarty, C. J. E., Douglas, P., & Green Revenga, C. (2005). Geospatial indicators of emerging water stress: An application to Africa. *Ambio, 34*(3), 230–236.

Wang, J. R., Broderick, A. J., & Barchowsky, A. (1999). Chromium Cr (VI) inhibits the transcriptional activity of nuclear factor-B by decreasing the interaction of p 65 with cAMP-responsive element binding protein. *The Journal of Biological Chemistry, 274*(51), 36207–36212.

Wang, Y., Yang, Z., Shen, Z., Tang, Z., Niu, J., & Gao, F. (2011). Assessment of heavy metals in sediments from a typical catchment of the Yangtze river, China. *Environmental Monitoring and Assessment, 172*, 407–417.

WHO. (2004). *Environmental chemistry* (2nd ed.). New York: WH Freeman and Co.

WHO (2007) Nitrate and nitrite in drinking-water. Background document for development of WHO guidelines for drinking-water quality.

Wosnie, A., & Wondie, A. (2014). Bahir Dar tannery effluent characterization and its impact on the head of Blue Nile River. *African Journal of Environmental Science and Technology, 8*(6), 312–318.

Wright, D. A., & Welbourn, P. (2002). *Environmental toxicology*. Cambridge: Cambridge University Press.

Yahya, A. N., Mohamed, S. K., & Mohamed, A. G. (2018). Environmental pollution by heavy metals in the aquatic ecosystems of Egypt. *Open Access Journal of Toxicology, 3*(1), 555603. https://doi.org/10.19080/OAJT.2018.03.555603.

Yonglong, L. U., Song, S., Wang, R., et al. (2015). Impacts of soil and water pollution on food safety and health risks in China. *Environment International, 77*, 5–15.

Yuan, C., Xu, Z., Fan, M., Liu, H., Xie, Y., & Zhu, T. (2014). Study on characteristics and harm of surfactants. *Journal of Chemical & Pharmaceutical Research, 6*, 2233–2237.

Yusuff, R. O., & Sonibare, J. A. (2004). Characterization of textile industries effluents in Kaduna, Nigeria and pollution implications. *Global NEST International Journal, 6*(3), 212–221.

Correlation Between Pollution Trends of Freshwater Bodies and Bacterial Disease of Fish Fauna

Gowhar Hamid Dar, Rouf Ahmad Bhat, Azra N. Kamili, Mohammad Z. Chishti, Humaira Qadri, Rubiya Dar, and Mohammad Aneesul Mehmood

Abstract Fish diseases are major challenges now-a-days which disrupt the stable supply of fishes around the world. Fish diseases are caused by bacterial and fungal infection and other environmental factors (poor water quality) are generally responsible for mass mortalities both in cultured as well as in wild fishes. Bacterial infection produces septicaemia, ulcerative and haemorrhagic diseases causing significant mortality in fishes of different habitats and affects the economy of the aquaculture sector. Polluted environs contains always disease spreading pathogens in addition to facultative microbes. The current review suggests that the incidences of bacterial infection in fishes have increased significantly, with new pathogens regularly recognized. Furthermore, the accounts of the whole genomes of various bacterial species over the years have allowed the identification of an important number of virulence genes that affect the pathogenic potential of these bacteria. The literature over review provides the most relevant information derived from the available bacterial genomes in relation to virulence and on the diverse virulence factors. Thus an attempt is made in the current review to portray the importance of profiling and evaluation of effect of the pathogenic bacteria in the fish fauna.

Keywords Fish · Pathogen · Bacteria · Profiling · Environmental pollution

---

G. H. Dar (✉) · R. A. Bhat · H. Qadri · M. A. Mehmood
Department of Environmental Sciences, School of Sciences, Sri Pratap College Campus, Cluster University of Srinagar, Srinagar, Jammu and Kashmir, India

A. N. Kamili · R. Dar
Centre of Research for Development, University of Kashmir, Srinagar, India

M. Z. Chishti
School of Life Sciences, Central University of Kashmir, Srinagar, India

H. Qadri et al. (eds.), *Fresh Water Pollution Dynamics and Remediation*,
https://doi.org/10.1007/978-981-13-8277-2_4

1 Introduction

Bacteria are found everywhere in the aquatic environment and the different diseases triggered by bacteria may result into heavy mortality in both cultured and wild fish. Most bacterial disease pathogens are generally believed to be part of the normal micro flora of the aquatic environment and are generally deemed as secondary or opportunistic pathogens in nature. Infections and diseases in the body of the host or in its physiology tend to occur only after onset of some key changes. So as to understand the different types of bacterial diseases in fish species, one must understand the relationship between the bacteria with its host and with their environment. To examine this question various studies have been undertaken throughout the world as different types of bacteria may respond either positively or negatively to various environmental stressors. Therefore, the contents of this chapter covers in detail the current status of knowledge regarding the effect of bacterial diseases in fishes and role of environmental conditions on bacterial communities of fish particularly with respect to trophic status of aquatic water body.

2 Bacterial Disease in Fish Fauna

The disease (red sore) has natural causative agent (*A. hydrophila*) which infects *Micropterus salmoides* (largemouth bass), due to this infection, the body of *Micropterus salmoides* shows the lesions affecting scales (few) indicating chronic ulcerations, dermal disorders, oedema (Huizinga et al. 1979). Liver and Kidneys are found to be affected by the production of toxic products from *A. hydrophila.* However, the pathological changes are not found too severe in spleen or heart and the red sore disease pathobiology in bass is hypothesized to be linked to elevated water temperature, increased metabolism, decreased body condition and stress. Developments in the control of bacterial kidney disease of salmonid fishes caused by *Renibacterium salmoninarum* highlight the knowledge gap in understanding this disease despite being reported more than 50 years ago (Elliot et al. 1989). The use of prophylactic or therapeutic feeding of sulphonamides and chemotherapy helped partially. It is now largely acknowledged that *R. salmoninarum* can be transmitted both vertically and horizontally and research prominence has shifted partly towards breaking the infection cycle of the disease so as to prevent it. The understanding of environmental stressors regarding the disease breakout in aquatic systems will surely prove beneficial. The distribution of bacterial flora in fishes isolated from skin, eggs, intestines, and gills had been described for a limited number of fish species (Cahill 1990) and the range of bacterial genera isolated from different fishes is found to be related to the aquatic habitat of the fish and varied with factors such as the salinity of water. The bacteria recovered from the skin and gills may be transients rather than residents on the fish surfaces. Microbiological studies conducted on fresh water fishes from the River ASA in Nigeria (Olayemi et al. 1990) indicate that the microbial flora of Skin, Lungs and Gastrointestinal tract of 65 fishes included 16 bacterial species, some with the likes of *Escherichia coli, Enterobacter aerogenes, Klebsiella pneumonia, Edwardsiella tarda, A. hydrophila, Acinetobacter, Staphylococcus* and *Micrococcus.*

3 Mechanism of Microbial Infections in Fishes

In the contemporary approach towards understanding communicable diseases, epidemiology or epizootiology plays an important role in attempts to explain relationships between the hosts, pathogens, environments, and outbreaks of diseases (Wedemeyer 1976). Outbreaks of bacterial diseases are influenced by the susceptibility of the host, virulence of the pathogens, and quality of the environment (Snieszko 1974) and stress is laid on the fact that control of diseases of fish and shellfish is primarily a managerial problem. The interrelationships between the pathogens, their host organisms and environmental factors are shown in Fig. 1. With the only normal fluctuations in ambient conditions in an unpolluted environment, there will be a natural balance between H, P and E, leading to sporadic outbreaks of disease (Snieszko 1974; Wedemeyer 1976). However, a reduction in the quality of E will lead to a marked increase in the frequency and severity of D, mainly by reducing the resistance of the host organisms to diseases (Wood 1974; Bohl 1989). Also, an increase in the population density of H will increase the risk of disease outbreaks, as shown in Table 1, as will an increase in the virulence of P. It is possible that adverse environmental conditions may decrease the ability of organisms to maintain an effective immunological response system, so that an increased susceptibility to different diseases might be expected to occur (Wedemeyer and McLeay

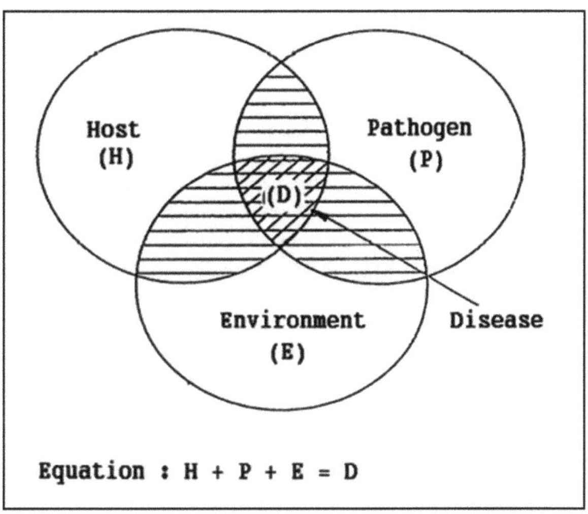

Fig. 1 Interactions between host, pathogen and environment, and the outbreak of diseases. (Wood 1974; Bohl 1989)

Table 1 Environmental factors which are harmful to warm and coldwater fish and increase their susceptibility to certain diseases (from Wedemeyer and McLeay 1981)

S.No.	Disease	Environmental stress factors predisposing to disease	References
1	Furunculosis *(Aeromonas salmonicida)*	Low oxygen (≈4 mg l^{-1}); crowding; handling in the presence of *A. salmonicida;* handling for up to a month prior to an expected epizootic	Koskivaara et al. (1991) and Koskivaara and Valtonen (1992)
2	Bacterial gill disease *(Myxobacteria* spp.)	Crowding; unfavourable environmental conditions such as chronic low oxygen (4 mg l^{-1}); elevated ammonia (0.02 mg l^{-1}unionized); particulate matter in water	Koskivaara et al. (1991) and Koskivaara and Valtonen (1992)
3	Columnaris *(Flexibacter columnaris)*	Crowding or handling during warm (15 °C) water periods if carrier fish are present in the water supply; temperature increase to about 30 °C, if the pathogen is present, even if not crowded or handled	Sovenyi and Szakolczai (1993) and Abowei and Briyai (2011)
4	Kidney disease *(Renibacterium salmoninarum)*	Water hardness less than about 100 mg l^{-1} (as $CaCo_3$); diets containing corn gluten or of less than about 30% moisture	Khan and Thulin (1991)
5	Hemorrhagic septicemia *(Aeromonas* and *Pseudomonas)*	Pre-existing protozoan infestations such as *Costia, Trichodina;* inadequate cleaning leading to increased bacterial load in water; particulate matter in water; handling; low oxygen; chronic sublethal exposure to heavy metals, pesticides or polychlorinated biphenyls (PCBs); for carp, handling after overwintering at low temperatures	
6	Vibriosis *(Vibrio anguillarum)*	Handling; dissolved oxygen lower than about 6 mg l^{-1}, especially at water temperatures of 10–15 °C; brackish water, of 10–15 per mille salinity	Toranzo et al. (2005) and Frans et al. (2011)
7	Parasite infestations *(Costia, Trichodina, Hexamita)*	Overcrowding of fry and fingerlings; low oxygen excessive size variation among fish in ponds	Pascoe and Cram (1977)
8	Spring viremia of carp and tail rot	Handling after overwintering at low temperatures. Crowding; improper temperatures; nutritional imbalances; chronic sublethal exposure to PCBs; or to suspended solids at 200–300 mg l^{-1}	Wedemeyer and McLeay (1981)
9	Coagulated yolk of eggs	Rough handling; malachite green containing and fry more than 0.08% zinc, gas supersaturation of 103% or more; mineral deficiency in incubation water	Wedemeyer and McLeay (1981)
10	"Hauling loss" (delayed mortality)	Hauling, stocking, handling in soft water (less than 100 mg l^{-1} total hardness); mineral additions not used; CO_2 above 20 mg l^{-1}	Wedemeyer and McLeay (1981)
11	Blue sac disease of eggs	Crowding; accumulation of nitrogenous metabolic wastes due to inadequate flow patterns	Wedemeyer and McLeay (1981)

1981). This certainly occurs in aquatic organisms, particularly fish, where acute and/or chronic pollution of surface waters (Bhat et al. 2017) can cause a reduction in the level of unspecific immunity to disease. Any marked change in surface water quality is reflected both directly and indirectly in the structure of the fish population. Indirect effects can occur from damage to the food web which consists of lower organisms in the aquatic environment. There is a wide range in the susceptibility of individual species of aquatic organisms to different pollutants. Organic pollution of water, followed by a decreased content of dissolved oxygen, creates a favourable environment for the growth of bacteria (Koskivaara et al. 1991; Koskivaara and Valtonen 1992). A direct relationship between the organic pollution of surface waters and outbreaks of furunculosis is well established, so that this disease may at times serve as a positive indicator of poor water quality; the causative agent, *Aeromonas salmonicida*, can survive for a maximum of 1 week in tap water, 12 weeks in stream water and as long as 6 months in organically polluted mud. Organic pollution of the aquatic environment is also an important factor in columnar is infection. Vibriosis occurs most frequently in brackish water, although in inland waters it can be found in localities receiving inputs of salt (Toranzo et al. 2005; Frans et al. 2011). Organic and even physical (e.g. inert suspended solids) pollution of water can be important factors in inducing flexibacteriosis in the gills of salmonids, by damaging the delicate gill respiratory epithelium.

Further, it is very important for aqua cultural managers to have a thorough understanding of biology, physiology, microbiology, immunology, ecology, and therapy, in order to make proper evaluations of the disease problem. The environmental circumstances not favorable for the fish may not necessarily be unfavourable for parasites and on the contrary, a lowering of the stressed conditions in fish may increase the likelihood of bacterial and parasitic infections (Schäperclaus 1991). A well-established fact is that contaminants and pollutants reach the aquatic ecosystem (Sved et al. 1997; Austin 1999), and have a strong tendency to get stored in various tissues of aquatic organisms (Han et al. 1997). Some of the contaminants and pollutants pose a serious threat to aquatic organisms (Lanno and Dixon 1996). However, limited research backup is available that pollutants may lead to the development of disease. Indeed, sufficient research supports the correlation between less disease incidences with the elevation of pollution level in aquatic ecosystems (Sandstrom 1994).

Many studies have reflected a higher incidence of diseased animals from polluted sites as compared to control sites (Vethaak 1992). Still there is deficit of standard qualitative and quantitative information about the contaminants in most of the research investigations. Nevertheless, there is lot of literature available which supports the process that some pollutants are harmful and lead to immunosuppression and finally may lead to fish death (Sovenyi and Szakolczai 1993) Curiously, some disease symptoms in gills and skin, are associated with many infectious diseases like gill disease and ulceration (Austin and Austin 1993; Sved et al. 1997). In this connection, it is very much relevant to highlight the increasing recognition of typical isolates of *Aer. salmoniczda* as a potential cause of skin lesions diseases and ulcerations in native marine fish (Wiklund and Bylund 1993; Nakatsugawa 1994; Pedersen et al. 1994; Wiklund et al. 1994; Wiklund 1995; Wiklund and Dalsgaard

1995; Larsen and Pedersen 1996; Austin et al. 1998). It still remains to be properly confirmed that whether or not this taxon, which has been related to the furunculosis disease of freshwater salmonids (Austin and Austin 1993), interacts mostly with marine fish that may have been already damaged by pollutants. Thus, from the above discussed issue, it seems that any environmental contamination whether through natural or anthropogenic interference occurring as a result of a natural phenomenon or through human influence, may act as a severe stressing factor and interferes with the microbial structure and their assemblage in a water source.

Fish disease along pollution gradient is monitored in order to quantify spatial patterns of 5 selected gross pathologies and the findings indicate that the most promising external disease for biological effect monitoring appeared to be epidermal hyperplasia/papilloma (Vethaak et al. 1992). There is strong evidential support that some pollutants and contaminants having immunosuppression effect lead to severe damage to fishes (Sovenyi and Szakolczai 1993) and the same fishes may die due to synergistic effect of bacteria diseases and other pollutants. *Flavobacterium branchiophilum* was found to be the most dominant bacteria in the biofilm of diseased gills when examined using monoclonal antibody probes (Speare et al. 1995). Likewise, when diagnostic tests for bacterial kidney (BKD) disease in *Oncorhynchus mykiss* was carried out and was confirmed with an experimentally infected fish kidney tissue by using enzyme linked immunosorbent assay (ELISA) test kit, fluorescent antibody (FA) testing, bacteriologic culture, and histopathology. In four groups, 75%of the samples of steelhead trout were found infected and ELISA, FA, and bacteriologic culture were positive for *R. salmoninarum* from the kidney tissue of the two groups. Though histopathology was not that much specific but all almost all infected fishes reflected proliferative histolytic interstitial nephritis (White et al. 1995). However, the investigation on the application of kidney biopsy of broodstock steelhead trout (*Oncorhynchus mykiss*) to confirm the status of BKD infection so as to decline it successfully, moreover the status of BKD of the broodstock should be determined and confirmed (White et al. 1996).

4 Impacts of Pollution Load on Bacterial Infection in Fish Fauna

The influence of pollution on fish populations and triggering of diseases in them state explicitly that anthropogenic factors contribute widely to the occurrence of disease in them as it alters the susceptibility of the host to pathogens that are integral and ubiquitous components of any habitat (Arkoosh et al. 1998) and the influence of water quality and temperature to adhesion of high and low virulence of *F. Columnare* strains to isolated gill arches suggest that high virulence strain adhered more readily than low virulence strain (Decostere et al. 1999). Adhesion of high virulence strain was enhanced by a number of factors including immersion of gill in bivalent, ion rich water, the presence of nitrite or organic matter and high temperatures. Bacterial fish pathogens monitored on rainbow trout, *Oncorhynchus mykiss* (Walbaum), in freshwater farms resulted in the isolation of 361 bacterial isolates and observation of enteric redmouth disease, furunculosis and rainbow trout fry syndrome/coldwater disease in them (Dalsgaard and Madsen 2000).

5 Isolation, Identification and Characterization of Pathogenic Bacteria in Fishes

The study on the identification and characterization of pathogenic *Aeromonas veronii* bacterial isolates associated with Epizootic Ulcerative Syndrome (EUS) in fish in Bangladesh were evaluated (Rahman et al. 2002) and 52 *Aeromonas* strains from EUS lesions in the fish were isolated. A clonal group of *A. veronii* biovar sobria is associated, and may be a causative agent of EUS in fish in Bangladesh. The characterization of non motile bacteria associated with freshwater fish spoilage, resembled phenotypically to *Psychrobacter* spp., (Garcia-Lopez et al. 2004). However, Oxidase-positive, nonmotile, nonfermenter Gram-negative rods isolated from freshwater fish can be wrongly ascribed to the genus *Psychrobacter*. The implications of the geographic distribution of some bacterial diseases such as vibriosis, "winter ulcer", photobacteriosis, furunculosis, flexibacteriosis, winter disease, streptococcosis, lactococcosis, BKD, mycobacteriosis and piscirickettsiosis and the main host species affected, together with the biochemical and antigenic diversity (Toranzo et al. 2005) describe the classical methods to isolate the microorganisms from their hosts as well as the serological and/or genetic tools for a rapid diagnosis of the diseases. The characterization of *Aeromonas sobria* isolated from fish Rohu (*Labeo rohita*) collected from polluted pond in J&K State laid emphasis on the fact that the infection in the pond could have been triggered by pollution levels, as the pond was affected by mismanagement practices, elevated pollution levels and anthropogenic activities (Dar et al. 2016). Fish disease and health management practices in aquaculture production laid importance on the fact that disease issues are of great concern in aquaculture and production of fish products is declining with the emergence of new types of bacterial diseases and the aquaculture industry is finding it more challenging to guarantee its sustainable development (Idowu et al. 2017). The pathogenic bacteria in *Oreochromis Niloticus* revealed the presence of 11 bacterial genera known as *Arthrobacter* sp., *Enterococcus* sp., *Staphylococcus* sp., *Micrococcus* sp., *Streptococcus* sp., *Aeromonas* sp., *Pseudomonas* sp., *Edwardsiella* sp., *Flexibacter* sp. and *Flavobacterium* sp., with a predominance of 55% Gram-negative bacilli in tilapia crops from body parts suffering from bleeding, ulceration, corneal opacity, and inflammation (Huicab-Pech et al. 2017).

The possible role of a plasmid in the pathogenesis of a fish disease caused by *Aeromonas hydrophila* by isolating pathogenic isolates of *Aeromonas hydrophila* from fish affected with ulcerative disease syndrome (UDS) and were found to be harboring a 21 kb plasmid (Majumdar et al. 2006). The incidences of *furunculosis* in *Schizothorax* spp. *(Schizothorax niger, S. esocinus, S. curvifrons* and *S. labiatus)* were reported from Wular Lake, Kashmir during summer and winter months (Chalkoo et al. 2007), however the percentage of the infection was maximum during winter with a mortality rate of 8–15%. The incidence of disease was highest (13.87%) in December, and lowest (0.40%) in May and October. *S. esocinus* exhibited the maximum (44.48%) percentage of infection, while as *S. labiatus* exhibited the minimum (14.28%) percentage of infection. Moreover, the relationship between bacterial fish diseases in a pond and the effect of application of livestock manures into it demonstrate the presence of higher population of *Aeromonas* and *Pseudomonas* in the pond with high manure supplication than the one which was not supplemented with manure (Chalkoo et al. 2007). Therefore, suggesting the possible impact of anthropogenic activities onto the health of fish species.

Pathogenic bacteria were isolated from the skin ulcerous symptomatic gourami (*Colisa lalia*) through 16S rDNA analysis and *Aeromonas* spp. infections in fishes is considered as the most common bacterial disease diagnosed in cultured warm water fish (Hossain 2008) among the Aeromonas spp., *Aeromonas veronii* and *Aeromonashydrophila* were frequently related to the haemorrhages at the base of the fins or on the skin, and gross ulcerative lesions. The coastal sewage discharge and also its impact on the fishes with reference to the antibiotic resistant enteric bacteria and enteric pathogens as the possible bio-indicators of pollution in the fish (colons and gills) and also in the sewage treated effluent (STE) in two marine sites in the Gulf of Oman, Muscat indicated that the sewage effluent causes contamination of marine wildlife along the coastal lines (Al-Bahry et al. 2009). Isolation and identification of pathogenic bacteria in edible fish from Fletcher Dam in, Zimbabwe was aimed at the isolation of human pathogenic bacteria in gills, intestines, mouth and skin of apparently healthy fishes, (*Tilapia rendali* and *Oreochromic mossambicus*) and it resulted in the separation and characterization of various human pathogenic bacteria and thus a potential hazard to humans (Sichewo et al. 2013). Similarly, the prevalence of pathogenic microflora in two major sea fish samples: Rupchanda (*Pampus chinensis*) and Surmai (*Scomberomorus guttatus*), collected from local market in Dhaka city (Noor et al. 2013) revealed the contamination of most of the fishes with a huge number of pathogens within a range of 2.0×10^2–1.9×10^9 cfu. mL^{-1} or cfu.g^{-1}. However, the insights of biosurfactant producing *Serratia marcescens* strain isolated from diseased tilapia fish and proposed that *Serratia marcescens* is an opportunistic bacterial pathogen with broad range of host ranging from vertebrates, invertebrates to plants (Chan et al. 2013).

Isolates of *Aeromonas hydrophila* isolated from various naturally infected fishes collected from both fresh and brackish water suggest that the pathogenicity assay proved that 33.3% of the tested *A. hydrophila* were pathogenic for tilapia in vitro with various levels of virulence (El-Barbary 2010). Microscopically, *A. hydrophila* toxins apparently were found to cause irreparable systemic damage to liver which leads to death. Further during the evaluation of probiotic effect of *Lactobacillus* isolates against bacterial pathogens in fresh water fishes a total of 59 Lactobacillus isolated from 5 different fresh water fishes including Cat fish (*Clariasorientalis*), Hari fish (*Anguilla* sp.), Rohu fish (*Labeo rohita*), Jillabe fish (Oreochromis sp.) and

Gende fish (Punitus carnaticus). A few selected *Lactobacillus* isolates were screened for antagonistic activity against *Aeromonas*, *Vibrio* sp. by agar diffusion assay and only, Lactobacilli RLD2 showed significant antagonistic activity against *Aeromonas* and *Vibrio* sp. alone (Dhanasekaran et al. 2010). It was also concluded that the Lactobacillus isolates could be used as probiotic bacteria in aquaculture, to manage *Aeromonasis*. The occurrence and characterization of the *Aeromonas* species prevalent among marine fishes report that *Aeromonas* are rod shaped bacteria that can be divided into eight to nine distinct species, including *A. caviae, A. eucrenophila, A. schubertii, A. sobria, A. veronii* and *A. hydrophilia* and among which *A. hydrophila* is a major cause of death in fishes and shellfishes (Aberoum and Jooyandeh 2010). The prevalent bacterial fish diseases of Latvia in seven state fish hatcheries and three private farms from examining a total of 3334 individual fish along with their pathological examination and by the bacteriological analysis of the material from ulcers of the surface, gills, heart, liver, kidney, spleen, muscles using commercial bacterial identification test strips, it was found that *Flexibacter* spp., *Aeromonas salmonicida*, and *Aeromonas hydrophila* are the pathogenic bacteria related to various fish diseases like aeromonosis and myxobacteriosis (Briede 2010).

The bacterial microflora associated with fresh Tilapia fish (*Oreochromis niloticus*) sold at Sokoto central market, Nigeria was studied by analysing sections of the skin, gills and intestine of 10 randomly selected fishes and some nine (9) bacterial species including six (6) gram positive (*Bacillus megatanium, Listeria monocytogenes, Bacillus Pumilus, Bacillus alvei, Bacillus licheniformis* and *Staphylococcus saprophyticus*) and three (03) gram negative bacteria (*Serratia mercescens, Providentia stuartii* and *Salmonella* spp.) were isolated (Shinkafi and Ukwaja 2010). Furthermore, the sharp tooth catfish (*Clarias gariepinus*) was evaluated for the effects of bacterial diseases on their skin and gills by evaluating 94 samples of live mature catfish (*Clarias gariepinus*) which were collected from the tributaries of River Nile. *Actinobacter lwoffii, Enterobacter amnigenus, Escherichia coli, Citrobacter amlonaticus, Serratia odorifera* and *Aeromonas jandaei, Staphylococcus epidermidis* were isolated from the infected body parts (El-Sayyad et al. 2010)

Some bacteria diseases in fish fauna like *Streptococcosis*, Infectious Abdominal Dropsy of Carp, *Furunculousis*, Motile Aeromonad Disease, Vibrosis, Columnaris disease, Bacteria kidney disease, Peduncle disease (fin rot), Bacteria gill disease, Pasteurellosis, Myxobacterial infections, gill and fin rot, Mycobacteriosis of fishes, Epitheliocystis suggest that understanding the life history, biology, diagnosis, pathology, epizootiology and control of these diseases might prove beneficial to the fish culturist (Abowei and Briyai 2011). *Vibrio anguillarum* also known as *Listonella anguillarum* is responsible for causing vibriosis which is a deadly haemorrhagic septicaemic disease affecting various marine and fresh water fish, bivalves and crustaceans (Frans et al. 2011). However, the bacterial taxonomy has progressed from reliance on highly artificial culture-dependent techniques involving the study of phenotype (including morphological, biochemical and physiological data) to the modern applications of molecular biology (mostly 16S rRNA gene sequencing) which gives an insight into evolutionary pathways i.e. (phylogenetics) over the years (Austin 2011).

The molecular detection of *Edwardsiella tarda* with *gyrB* gene isolated from *Arapaima gigas*, show that the bacteria isolated from different organs of the fish characterized by Vitek System®2 exhibit 98% probability to *Edwardsiella tarda* which was characterized by PCR (Choresca Jr. et al. 2011). The Coldwater disease (CWD) of fishes caused by bacteria *Flavobacterium psychrophilum,*) is a bacterial disease that affects a broad host-species range of fishes inhabiting the cold, fresh waters and the disease occurs predominately at a temperature of 16 °C and below, and is most prevalent and severe at 10 °C and below (Starliper 2011). It is further reported that *Flavobacterium psychrophilum*is a Gram-negative bacterium and can be recovered from affected host tissues and can be characterized using standard biochemical techniques.

6 Correlation of Pathogenic Bacteria with Different Tissues of Fish Fauna

There is a clear cut relationship between *Aeromonas* species and/or serogroups and specific disease symptoms in common carp (*Cyprinus carpio* L.) and rainbow trout (*Oncorhynchus mykiss* Walbaum) as the adhesion of *Aeromonas* strains to various tissues to evaluate the disease spectrum and found that adhesion intensity of *Aeromonas* strains was significantly higher to tissues (Kozinska and Pekala 2012). The strains of *A. hydrophila*have tendency to cause skin ulcers as well as septicaemia in both carp as well as trout while it was seen that the other strains were able to cause only skin ulcers or some specific internal lesions with or without septicaemia depending upon which species and group they represent. *Pseudomonas* species from samples of different species of cultured fishes was collected from different fish farms at different localities in Egypt and four *Pseudomonas* species were recovered from naturally infected fishes; *P. fluorescence*, *P. putida, P. aeruginosa* and *P. anguilliseptica* (El-Hady and Samy 2011). The in-vitro sensitivity test of isolated *Pseudomonas* strains was conducted to different chemotherapeutic agents and on molecular typing of the isolated *Pseudomonas* spp. by plasmid profile analysis as well as by SDS-PAGE, it was found that most of the isolates showed a degree of variation in plasmid number (1 up to 4 plasmids) and SDS-PAGE revealed that one isolate from *Pseudomonas fluores*cence, *Pseudomonasaeruginosa* and *Pseudomonas anguilliseptica* species shared one band that was present at 144.8 kDa. The isolation of some emergent bacterial pathogens from capture and culture fisheries in Bangladesh were studied (Hossian et al. 2011). Twenty five (25) 25 bacterial strains were successfully isolated, mostly belonging to the groups, *Aeromonas* and *Pseudomonas* from different sources and among these strains *P. fluorescens* strains were found to be highly virulent, followed by *A. hydrophila* and *A. veronii biovar sobria.*

A study on the virulence of *Pseudomonas* and *Aeromonas* bacteria (Shayo et al. 2012) recovered from *Oreochromis niloticus* (Perege) resulted in the isolation of several bacterial species including *Aeromonas hydrophila*, *A. veronii* and

Pseudomonas aeruginosa from normal and ulcerative affected Tilapia species. On checking the virulence of these bacterial species on healthy fish, it was found that 112 out of 180 infected fish developed clinical abnormalities like skin darkness, scales detachment, blindness and large irregular haemorrhages on the body surface, fin necrosis, exophthalmia and eye cataract/trachoma within 4 days with mortality rate reaching up to 95%. Correlation between microbiological water quality and tissue lesions in gills from Nile tilapia (*Oreochromis niloticus*) and hybrid tambacu (*Colossoma macropomum* female × *Piaractusmesopotamicus* male), showed that the water samples reflect the presence of total coliforms (*Escherichia coli* and heterotrophic bacteria) and the gills, were dominated by Gram positive and Gram negative bacteria diversity. The tissue lesions were exhibiting lamellar fusion, interlamellar hyperplasia, sub-epithelial edema and telangiectasia (Santos et al. 2012). Surveillance of *Aeromonas sobria* and *Aeromonas hydrophila* from commercial food stuffs and environmental sources was taken up and a total of 71 suspected *Aeromonas* strains were isolated from 154 commercial food and environmental samples and upon biochemical characterization of these isolates, 56 were identified as *A. sobria* and remaining 15isolates were identified as *A. hydrophila* (Das et al. 2012). Additionally, the mass mortality of *Garra rufa* from a fish hatchery farm in Slovakia was investigated and upon the clinical observation, it was found that the fish was characterized by abnormal swimming behaviour, bleeding from skin lesions and haemorrhages showed the presence of *A. Sobria,* the main reason for the mortality in fishes (Majtan et al. 2012).

7 Effects of Water Quality Characteristics on Bacterial Pathogencity in Fish Fauna

The effects of temperature and pH on the inactivation of fish viral and bacterial pathogens by conducting laboratory studies on the inactivation of a number of fish pathogenic viruses and bacteria at 60 °C, pH 4.0 and pH 12.0 and found that the most resistant bacterium to 60 °C, pH 4.0 as well as pH 12.0 was *Lactococcus garvieae* (Dixon et al. 2012). *Vibrio* spp. in diseased *Channa punctatus* from aquaculture farm contaminated by agricultural wastewater were isolated and identified and the presence of *V. harveyi, V. parahaemolyticus, V. fischeri* and *V. cholera* in the fishes was established which is a cause of concern (Sankar et al. 2012). *C. Punctatus* was found to be mostly infected by *V. harveyi, V. parahaemolyticus* and *V. cholera.* Septicaemia, ulcerative and haemorrhagic diseases in fish caused by *Aeromonas* infections suggest that these types of infections cause significant mortality in both wild and farmed freshwater and marine fish species resulting the damage to the economics of the aquaculture sector (Beaz-Hidalgo and Figueras 2013). It was further proposed that the high number of plasmids, IS and pseudogenes revealed by the genome of *A. salmonicida* is a result of evolution and adaptation of this species to infect fish. In contrast, the genome sequence of *A. hydrophila*, which has few pseudogenes, shows a wider metabolic capacity to infect a wider range of host organisms. *Aeromonas* spp. was thus considered as important fish pathogens and a persistent threat to the aquaculture sector. Similarly, Columnaris disease in fish as evaluated and by analysing different aspects of this disease state that *Flavobacterium columnare* is the causative agent of columnaris disease and this bacterium affects both cultured and wild freshwater fish including many susceptible commercially important fish species. *F. Columnare* can cause widespread infections including skin lesions, fin erosion and gill necrosis, with a high degree of mortality, leading to severe economic losses (Declercq et al. 2013)

Studies on *Aeromonas hydrophila* in cultured *Oreochromis niloticus* in Egypt were carried out and some ten isolates of *A. hydrophila* strains from 40 cultured *O. niloticus* collected randomly from the fish farm with a prevalence rate of 25% were recovered (Deen et al. 2014). The clinical symptoms of the collected fish exhibited loss of escape reflex; skin darkness; bilateral exophthalmia and ulcers with postmortem examinations revealing varied degrees of congestion, hemorrhage and enlargement in internal organs. Similarly, a study on the isolation and identification of bacteria associated with fresh and smoked Fish (*Clarias gariepinus*) in Minna Metropolis, Nigeria under in-vitro assay conditions revealing the contamination of the sample by many bacterial species (Ibrahim et al. 2014). Comparing the effects of *Flavobacterium branchiophilum* and *F. Succinicans* associated with bacterial gill disease in rainbow trout (*Oncorhynchus mykiss* Walbaum) in water recirculation aquaculture systems, it was found that *F. branchiophilum* is the dominant bacterial species present on the gills of rainbow trout affected by natural and environmentally induced BGD (Good et al. 2015). The detection and characterization of potentially pathogenic *Aeromonas sobria* from fish silver carp (*Hypophthalmichthys molitrix*) collected from a fish pond at District Poonch of Jammu and Kashmir State was taken for analysis (Dar et al. 2015). A total of 33 colonies of *A. sobria* strain were isolated from the fish. Histopathology of Gills, showed haemorrhagic gill epithelia and epithelial hyperplasia. Lamelar epithelial hypertrophy and hyperplasia with degenerative changes of the epithelium and hypertrophic epitheliocystis infected cells on gills was also observed.

8 Flavobacterial Infection as a Most Common Threat to Fish Industry

Emerging flavobacterial infections in fish, pointed out a fact that Flavobacterial diseases in fish are caused by multiple bacterial species within the family Flavobacteriaceae and are responsible for devastating losses in wild and farmed fish stocks around the world (Loch and Faisal 2015). It was further pointed out that in addition to directly imposing negative economic and ecological effects, flavobacterial disease outbreaks are also notoriously difficult to prevent and control despite nearly 100 years of scientific research. Various modern trends to control and manage *Aeromonas* infection in fish species were studied and it was pointed out that *Aeromonas hydrophila,* is a ubiquitous, free-living, Gram-negative bacterium, is prevalent in aquatic habitats with cosmopolitan distribution that has resulted in heavy mortalities in farmed and feral fishes (Harikrishnan and Balasundaram 2015).

9 Conclusion

Perusal of the literature suggests that a number of studies have been carried out related to the fish diseases using some most modern techniques like molecular typing, SDS PAGE, PCR, enzyme linked immunosorbent assay (ELISA), Flourescent Antibody (FA) technique, bacteriological culture and histopathology. The studies are mostly confined to some typical bacterial or fungal diseases like bacterial kidney diseases (BKD's), bacterial gill diseases (BGD's), cold water disease (CWD's), Ulcerative disease syndrome (UDS), epizootic Ulcerative Syndrome (EUS). However, some major areas in the fish pathology involving the impact of pathogenic bacteria on fish, susceptibility of the fishes to the varying environmental conditions and influence of trophic studies are still missing in the scientific literature. The horizons of the new studies should be broadened to involve the most modern techniques like VITEK 2, specific gene targeting, microbial culture, classical morphology and histopathology to get an idea about the susceptibility of the fishes both in lotic and lentic systems to different bacterial diseases.

References

Aberoum, A., & Jooyandeh, H. (2010). A review on occurrence and characterization of the *Aeromonas* species from marine fishes. *World Journal of Fish and Marine Sciences, 2*(6), 519–523.

Abowei, J. F. N., & Briyai, O. F. (2011). A review of some bacteria diseases in Africa culture fisheries. *Asian Journal of Medical Sciences, 3*(5), 206–217.

Al-Bahry, S. N., Mahmoud, I. Y., Al-Belushi, K. I. A., Elshafie, A. E., Al-Harthy, A., & Bakheit, C. K. (2009). Coastal sewage discharge and its impact on fish with reference to antibiotic resistant enteric bacteria and enteric pathogens as bio-indicators of pollution. *Chemosphere, 77*, 1534–1539.

Arkoosh, M. R., Casillas, E., Clemons, E., Kagley, A. N., Olson, R., Reno, P., & Stein, J. E. (1998). Effect of pollution on fish diseases: Potential impacts on salmonid populations. *Journal of Aquatic Animal Health, 1*, 182–190.

Austin, B. (1999). The effects of pollution on fish health. *Journal of Applied Microbiology, Symposium Supplement, 85*, 234–242.

Austin, B. (2011). Taxonomy of bacterial fish pathogens. *Austin Veterinary Research, 42*, 20p. http://www.veterinaryresearch.org/content/42/1/20.

Austin, B., & Austin, D. A. (1993). *Bacterial fish pathogens. Disease in farmed fish and wild fish* (2nd ed.). Chichester: Ellis Horwood.

Austin, B., Austin, D. A., Dalsgaard, I., et al. (1998). Characterization of a typical *Aerumonus sulmunrcidu* by different methods. *Systemuri und Applied Microbiulogy, 21*, 50–64.

Beaz-Hidalgo, R., & Figueras, M. J. (2013). *Aeromonas* spp. whole genomes and virulence factors implicated in fish disease. *Journal of Fish Diseases, 36*, 371–388.

Bhat, R. A., Shafiq-ur-Rehman, D. M. A., Mushtaq, N., Bhat, J. I. A., & Dar, G. H. (2017). Current status of nutrient load in Dal Lake of Kashmir Himalaya. *Journal of Pharmacognosy and Phytochemistry, 6*(6), 165–169.

Bohl, M. (1989). *Optimal water quality-basis of fish health and economical production. Current trends in fish therapy* (pp. 18–32). Giessen: Deutsche Veterinärmedizinische Gesellschaft e. V..

Briede, I. (2010). The prevalent bacterial fish diseases in fish hatcheries of Latvia. *Environmental and Experimental Biology, 8*, 103–106.

Cahill, M. M. (1990). Bacterial flora of fishes: A review. *Microbial Ecology, 19*, 21–41.

Chalkoo, S. R., Najar, A. M., Qureshit, T. A., & Shafit, A. (2007). *Furunculosis* in snow trout (Schizothoracinae) in Kashmir: First report. *Journal of Indian Fisheries Association, 34*, 59–73.

Chan, X. Y., Chang, C. Y., Hong, K. W., Tee, K. K., Yin, W. F., & Chan, K. G. (2013). Insights of biosurfactant producing *Serratia marcescens* strain W2.3 isolated from diseased tilapia fish: A draft genome analysis. *Gut Pathogens, 5*, 29p. http://www.gutpathogens.com/content/5/1/29.

Choresca-Jr, S. H., Gomez, D. K., Shin, S. P., Kim, J. H., Han, J. E., Jun, J. W., & Park, S. C. (2011). Molecular detection of *Edwardsiella tarda* with gyrB gene isolated from pirarucu, *Arapaima gigas* which is exhibited in an indoor private commercial aquarium. *African Journal of Biotechnology, 10*(5), 848–850.

Dalsgaard, I., & Madsen, L. (2000). Bacterial pathogens in rainbow trout, *Oncorhynchus mykiss* (Walbaum), reared at Danish freshwater farms. *Journal of Fish Diseases, 23*, 199–209.

Dar, G. H., Dar, S. A., Kamili, A. N., Chishti, M. Z., & Ahmad, F. (2015). Detection and characterization of potentially pathogenic *Aeromonas sobria* isolated from fish *Hypophthalmichthys molitrix* (Cypriniformes: Cyprinidae). *Microbial Pathogenesis, 91*, 136–140.

Dar, G. H., Kamili, A. N., Chishti, M. Z., Dar, S. A., Tantry, T. A., & Ahmad, F. (2016). Characterization of *Aeromonas sobria* isolated from fish Rohu (*Labeo rohita*) collected from polluted pond. *Journal of Bacteriology and Parasitology, 7*, 273p. https://doi.org/10.4172/2155-9597.1000273.

Das, A., Vinayasree, V., Santhosh, C. R., & Hari, S. S. (2012). Surveillance of *Aeromonas sobria* and *Aeromonas hydrophila* from commercial food stuffs and environmental sources. *Journal of Experimental Sciences, 3*(9), 36–42.

Declercq, M. A., Haesebrouck, F., Den-Broeck, W. V., Bossier, P., & Decostere, A. (2013). *Columnaris* disease in fish: A review with emphasis on bacterium-host interactions. *Veterinary Research, 44*, 27.

Decostere, A., Haesebrouck, F., Turnbull, J. F., & Charlier, G. (1999). Influence of water quality and temperature on adhesion of high and low virulence *Flavobacterium columnare* strains to isolated Gill arches. *Journal of Fish Diseases, 22*, 1–11.

Deen, A. E. N. E., Dorgham, S. M., Hassan, A. H. M., & Hakim, A. S. (2014). Studies on *Aeromonas hydrophila* in cultured *Oreochromis niloticus* at Kafr El Sheikh Governorate, Egypt with reference to histopathological alterations in some vital organs. *World Journal of Fish and Marine Sciences, 6*(3), 233–240.

Dhanasekaran, D., Saha, S., Thajuddin, N., Rajalakshmi, M., & Panneerselvam, A. (2010). Probiotic effect of *Lactobacillus isolates* against bacterial pathogens in fresh water fish. *Journal of Coastal Development, 13*, 103–112.

Dixon, P. F., Smail, D. A., Algot, M., Hastings, T. S., Bayley, A., Byrne, H., Dodge, M., Garden, A., Joiner, C., Roberts, E., Verner-Jeffreys, D., & Thompson, F. (2012). Studies on the effect of temperature and pH on the inactivation of fish viral and bacterial pathogens. *Journal of Fish Diseases, 35*, 51–64.

El-Barbary, M. I. (2010). Some clinical, microbiological and molecular characteristics of *Aeromonas hydrophila* isolated from various naturally infected fishes. *Aquaculture International, 18*, 943–954.

El-Hady, M. A., & Samy, A. A. (2011). Molecular typing of *Pseudomonas* species isolated from some cultured fishes in Egypt. *Global Veterinaria, 7*(6), 576–580.

Elliot, D. G., Pascho, R. J., & Bullock, G. L. (1989). Developments in the control of bacterial kidney disease of salmonid fishes. *Diseases of Aquatic Organisms, 6*, 201–215.

El-Sayyad, H. I., Zaki, V. H., El-Shebly, A. M., & El-Badry, D. A. (2010). Studies on the effects of bacterial diseases on skin and gill structure of *Clarias gariepinus* in Dakahlia Province, Egypt. *Annals of Biological Research, 1*(4), 106–118.

Frans, I., Michiels, C. W., Bossier, P., Willems, K. A., Lievens, B., & Zediers, H. (2011). *Vibrio anguillarum* as a fish pathogen: Virulence factors, diagnosis and prevention. *Journal of Fish Diseases, 34*, 643–661.

Garcia-Lopez, I., Otero, A., Garcia-Lopez, M. L., & Santos, J. A. (2004). Molecular and phenotypic characterization of nonmotile Gram negative bacteria associated with spoilage of freshwater fish. *Journal of Applied Microbiology, 96*, 878–886.

Good, C., Davidson, J., Wiens, J. D., Welch, T. J., & Summerfelt, S. (2015). *Flavobacterium branchiophilum* and *F. succinicans* associated with bacterial gill disease in rainbow trout *Oncorhynchus mykiss* (Walbaum) in water recirculation aquaculture systems. *Journal of Fish Diseases, 38*, 409–413.

Han, B. C., Jeng, W. L., Jeng, I. L. I. S., Kao, L. T., Hleng, P. J., & Huang, Y. L. (1997). Rock-shells (*Thair clurigera*) as an indicator of As, Cu, and Zn contamination on the Putai Coast of the Black-Foot disease area in Taiwan. *Archives of Environmental Contamination and Toxicolgy, 32*, 456–461.

Harikrishnan, R., & Balasundaram, C. (2015). Modern trends in *Aeromonas hydrophila* disease management with fish. *Reviews in Fisheries Science, 13*, 281–320.

Hossain, M. (2008). Isolation of pathogenic bacteria from the skin ulcerous symptomatic gourami (*Colisa lalia*) through 16S rDNA analysis. *University Journal of Zoology Rajshahi University, 27*, 21–24.

Hossian, M. M. M., Rahman, M. A., Mondal, S., Mondal, A. S. M. S., & Chowdhury, M. B. R. (2011). Isolation of some emergent bacterial pathogens recovered from capture and culture fisheries in Bangladesh. *Bangladesh Research Publication, 6*, 77–90.

Huicab-Pech, Z. G., Chavez, C. M. R., & Reynoso, F. L. (2017). Pathogenic bacteria in *Oreochromis Niloticus* Var. stirling tilapia culture. *Fisheries and Aquaculture Journal, 8*, 197. https://doi.org/10.4172/2150-3508.1000197.

Huizinga, H. W., Esch, G. W., & Hazen, T. C. (1979). Histopathology of red-sore disease (*Aeromonas hydrophila*) in naturally and experimentally infected largemouth bass *Micropterus salmoides* (Lacepede). *Journal of Fish Diseases, 2*, 263–277.

Ibrahim, B. U., Baba, J., & Sheshi, M. S. (2014). Isolation and identification of bacteria associated with fresh and Smoked Fish (*Clarias gariepinus*), in Minna Metropolis, Niger State. Nigeria. *Journal of Applied and Environmental Microbiology, 2*, 81–85.

Idowu, T. A., Adedeji, H. A., & Sogbesan, O. A. (2017). Fish disease and health management in aquaculture production. *International Journal Environmental & Agricultural Science, 1*, 2–6.

Khan, R. A., & Thulin, J. (1991). Influence of pollution on parasites of aquatic animals. *Advances in Parasitology, 30*, 202–238.

Koskivaara, M., & Valtonen, E. T. (1992). *Dactylogyrus* (Monogenea) communities in three lakes in central Finland. *Parasitology, 104*, 263–272.

Koskivaara, M., Valtonen, E. T., & Prost, M. (1991). Seasonal occurrence of gyrodactylid monogeneans on the roach (*Rutilus rutilus*) and variations between four lakes of differing water quality in Finland. *Aqua Fennica, 21*, 47–55.

Kozinska, A., & Pekala, A. (2012). Characteristics of disease spectrum in relation to species, serogroups, and adhesion ability of motile aeromonads in fish. *The Scientific World Journal*. https://doi.org/10.1100/2012/949358.

Lanno, R. P., & Dixon, D. G. (1996). Chronic toxicity of waterborne thiocyanate to rainbow trout (*Oncorhynchus mykiss*). *Canadian Journal of Fisheries and Aquatic Sciences, 53*, 2137–2146.

Larsen, J. L., & Pedersen, K. (1996). Atypical Aeromonas salmonicida isolated from diseased turbot (*Scophthalmus maximus*). *Acta Veterinaria Scandinavia, 37*, 139–146.

Loch, T. P., & Faisal, M. (2015). Emerging flavobacterial infections in fish: A review. *Journal of Advanced Research, 6*, 283–300.
Majtán, J., Černy, J., Ofúkaná, A., Takáč, P., & Kozánek, M. (2012). Mortality of therapeutic fish *Garra rufa* caused by *Aeromonas sobria*. *Asian Pacific Journal of Tropical Biomedicine, 2*(2), 85–87.
Majumdar, T., Ghosh, S., Pal, J., & Mazumder, S. (2006). Possible role of a plasmid in the pathogenesis of a fish disease caused by *Aeromonas hydrophila*. *Aquaculture, 256*, 95–104.
Nakatsugawa, T. (1994). Atypical *Aeromonas salmonrcida* isolated from cultured shotted halibut. *Fish Pathology, 29*, 193–198. New Delhi.
Noor, R., Acharjee, M., Ahmed, T., Das, K. K., & Paul, L. (2013). Microbiological study of major sea fish available in local markets of Dhaka city, Bangladesh. *Journal of Microbiology, Biotechnology and Food Sciences, 2*(4), 2420–2430.
Olayemi, A. B., Adedayo, O., & Ojo, A. O. (1990). Microbiological studies on freshwater fishes from the Asa River, Ilorin, Nigeria. *Journal of Aquaculture in the Tropics, 5*, 139–139.
Pascoe, D., & Cram, P. (1977). The effect of parasitism on the toxicity of cadmium to the three-spined stickleback, *Gasterosteus aculeatus* L. *Journal of Fish Biology, 10*, 467–472.
Pedersen, K., Kofod, H., Dalsgaard, I., & Larsen, J. L. (1994). Isolation of oxidase-negative *Aeromonas sulmonrrida* from diseased turbot *Scophthalmus maximus*. *Diseases of Aquatic Organisms, 18*, 149–154.
Rahman, M., Colque-Navarro, P., Kuhn, I., Huys, G., Swings, J., & Mollby, R. (2002). Identification and characterization of pathogenic *Aeromonas veronii* Biovar Sobria associated with Epizootic Ulcerative Syndromein Fish in Bangladesh. *Applied and Environmental Microbiology, 68*(2), 650–655.
Sandstrom, O. (1994). Incomplete recovery in a coastal fish community exposed to effluent from a modernized Swedish kraft mill. *Canadian Journal of Fisheries, 51*, 2195–2202.
Sankar, G. P., Saravanan, J., Krishnamurthy, P., Chandrakala, N., & Rajendran, K. (2012). Isolation and identification of *Vibrio* spp. in diseased *Channa punctatus* from aquaculture fish farm. *Indian Journal of Geo-Marine Sciences, 41*(2), 159–163.
Santos, D. M. S., Cruz, C. F., Pereira, D. P., Alves, L. M. C., & de-Moraes, F. R. (2012). Microbiological water quality and gill histopathology of fish from fish farming in Itapecuru-Mirim County, Maranhão State. *Maringá, 34*, 199–205.
Schäperclaus, W. (1991). *Fish diseases* (Vol. 1 and 2). New Delhi: Oxonian Press Pvt. Ltd.
Shayo, S. D., Mwita, C. J., & Hosea, K. (2012). Ulcerative Aeromonas infections in *Tilapia* (*Cichlidae: Tilapiini*) from Mtera Hydropower Dam, Tanzania. *Open Access Scientific Reports, 1*, 1–4.
Shinkafi, S. A., & Ukwaja, V. C. (2010). Bacteria associated with fresh tilapia fish (*Oreochromis niloticus*) sold at Sokoto Central Market in Sokoto, Nigeria. *Nigerian Journal of Basic and Applied Science, 18*(2), 217–221.
Sichewo, P. R., Gono, R. K., Muzvondiwa, J. V., & Sizanobuhle, N. (2013). Isolation and identification of pathogenic bacteria in edible fish: A case study of Fletcher Dam in Gweru, Zimbabwe. *International Journal of Science and Research, 9*, 269–273.
Snieszko, S. F. (1974). The effects of environmental stress on outbreaks of infectious diseases of fish. *Journal of Fish Biology, 6*, 197–208.
Sovenyi, J., & Szakolczai, J. (1993). Studies on the toxic and immunosuppressive effects of cadmium on the common carp. *Acta Veterinaria Hungaricu, 41*, 415–426.
Speare, D. J., Markham, R. J. F., Despres, B., Whitman, K., & MacNair, N. (1995). Examination of gills from salmonids with bacterial gill disease using monoclonal antibody probes for *Flavobacterium branchiophilum* and *Cytophaga columnaris*. *Journal of Veterinary Diagnostic Investigation, 7*, 500–505.
Starliper, C. E. (2011). Bacterial Coldwater disease of fishes caused by *Flavobacterium psychrophilum*. *Journal of Advanced Research, 2*, 97–108.
Sved, D. W., Roberts, M. H., Jr., & Van, V. P. A. (1997). Toxicity of sediments contaminated with fractions of creosote. *Water Research, 31*, 294–300.

Toranzo, E., Magarinos, B., & Romalde, L. (2005). A review of the main bacterial fish diseases in mariculture systems. *Aquaculture, 246*(4), 37–61.

Vethaak, A. D. (1992). Diseases of flounder (*Platichthys flesus* L.) in the Dutch Wadden Sea, and their relation to stress factors. *Netherlands Journal of Sea Research, 29*, 257–272.

Vethaak, A. D., Bucke, D., Lang, T., Wester, P., Johl, J., & Carr, M. (1992). Fish disease monitoring along a pollution transect: A case study using dab *Limanda limanda* in the German Bight, North Sea. *Marine Ecology Progress Series, 91*, 173–192.

Wedemeyer, G. A. (1976). *Physiology of fish in intensive culture systems*. London: Chapman and Hall.

Wedemeyer, G. A., & McLeay, O. J. (1981). Methods for determining the tolerance of fishes to enviromental stressors. In *Pickering: Stress and fish* (pp. 247–275). London/Toronto/Sydney/San Francisco: Academic Press.

White, M. R., Wu, C. C., & Albregts, S. R. (1995). Comparison of diagnostic tests for bacterial kidney disease in juvenile steelhead trout (*Oncorhynchus mykiss*). *Journal of Veterinary Diagnostic Investigation, 7*, 494–499.

White, M. R., Albregts, S. R., Wu, C. C., & Breidert, B. (1996). The use of kidney biopsy of broodstock steelhead trout (*Oncorhyncus mykiss*) to determine the status of bacterial kidney disease infection. *Journal of Veterinary Diagnostic Investigation, 8*, 519–522.

Wiklund, T. (1995). Virulence of 'atypical' *Aeromonas salmonicida* isolated from ulcerated flounder *Platichthys Jesus*. *Diseases of Aquatic Organisms, 21*, 145–150.

Wiklund, T., & Bylund, G. (1993). Skin ulcer disease of flounder *Platichthys Jesus* in the northern Baltic Sea. *Diseases of Aquatic Organisms, 17*, 165–174.

Wiklund, T., & Dalsgaard, I. (1995). Atypical *Aeromonas salmonicida* associated with ulcerated flatfish species in the Baltic Sea and the North Sea. *Journal of Aquatic Animal Health, 7*, 218–224.

Wiklund, T., Dalsgaard, I., Erola, E., & Olivier, G. (1994). Characteristics of 'atypical', cytochrome oxidase-negative *Aeromonas salmonicida* isolated from ulcerated flounders (*Platichthyspesus* (L.)). *Journal of Applied Bacteriology, 76*, 511–520.

Wood, J. (1974). *Diseases of pacific Salmon, their prevention and treatment* (2nd ed.). Olympia: State of Washington, Dept. of. Fisheries.

Impact of Pollution on Quality of Freshwater Ecosystems

T R Kumaraswamy, Sabiha Javeed, Maheen Javaid, and Kumara Naika

Abstract Pollution on freshwater ecosystems is one of the major threats to the biota worldwide. Population growth and human activities are immensely contributing to the degradation and pollution of the freshwater. Freshwater sources, both surface and groundwater are contaminated with different kinds of pollutants (toxic metals and pesticides) discharged from different sources. Various human activities are the main reasons contributing to the decline in quality of freshwaters. The toxic pollutants have adverse effects on aquatic ecosystems and are responsible exclusively or in combination for causing lethal diseases to humans. This chapter tries to discuss the effects of freshwater pollution on its ecosystem.

Keywords Nutrient stress · Microorganisms · Freshwater · Toxic metals · Pollution load · Industrial effluent

1 Introduction

Freshwater environments are among the most beneficial biological systems on the world (Ghermandi et al. 2008) and accounts for many significant services to human culture. They are also very ecologically sensitive and adaptive ecosystems (Turner et al. 2000). Freshwater ecosystems show enormous diversity according to their origin, physical location, water management and chemistry, prevailing species, and soil and sediment physiognomies. One of the primary generally utilized freshwater classifications systems categorizes freshwater systems into lacustarine (lakes), riverine (along waterways and streams), and palustarine (marshes and lowlands) in view of their hydrological, natural and land attributes. However, the Ramsar Convention on wetland biological communities, which is a worldwide arrangement

T. R. Kumaraswamy (✉) · K. Naika
Department of Environmental Science, Bangalore University, Bengaluru, Karnataka, India

S. Javeed · M. Javaid
Department of Environmental Science, University of Kashmir, Srinagar, Jammu and Kashmir, India

H. Qadri et al. (eds.), *Fresh Water Pollution Dynamics and Remediation*,
https://doi.org/10.1007/978-981-13-8277-2_5

marked in 1971 for national activity and universal participation for the preservation and savvy utilization of Freshwater environments and their assets, characterizes Freshwater ecosystems as "areas of marsh, fen, peatland or water, whether natural or artificial, permanent or temporary, with water that is static or flowing". According to the Ramsar Convention definition, the vast majority of normal water bodies, (for example, waterways and lakes) and manmade Freshwater ecosystems, (for example, lakes, cultivate lakes, flooded fields, hallowed forests, supplies, rock pits and channel). Numerous Freshwater ecosystems which delivers important critical functions have continually ignored in the policy development. Thus, numerous freshwater environments, biological communities are debilitated, and many are now debased and lost because of urbanization, population increase, and expanded financial exercises.

The undesirable economic, social, and environmental consequences of diminishing water quality in the freshwater ecosystems are a concern for the world. The problem of deteriorating water quality is particularly more alarming in the case of little water bodies, for example, lakes, tanks and lakes in the past, these water sources performed several economic (fisheries, livestock and forestry), social (water supply), and ecological functions (groundwater recharge, nutrient recycling, and biodiversity maintenance). Regardless of every one of these advantages, numerous leaders and even a large number of the 'essential partners' consider them 'badlands'. Every stake holder asserts a stake in them, as they are in de-restriction, but seldom are enthusiastic to recompense for this extractive use (Verma et al. 2001). These freshwater bodies are frequently subjected to changes in land use in their catchments driving to decrease in inflows and weakening nature of the "runoff" navigating through agricultural fields and urban zones. Then again, huge numbers of them become the "sink" for unprocessed sewages from urban focuses and businesses. Infringement of a supply zone for urban advancement, an unnecessary preoccupation of water for agribusiness is one more significant issue (Verma et al. 2001). The absence of congruity among government approaches in the zones of financial aspects, condition, landscape protection, improvement arranging is one explanation behind the disintegration of these water bodies (Turner et al. 2000). The absence of good administration and administration are additionally real reasons (Kumar et al. 2012) for the decline in freshwater systems.

2 Sources of Pollution in Freshwater Ecosystems

Contaminants enter surface water from a definite, recognizable supply or from an intensive, poorly outlined region. Contaminants that originates at a distinct location, such as a pipe, ditch, tank, or sewer, are examples of point source pollution. Point sources are simple to spot and so are comparatively simple to block. Point source pollutants can go into the water directly. Vessels on large lagoons may leak lubricant or dump waste (Bhat et al. 2018), from time to time unintentionally. Routine vessel operations such as discharging ballast water (water that is used to stabilize a ship)

Table 1 Characteristics of the point and nonpoint sources of elements to receiving waters

Point Source water pollution	Non-point source water pollution
Municipal and industrial effluent	Runoff from agriculture land
Runoff and leachate from waste dumping locations	Runoff from meadow and range
Runoff from animal pens and stable	Runoff from unsewered and sewered areas
Runoff from mining sites, oil refineries, industrial locates	Putrefying tank leachate
Overflow from storm water sewers	Runoff from building establishment
Overflows from drainage and human waste sewers	Runoff from abandoned mines
Runoff from construction locates	Atmospheric deposition over a water surface
	Activities on land that generate impurities, such as logging, wetland alteration, construction, and development of land or waterways

can transport invasive plants and animals and cause the demise to native species. Contaminants that come from a grander area, such as agricultural field, livestock pens, atmosphere, are examples of non-point source pollution. Since non-point source pollution comes from many contaminators, it is much more difficult to control than is point source pollution (Carpenter et al. 1998) (Table 1).

2.1 The Atmosphere

The atmosphere is a vital part of the water sequence and, the position of the earth's weather. This layer of life-giving gases is also the basin for the vaporous discarded products of current human culture. These wastes create air pollution, the contamination of the atmosphere by vapors and particles in amounts that may be injurious to human well-being and the environment. Air impurities have a range of unpleasant effects, from raising the global temperature and destroying natural atmospheric processes to causing damage to the environment and human health (Goolsby and Battaglin 2001). The combustion of fossil fuel releases into the air massive amounts of pollutants such as nitrogen dioxide, sulphur dioxide, carbon monoxide and hydrocarbons. These contaminants drift through the atmosphere or are washed away by rain are of specific concern to humans because heavy metals such as mercury and lead, which are wastes emitted from the burning of coal and other constituents. Sulfur and nitrogen from coal ignition produces the acids that fall as acid rain and create acid streams and lakes. Acid rain is significantly a lot of acidic than traditional rain, it has a pH of around 5.6. Nitrogen wastes in the atmosphere create nitrates, which act as nutrients in the water (Carpenter et al. 1998).

2.2 Domestic Sewage

Everything that is washed down a toilet descends through a sink, or enters a sewer drain in the road becomes sewage, the waste matter that passes through sewers. Sewage is 95% water. The remaining 5% is mostly human waste but also includes oil, toxic chemicals, fertilizers, pharmaceuticals (drugs), pesticides, pathogens, and trash. The organic material is biodegradable, which means it can be broken down by bacteria into stable, nontoxic inorganic compounds, such as carbon dioxide, water and ammonia. Pathogens, synthetic chemicals, and most trash are not biodegradable (Adetunde and Glover 2010). Sewage might run raw into lakes and streams, or it may be treated. In industrialized nations, sewage goes through a sewage treatment plant before it is released into the environment. But even where there are sewage treatment plants, the sewage is not always thoroughly cleansed. The sewage systems of many bigmetropolises are now old and overstretched. Storms cause wastewater to overflow so that sewage is dumped directly into streams and lakes. Some pollutants, such as pest spawns, nutrients, and synthetic organic chemicals, are not removed by the treatment procedure. Untreated sewage fouls the waters of many developed nations. In developing countries, sewage treatment costs are too high, and 90% of sewage enters inland waterways untreated. Large cities annually release voluminous amounts of raw waste matter into native waterways. Drinking or swimming in contaminated water results in hundreds of millions of cases of intestinal diseases each year (Mishima and Gage 1992).

2.3 Runoff

Water that flows across roads, roofs, landfills and contaminated soil often drains directly into streams or lakes. This runoff can be polluted with oil, with impurities that were applied as pesticides or fertilizers, with chemicals from improperly maintained landfills, with pathogens from pet waste, with road salts, and with heavy metals from mines and other sources. More metropolises are diverting roadway runoff to waste water treatment plants. However, the plants are unable to remove some kinds of pollutants, specifically fertilizers and other chemical compounds, which wind up running off into surface waters anyway. Animal wastes enter surface water as runoff, primarily from animal feeding operations at farms (Baig et al. 2009; Goolsby and Battaglin 2001).

2.4 Industrial Waste

Industrial and municipal solid waste may be piped into surface water directly (Bhat et al. 2012, 2014), or it may be stored in ponds and contaminated waste sites (Bhat et al. 2013). Water trickling through disposal areas brings contaminants to streams, lakes, ponds, and groundwater. Many waste disposal sites were built before regulations were in place. Others are improperly maintained, and there is often little enforcement (Carpenter et al. 1998).

3 Impacts of Pollution

3.1 Impacts of Pollution on the Quality of Bottom Sediments in Freshwater Ecosystems

The runoff of clay, silt, and sand from the land into streams and lakes is a natural and necessary process. Sediments build up floodplains and wetlands, and they bring nutrients with them. However, logging, plowing, and other disturbances to the land caused by the construction of roads or buildings can cause excess sediment runoff. Sediments become pollutants if they muddy a stream or lake and hinder photosynthesis, bury aquatic ecosystems, and clog the feeding apparatus or gills of animals. Also, sediments may have toxic chemicals attached to them, thus providing the chemicals with a mode of transportation through the ecosystem (Goolsby and Battaglin 2001).

3.2 Impacts of Pollution on the Quality of Water in Freshwater Ecosystems

3.2.1 Acid Rain

Lakes and streams with low pH support a lower quantity and variety of life. Most marine plants grow best in water with a pH of 7.0–9.2. As pH decreases, plant population also declines, which decreases nourishment for some aquatic birds. If pH continues to lower, the numbers of freshwater shrimp, crayfish, clams, and some fish start to decrease. At pH 5.5, the bacteria that decompose leaf litter and other wreckage begin to perish, cutting off the nutrient source for plankton. If acid runoff brings aluminum into the lake in great quantities, fish populations experience additional stress, leading to lower body weight and smaller size. Native fish are then less able to compete with alien species for food and habitat (Adetunde and Glover 2010). As their environment becomes intolerable, fish populations decline. Young fish that hatch into acidic, metal-rich waters will not survive into maturity, or they may be deformed or stunted in their growth. If the pH goes below 5, females will not spawn, and fish spawns will not hatch. These stresses make the fish more vulnerable to disease and other problems. With a pH below 5, adult fish die. Lake water with a pH less than 4.5 becomes entirely uninhabited of fish. Loss of bacteria causes organic material to lie undecayed on the bottom of a lake while allowing moss to cover its shores (Mishima and Gage 1992). Melting ice or heavy rainstorms can bring in the excess acidic runoff, gradually and briefly lowering the pH of streams and lakes. Waterways that already have low pH are at risk of serious damage from these temporary increases in acidity. Temporary acidification can totally distress an ecosystem and result in considerable fish eradication (Goolsby and Battaglin 2001).

Although frogs can tolerate lower acidity than fish, they cannot live without nourishment. Thus, the loss of food due to increased acidity of the water causes their populations to decline. Birds and mammals that depend on the lake for fish or plants also perish off or leave the area. However, some organisms, plants and mosses, and black fly larvae endure or even flourish in an acidic environment. In fact, acid rain can invert whole freshwater ecosystems, allowing lakes to be taken over by alien organisms and leaving few, if any, native organisms (Kumar et al. 2005).

3.2.2 Heavy Metals

Most heavy metals are a natural part of the environment in low concentrations. Iron and aluminum, for example, are significant constituents of numerous kinds of rocks, and mercury and lead are discharged out by volcanoes. Plants and animals need tiny quantities of some essential metals in limited quantities to carry out their life processes. For example, hemoglobin, the molecule that transferences oxygen in the blood, consumes iron. Many enzymes contain zinc. Other heavy metals essential for life processes include copper, vanadium, and cobalt. All heavy metals are poisonous to organisms in some amount (Mehmood et al. 2019). Mercury, lead, and cadmium are not used by plants or animals and are toxic even in minute quantities. Heavy metals that are biologically useful in small amounts are poisonous in larger quantities. Because these metals bio accumulate, they are especially dangerous to animals that feed high on the food chain (Goolsby and Battaglin 2001; Mehmood et al. 2019). Anthropological activities release heavy metals into the environment. Combustion of coal, fuel oils, fuel additives and waste release heavy metals into the air, as does steel and iron manufacturing (Singh et al. 2018). The metals eventually descent or are rained out of the atmosphere. Runoff from the land carries heavy metals from atmospheric fallout, mines and metal refineries, urban areas, human waste, landfills, and contaminated sediments into inland waters. Large storms cleanse areas that are isolated. Storm waters are particularly laden with heavy metals. Mercury is probably the most damaging heavy metal and is most hazardous when released into the atmosphere by the combustion of coal. In addition, the burning of municipal and medical wastes discharges large quantities of mercury into the atmosphere. Once released, the metal cools and turns into aerosol droplets. These droplets may travel hundreds of miles through the atmosphere but will eventually fall to the ground or into the water to be deposited into the sediment. Bacteria then convert this mercury into organic mercury, usually the dangerous methyl mercury (Adetunde and Glover 2010).

The methyl mercury is easily absorbed through the skin, lungs, and the gut of animals. The compound is tremendously lethal and is toxic to some algae and to the larvae of some small invertebrates. Methyl mercury bio accumulates in top predator's including fish, such as tuna, that people consume. Humans are very tantalizing to methyl mercury: It causes brain, liver, and kidney damage. Recognition of the hazards of mercury to human health resulted in a great reduction in the global mercury manufacture beginning in 1990. Enormous amounts of lead are produced each year, and it is the most common toxic material found in humans. In the form of tetraethyl lead, it has been used as a gasoline additive or as a component of paints. The metal is ubiquitous in computer screens, electronics, batteries, medical equipment, and a myriad of other modern devices. Lead enters the water in the fallout from the atmosphere, in industrial waste, from landfills, and in gasoline residue (Singh et al. 2018). The metal is not toxic to lower organisms and does not seem to bio accumulate. People obtain most of their lead by breathing it from the atmosphere or by ingesting it in paint flakes. Lead leads to the nervous system, brain, and blood disorders (Rashid et al. 2019), especially in children (Howarth et al. 1996).

3.2.3 Pharmaceuticals

Pharmaceutical pollutants include human and veterinary drugs and even illicit (recreational) drugs. Synthetic hormones such as those in birth control pills are a major problem in the waterways, but they are not the only pharmaceuticals that find their way into the environment. The amount of pharmaceuticals and personal care products (shampoos, suntan lotions, perfumes, and soaps, for example) that finds its way into the environment each year is equal to the amount of pesticides used annually (Mian et al. 2010). Other drugs are found in waterways as well. Antibiotics (and antibiotic soaps) are produced to exterminate bacteria. Yet numerous species of bacteria are favorable to the environment, and antibiotics are not always selective about which bacteria they kill. Bacteria have been known to develop resistance to antibiotics, and even low-level concentrations of these substances in the environment could increase the number of disease-causing bacteria that are able to resist these medications (Adetunde and Glover 2010; Mishima and Gage 1992). Proper disposal of pharmaceutical waste is a complex issue, and consistent guidelines are not yet in place. Because pharmaceuticals are not regulated as pollutants, people in households can dispose of them however they choose, although medical care facilities do have guidelines for the disposal of unused medications.

3.2.4 Thermal Pollution

To cool a thermal plant, water is drained from the watercourse, channeled through the area that essentials to be cooled, and then reverted to the river. Approximations are that nearly half of all water used in the United States each year is for cooling electric power plants. A 1000-megawatt power plant heats more than 2 million gallons (10 million l) of water by 95 °F (35 °C) every hour. The water adjoining a plant may be as much as 18 °F (10 °C) higher than the water farther away (Wetzel 2001). Heated water has numerous properties on the nearby ecosystem. Warmer temperatures upsurge the capability of plants to photosynthesize, which may outgrowth an algal bloom. Warmer temperatures also increase the stress on plants and animals in the water. Warm water holds less dissolved oxygen than cool water, making it more difficult for aquatic animals to breathe. Some species suffocate in water temperatures greater than 95 °F (35 °C). Higher temperatures and lower oxygen may increase the animals' vulnerability to complications from pathogens and toxic substances. As a result, the biodiversity, the number of different species in the ecosystem may decrease. Studies have shown that phytoplankton diversity decreases at thermal waste sites (Xiaolong et al. 2010). Temperature changes may harm fish and other aquatic organisms in other ways. Fish and invertebrates are ectotherms. Their body temperatures are the same as the surrounding water. These "coldblooded" animals are slow moving and slow growing and are adapted to specific water temperature. Warmer temperatures speed their metabolism; for example, their heart rate doubles with every 18 °F (10 °C) rise in water temperature, which harms their ability to survive and reproduce. Native fish that like cooler water, such as trout, may lose ground while nonnative species, algae, and bacteria may increase and thrive. By contrast, endotherms are "warm-blooded" animals that retain their body temperatures almost constant, independent of the temperature of their surroundings. Endotherms fuel their heat by consumption of lot of food and maintain their body temperatures with insulation such as fur, feathers, and blubber (Adetunde and Glover 2010).

Power plants have changed categories of cooling systems. The easiest and cheapest cooling method are the ones through the system. Cool water is reserved from a nearby water body, and hot water is returned to the same water body. The once-through system is by far the most environmentally destructive. Closed-cycle cooling reuses the cooling water so that the waste heat does not leave the plant. A favored type of closed-cycle cooling, which is expensive to build and operate, pumps the hot water into towers, where the excess heat is released into the atmosphere (Steiner et al. 2006).

Two-thirds of the energy produced by the plant becomes not electricity, but unusable heat. Plants, eggs, larvae, and juvenile fish, including about 35% of the young striped bass in that portion of the river, are lost as the cooling water runs through the plant. Some larger fish are killed or injured when they are trapped on the screens that prevent them from being sucked into the cooling system. In contrast, closed cycle technology not only drastically limits waste heat. It also reduces fish kills by up to 97% (Ramakrishnaiah et al. 2009). Dramatic and voluminous thermal pollution that enters the water as runoff from paved surfaces is a grave concern. Because pavement absorbs heat better than natural surfaces, especially in the summer, the temperature of water flowing from a parking lot may be a few degrees higher than water flowing off a natural surface. Even small differences in temperature can alter the environment enough to stress the native fish and plant species (Baig et al. 2009).

3.2.5 Nutrient Load

Nutrients enter streams and lakes from runoff from land, fallout from the atmosphere, and recycling of plant and animal tissue within the aquatic environment (Bhat et al. 2017). Without nutrients, plants and animals could not grow, replenish their bodies, or have energy for living. Because they are essential for life, it seems impossible that nutrients could be a pollutant, yet nutrients are the most serious pollutants entering freshwater systems today. Recent escalations in nutrient input are having a dire effect on freshwater ecosystems (Bu et al. 2010). Nutrients themselves are occasionally deadly. Depending on temperature and pH, ammonia can be venomous to fish and other aquatic creatures. Excess nutrients come chiefly from human and animal wastes, detergents, and fertilizers. Sewage and runoff that enters the water directly, either from a deliberate act or from leaks and spills from wastewater lagoons, carry an enormous amount of nutrient pollution which is lethal for fish survival.

Although this did not occur in any of the incidents mentioned, high nitrate concentrations in drinking water can result in the illness in infants known as methemoglobinemia or "blue baby" syndrome. In this illness, the baby's digestive system acclimates the nitrates to nitrite, a process that interferes with the blood's ability to carry oxygen. The baby's tissues, deprived of oxygen, turn blue (Steiner et al. 2006). Smaller quantities of nutrients in a lake do not cause an immediate fish kill but accelerate eutrophication. Eutrophication is part of a lake's natural aging process as it progresses from oligotrophic to mesotrophic to eutrophic. Excess nutrients drastically reduce the natural pace of this process from thousands of years to just a few years. During eutrophication, excess nutrients act as fertilizer for algae and aquatic plants and bring on what is called a bloom. When the plants die, bacteria populations expand to consume the tissue. These aerobic bacteria need oxygen, and an enormous number of them deplete the gas from the water. The decaying tissue also warms the lake's water, causing oxygen and other gases to bubble out into the atmosphere. Oxygen poor water is called hypoxic; fish and most other animals cannot survive in it. Hypoxic waters then become lifeless zones, regions that are unreceptive to most forms of life. As native species die off or leave the area, different species begin to appear (Adetunde and Glover 2010).

3.2.6 Invasive Species

Invasive species is a gargantuan problem in some zones. Over the past 200 years, the rate of species invasion has mounted exponentially as people have moved more spontaneously around the earth. For an alien species, the main route to a new marine ecosystem is via vessel. For instance, the ballast water drawn into tanks in ships to stabilize the vessels contains the organisms that happened to be swimming in the water when it was pumped in, such as plankton, jellies, larval mollusks, and crustaceans. When the ballast water is discarded, the non-native species (often along with pollutants) are ejected with it. Species can also drift to new environments while attached to vessels and propellers. They can also travel packed with bait worms and other consignment (Ramakrishnaiah et al. 2009). Aquarium discarding is another common method for a non-native species to annex a new aquatic system. When people are tired of their aquariums, or when their aquatic animals become difficult to manage, they dump the organisms into the nearest stream, lake, or pond.

Freshwater Asian clams (Corbicula fluminea), used in aquariums, are speedily diffusing and dislodging native species in the United States. Thousands of water bodies are infested with aquarium plants that were dumped from tanks by their holders (Goolsby and Battaglin 2001). After a species is familiarized to a new environment, numerous possible consequences may result. In most circumstances, environments for survival are not right, and the alien organism succumbs. Sometimes, an alien species may be compatible with the natives and contribute to the biodiversity of the ecosystem. Rarely do invasive species thrive; but when they do, it may be because they have no predators and so are able to out-compete native species for nourishment and living space. Exploding populaces of alien organisms significantly decline the diversity of an ecosystem by shifting the habitat to the degree that it becomes unsuitable for the native species, thereby driving the native species toward annihilation (Steiner et al. 2006).

3.2.7 Human Beings

Manmade organic compounds contained in pesticides, flame retardants, industrial solvents, and cleaning fluids are commonly found in aquatic environments (Folke et al. 2002; Gurung et al. 2012). Although these persistent organic pollutants (POPs) have many usages, some are deadly even in minute quantities. POPs do not biodegrade or disintegrate in the environment. POPs are exceptionally effective at bioaccumulation. Animals consume tissues and shells with adsorbed POPs as they descend through the water column, or after they accumulate in bottom sediments. As a result, concentrations are especially high in top predators that consume aquatic organisms. Examples of these predators include mammals such as otters, whales, and dolphins, and fish eating birds such as eagles, ospreys, and gulls (Fischer and Young 2007; Loyau and Schmeller 2017). One compound was found to be 71 times more concentrated in polar bears than in the seals they eat. Many POPs are toxic, some even in small amounts (Ramakrishnaiah et al. 2009). They evaporate and enter the atmosphere, particularly in warm conditions. Once in the air, they travel everywhere, even thousands of miles (km) from where they were used. The compounds then rain out of the atmosphere or attach to dust particles and are blown into lakes and streams. POPs are found in high levels even in animals and people that inhabit the remote reaches of the planet. Researchers in the 1980s were shocked to discover that the breast milk of native women in the Canadian Arctic, which they analyzed because they needed and assumed POP quantities would be minimal, had high concentrations of these toxins (Xiaolong et al. 2010). Most POPs are insoluble in water, but are soluble in fats. Polar bears rely on stored body fat for part of the year, making them particularly vulnerable to the effects of POPs. People, too, metabolize the chemicals when they lose weight or pass nutrients on to a developing fetus or in breast milk (Paul and Meyer 2001).

Polychlorinated biphenyls (PCBs) are extremely stable, water-soluble compounds that were once used as flame retardants; to cool and insulate electrical devices; to manufacture paints, plastics, adhesives, and other materials; and to strengthen wood and concrete. PCBs were never supposed to be released into the environment, but they leaked from equipment and waste disposal locations. Their stability allows them to persist in the soil and water for many years. Although PCBs have been banned in industrialized nations for decades, they are still everywhere in the environment, particularly in the animals at the top of the food web. Fortunately, concentrations of PCBs are dropping as they become attached to sediments, fall to lake bottoms, and are buried (Ramakrishnaiah et al. 2009). The damage brought by PCBs is multifaceted. They are extremely toxic to fish and invertebrates, even in small concentrations, and are strong endocrine disruptors. They interfere with reproduction and development in birds and mammals, reducing the number and survival rates of their offspring. In mammals, PCBs interfere with the metabolism of thyroid hormones, which regulate a diversity of physiological processes, including brain development and metabolism. PCBs also reduce immune system function: Polar bears are losing their ability to fight common infections and are also beginning to show some endocrine effects, such as masculinization in some females (Okeke and Igboanua 2003). In humans, PCB exposure has been linked to developmental neurological problems in children (Goolsby and Battaglin 2001), liver dysfunction, skin and respiratory problems, dizziness, and possibly cancer. One recent study found that levels of three chemicals (PCBs, hexachlorobenzene, and chlordane) were higher in the mothers of men with testicular cancer than in a control group, suggesting that the cancer was initiated in Utero (Carpenter et al. 1998).

4 Conclusion

Pollution worsens the quality of freshwater ecosystems due to addition of foreign bodies lethal to aquatic biota and human society. The persistent pollutants and potentially harmful toxic metals, passed to distant zones by climate extremes, intermingling with local and introduced biota can alter freshwater ecosystems intensely. Such altered freshwater ecosystems might be less stable, less healthy and non-functional most of the period to provide ecosystem services to human society and other forms of life. Despite the prominence of freshwater ecosystems to the livelihood of biodiversity and mankind, research efforts are reduced by monetary restraints and the absence of wide-ranging, multidisciplinary methods to capture all facets of change. The focus should be given to the better understandings, mechanisms of pollutants arrival into freshwaters, both temporally and spatially. Recent distribution of pollutants with sufficient seasonal resolution needs to be complemented by changes in freshwater biodiversity and human land use regimes.

References

Adetunde, L. A., & Glover, R. L. K. (2010). Bacteriological quality of borehole water used by students' of university for development studies, Navrongo campus in upper-east region of Ghana. *Current Research Journal of Biological Sciences, 2*(6), 361–364.

Baig, J. A., Kazi, T. G., Arain, M. B., Afridi, H. I., Kandhro, G. A., Sarfraz, R. A., et al. (2009). Evaluation of arsenic and other physico-chemical parameters of surface and ground water of Jamshoro, Pakistan. *Journal of Hazardous Materials, 166*(2–3), 662–669.

Bhat, R. A., Dar, G. H., Jehangir, A., Bhat, B. A., & Yousuf, A. R. (2012). Municipal solid waste generation and present scenario of waste management during Yatra season in Pahalgam: A tourist health resort of Kashmir valley. *International Journal of Current Research, 4*, 004–009.

Bhat, R. A., Kamili, A. N., & Bandh, S. A. (2013). Characterisation and composition of Municipal Solid Waste (MSW) generated in Yusmarg: A health resort of Kashmir valley: A Glance at the World. *Waste Management, 33*, 774–777.

Bhat, R. A., Nazir, R., Ashraf, S., Ali, M., Bandh, S. A., et al. (2014). Municipal solid waste generation rates and its management at Yusmarg forest ecosystem, a tourist resort in Kashmir. *Waste Management and Research, 32*, 165–169.

Bhat, R. A., Shafiq-ur-Rehman, D. M. A., Mushtaq, N., Bhat, J. I. A., & Dar, G. H. (2017). Current status of nutrient load in Dal Lake of Kashmir Himalaya. *Journal of Pharmacognosy and Phytochemistry, 6*(6), 165–169.

Bhat, R. A., Dervash, M. A., Mehmood, M. A., & Hakeem, K. R. (2018). Municipal solid waste generation and its management, a growing threat to fragile ecosystem in Kashmir 756 Himalaya. *American Journal of Environmental Sciences, 2*(11), 145–167.

Bu, H., Tan, X., Li, S., & Zhang, Q. (2010). Water quality assessment of the Jinshui River (China) using multivariate statistical techniques. *Environmental Earth Sciences, 60*(8), 1631–1639.

Carpenter, S. R., Caraco, N. F., Correll, D. L., Howarth, R. W., Sharpley, A. N., & Smith, V. H. (1998). Nonpoint pollution of surface waters with phosphorus and nitrogen. *Ecological Applications, 8*(3), 559–568.

Fischer, A., & Young, J. C. (2007). Understanding mental constructs of biodiversity: Implications for biodiversity management and conservation. *Biological Conservation, 136*(2), 271–282.

Folke, C., Carpenter, S., Elmqvist, T., Gunderson, L., Holling, C. S., & Walker, B. (2002). Resilience and sustainable development: Building adaptive capacity in a world of transformations. *Ambio: A Journal of the Human Environment, 31*(5), 437–440.

Ghermandi, A., van den Bergh, J. C., Brander, L. M., de Groot, H. L., & Nunes, P. A. (2008). The economic value of wetland conservation and creation: A meta-analysis. *Water Resources Research, 46*(12).

Goolsby, D. A., & Battaglin, W. A. (2001). Long-term changes in concentrations and flux of nitrogen in the Mississippi River Basin, USA. *Hydrological Processes, 15*(7), 1209–1226.

Gurung, A. B., von Dach, S. W., Price, M. F., Aspinall, R., Balsiger, J., Baron, J. S., & Kohler, T. (2012). Global change and the world's mountains—Research needs and emerging themes for sustainable development: A synthesis from the 2010 Perth II Conference. *Mountain Research and Development, 32*(S1), S47–S54.

Howarth, R. W., Billen, G., Swaney, D., Townsend, A., Jaworski, N., Lajtha, K., & Berendse, F. (1996). Regional nitrogen budgets and riverine N & P fluxes for the drainages to the North Atlantic Ocean: Natural and human influences. In *Nitrogen cycling in the North Atlantic Ocean and its watersheds* (pp. 75–139). Dordrecht: Springer.

Kumar, R., Singh, R. D., & Sharma, K. D. (2005). Water resources of India. *Current Science, 89*(5), 794–811.

Kumar, M. D., Sivamohan, M. V. K., & Bassi, N. (2012). Technology choices and institutions for improving the economic and livelihood benefits from multiple-use tanks in western Odisha. In *Water management, food security and sustainable agriculture in developing economies* (pp. 158–183). Oxon: Routledge.

Loyau, A., & Schmeller, D. S. (2017). Positive sentiment and knowledge increase tolerance towards conservation actions. *Biodiversity and Conservation, 26*(2), 461–478.

Mehmood, M. A., Qadri, H., Bhat, R. A., Rashid, A., Ganie, S. A., Dar, G. H., & Shafiq-ur-Rehman. (2019). Heavy metal contamination in two commercial fish species of a trans-Himalayan freshwater ecosystem. *Environmental Monitoring and Assessment Environment, 191*, 104. https://doi.org/10.1007/s10661-019-7245-2.

Mian, I. A., Begum, S., Riaz, M., Ridealgh, M., McClean, C. J., & Cresser, M. S. (2010). Spatial and temporal trends in nitrate concentrations in the River Derwent, North Yorkshire, and its need for NVZ status. *Science of the Total Environment, 408*(4), 702–712.

Mishima, A., & Gage, R. L. (1992). *Bitter Sea: The human cost of Minamata disease*. Tokyo: Kosei Publishing Company.

Okeke, C. O., & Igboanua, A. H. (2003). Characteristics and quality assessment of surface water and groundwater recourses of Akwa Town, Southeast, Nigeria. *Journal of Nigerian Association of Hydrogeologists Geology, 14*, 71–77.

Paul, M. J., & Meyer, J. L. (2001). Streams in the urban environment. *Annual Review of Ecology and Systematics, 32*, 333–365.

Ramakrishnaiah, C. R., Sadashivaiah, C., & Ranganna, G. (2009). Assessment of water quality index for the groundwater in Tumkur Taluk, Karnataka State, India. *Journal of Chemistry, 6*(2), 523–530.

Rashid, A., Bhat, R. A., Qadri, H., & Mehmood, M. A. (2019). Environmental and socioeconomic factors induced blood lead in children: An investigation from Kashmir, India. *Environmental Monitoring and Assessment, 191*(2), 76. https://doi.org/10.1007/s10661-019-7220-y.

Singh, D. V., Bhat, J. I. A., Bhat, R. A., Dervash, M. A., & Ganei, S. A. (2018). Vehicular stress a cause for heavy metal accumulation and change in physico-chemical characteristics of road side soils in Pahalgam. *Environmental Monitoring and Assessment, 190*(6), 353. https://doi.org/10.1007/s10661-018-6731-2.

Steiner, T. S., Samie, A., & Guerrant, R. L. (2006). Infectious diarrhea: New pathogens and new challenges in developed and developing areas. *Clinical Infectious Diseases, 43*, 408–410.

Turner, R. K., Van Den Bergh, J. C., Söderqvist, T., Barendregt, A., Van Der Straaten, J., Maltby, E., & Van Ierland, E. C. (2000). Ecological-economic analysis of wetlands: Scientific integration for management and policy. *Ecological Economics, 35*(1), 7–23.

Verma, M., Bakshi, N., & Nair, R. P. (2001). *Economic valuation of Bhoj Wetland for sustainable use. Unpublished project report for World Bank assistance to Government of India, Environmental Management Capacity-Building* (p. 35). Bhopal: Indian Institute of Forest Management.

Wetzel, R. G. (2001). *Limnology: Lake and river ecosystems*. Oxford: Gulf Professional Publishing.

Xiaolong, W., Jingyi, H., Ligang, X., & Qi, Z. (2010). Spatial and seasonal variations of the contamination within water body of the Grand Canal, China. *Environmental Pollution, 158*(5), 1513–1520.

Heavy Metal Intrusion and Accumulation in Aquatic Ecosystems

Khursheed Ahmad Wani, Javid Manzoor, Ashaq Ahmad Dar, Razia Shuab, and Rafiq Lone

Abstract Most of the heavy metals have deleterious impacts on the growth and productivity of the plants, and also alters the general physiological characteristics of plants. But, some plants cope with the pollution stress and accumulate more and more toxic and heavy metals in their modified tissues. These toxic substances not only affect the plant physiology, but have toxic impacts on soil health. The toxic metals, in context to many organic compounds are not decomposed by the microbiological activity. Toxic levels of Lead (Pb), Cadmium (Cd) and Mercury (Hg) affects plant processes at physiological and biochemical levels some of the heavy metals are accumulated in aquatic environment and most of them get absorbed in the aquatic plants. Theretofore, a miscellaneous positive correlation among selected aquatic plants and specific heavy metals was reported, however mechanism of a particular model species is still vague. The intrusion of heavy metals may also change the nutrient pool of the aquatic ecosystem that may affect the overall productivity of the system. The work will review some of the important heavy metals, the plants that are useful to reduce the concentration of these metals from different ecosystems.

Keywords Heavy metals · Phytoremediation · Tolerance · Growth · Productivity · Toxicity

K. A. Wani (✉)
Department of Environmental Science, Govt Degree College, Bijbehara, Jammu and Kashmir, India

J. Manzoor
Department of Environmental Science, Jiwaji University Gwalior, Gwalior, India

A. A. Dar
Department of Ecology and Environmental Science, Pondicherry University, Puducherry, India

R. Shuab
Department of Botany, University of Kashmir, Srinagar, India

R. Lone
Department of Botany, SBBS University, Jalandhar, Punjab, India

H. Qadri et al. (eds.), *Fresh Water Pollution Dynamics and Remediation*,
https://doi.org/10.1007/978-981-13-8277-2_6

1 Introduction

Heavy metals are the high density metallic compound components and are among the essential class of contaminants in the earth. The concentration of heavy metals in nature has expanded because of different anthropogenic activities like burning of petroleum derivatives, release of waste, utilization of fertilizers and pesticides and so forth (Mehmood et al. 2019). Heavy metals are characterized as metallic components that have a moderately high thickness contrasted with water. Heavy Metals are characterized as high-density metallic components with atomic no. >20. Heavy metals contaminants that are regularly found in nature are cadmium (Cd), chromium (Cr), copper (Cu), mercury (Hg), lead (Pb), nickel (Ni) and zinc (Zn). Presumably some of these are important for plant development and are referred to as micronutrients, for example, Zn, Cu, Mn, Ni and Co, while others (Cd, Pb and Hg) have obscure organic capacities (Gaur and Adholeya 2004). Natural ecosystems are influenced by the metals and don't experience biodegradation yet can be aggregated in living life forms, subsequently causing different sicknesses and disarranges even in moderately bring down concentration (Clark 1993). The mushroom growth of industries, rapid urbanization and ever increasing population is one of the common causes of environmental degradation and pollution (Bhat et al. 2017). Phytoremediation is an environmentally friendly technique that is ecologically sound and economically effective is a smart way to the current cleanup methods that are very expensive. This technology involves proficient use of plants to eliminate, detoxify or immobilize heavy metals in reasonable means. Recently, phytoremediation as a cost effective and environmentally friendly technology has been developed by scientists and engineers in which biomass/microorganisms or live plants are utilized to remediate the contaminated regions. It very well may be sorted into different applications, including phytofiltration, phytostabilization, phytoextraction, and phytodegradation (Ahmadpour et al. 2012). Different macrophyte species *Phragmites australis* (Cav.) Trin., *Typha orientalis Presl, Lythrum salicaria Linn., Arundo donax Linn. var. versicolor Stokes, Typha minima Funk, Juncus effusus L., Pontederia cordata L., Cyperus alternifolius Linn.* subsp. *flabelliformis (Rottb.) Kükenth., Acorus calamus Linn.*, and *Iris pseudacorus Linn.* were investigated and compared for their shapes, biomass, roots, and ability to accumulate heavy metals. *Acorus calamus Linn*, *T. orientalis Presl*, *P. australis (Cav.) Trin.*, *T. minima Funk*, and *L. salicaria Linn.* displayed the maximum concentration of metal tolerance, whereas *P. cordata L.*, *I. pseudacorus Linn.*, and *C. alternifolius Linn.* sub sp. *flabelliformis (Rottb.)* Kükenth. had the minimum. The concentration of different metals ranges from minimum to maximum, such as *T. minima Funk, P. australis (Cav.) Trin., L. salicaria Linn., A. donax Linn. var. versicolor Stokes, P. cordata L., and A. calamus Linn.*, whereas T. *orientalis Presl and C. alternifolius Linn.* sub sp. *flabelliformis (Rottb.)* Kükenth had poor capacity to gather heavy metals (Sun et al. 2013).

This expansion of overwhelming metal fixation is a noteworthy worry to the people and environment (Kabata-Pendias 2011), in light of their non-biodegradable nature. Instant and necessary measures are required to remediate such polluted systems. Of all the remediation advances, phytoremediation has been favored due to its cost-viability, ecofriendly nature (Uqab et al. 2016) and easy maintenance (Kamran et al. 2014).

Phytoremediation is a novel methodology and a coordinated multidisciplinary approach which gives an incredible potential to treat such polluted systems utilizing plants (Uqab et al. 2016: Jadia and Fulekar 2009: Sarma 2011). Numerous regular strategies are extremely costly, relentless and don't give the adequate outcomes. Phytoremediation, serves a biological option, has increased expanding consideration due to its cost-effective nature.

Plants are of special interest, as most of the plants accumulate dangerous metals and supplements in huge amounts in contrast with to terrestrial plants (Pratas et al. 2012). Also, in view of biochemical arrangement, propensity, species, bounty and condition, these macrophytes has been found to assimilate these toxins at various rates and efficiencies. Studies have discovered that amid the metal-binding cysteine-rich peptides (phytochelatins), which detoxify heavy metals by forming complexes with them (Kinnersely 1993). Plants are equipped for expelling the metal pollution from water and additionally from soil. Aquatic plants of different kinds whether free gliding, submerged all are known for evacuating heavy metals. Phytoremediation is an ecofriendly innovation that utilizations characteristic or genetically modified plants, with their related rhizospheric microorganisms which invigorate plant development and purify soil and water in blend with the plants (Uqab et al. 2016). Phytoremediation is an all around arranged cleanup innovation for an assortment of natural and inorganic toxins. Plants extricate metals, hydrocarbon mixes and man-made synthetic compounds, for example, herbicides, fungicides, pesticides and anti-infection agents. Phytoremediation is an environmental friendly, cheap, efficient and most reliable as it helps to remove the contamination. Plants have and utilize an assortment of instruments to manage the contaminations particularly heavy metals, hydrocarbon mixes and manmade synthetic compounds, for example, herbicides, fungicides, pesticides. Plants sequester them in their cell dividers. Plants chelate these contaminants in the soil in inert structures or complex those in their tissues and can store them in vacuoles, far from the delicate cell cytoplasm where most metabolic procedures happen. Organics might be debased in the root zone contingent upon their properties of plants or taken up, trailed by corruption, sequestration, or volatilization. Effectively phytoremediated natural contaminations incorporate natural solvents, for example, TCE (the most widely recognized toxin of groundwater) (Newman and Reynolds 2004), herbicides, for example, atrazine (Burken and Schnoor 1997). Explosives, for example, TNT (Hughes et al. 1997), petroleum hydrocarbons, and the fuel added substance MTBE (Davis et al. 2003) and polychlorinated biphenyls (PCBs). Phytoremediation is a developing innovation that utilizations plants to expel contaminants from soil and water (Bhadra et al. 1999). Phytoremediation technologies which are used for the uptake of heavy metals include mechanisms of phytoextraction, phytostabilisation, rhizofilltration, and phytovolatilization (Fig. 1).

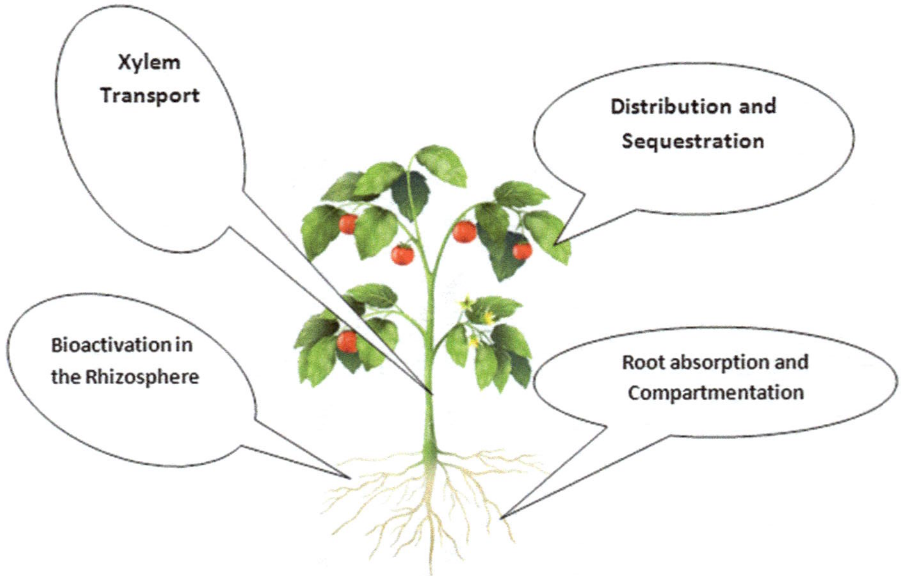

Fig. 1 Heavy metal hyperaccumulation by plants

The fundamental focal point of this chapter is to examine the capability of phytoremediation procedure to treat substantial metal polluted destinations, to give data about the components embraced by plants for overwhelming metal take-up and furthermore to give a concise rundown of sea-going plants productive for remediation of different metals.

2 Methodology

The search for relevant literature was approached with a rather broad perspective. Keywords were heavy metals, aquatic plants, phytoremediation, cadmium, phytotoxicity, metal stress, Hyperaccumulator, hyperaccumulation, contamination, toxicity, metal stress, nutrient pool, nutrient dynamics (Bhat et al. 2017), photosynthetic rate, growth, yield, multiple pollution, soil enzyme, biomass accumulation, sodium chloride, nitrate reductase activity, fast growing plant, phytoextractor, antioxidant enzyme, oxidative stress, hydrogen peroxide, NADP oxidase, phytochelation and their synonyms.

International data base was searched; resulting in a total of 1000 references, out of these more than 600 came from Elsevier and Science direct. ISI web base of knowledge was also used to select the most of the references. The rest came from smaller databases. The references were sorted in two rounds, in the first round irrelevant references were excluded. The quality of reference was assessed by using the criteria such as contribution of new knowledge, originality of empirical findings, use of theory in design and analysis and finally, whether, the reference took the special characteristics of environment toxicant such as lead, mercury, cadmium into consideration.

3 Heavy Metals and Metal Accumulation

3.1 Lead (Pb)

The orderly investigation on physiological impacts of lead on plant kingdom and other living ecosystems, work began at the end of 1960's, although the dangerous impacts of the metals were known for long time. Moreover, in contrast to dicots, monocot response to heavy metals was reported in detail (Broyer et al. 1972). In spite of the fact that lead is thought to be lethal to the plants, metal sensitivity and reaction of various responses depend upon physiology and genetic make-up of plant. The pattern of impacts and capability of imperative product plants to tolerate the metal toxicity need a careful examination. Effect of lead on physiology and biochemistry on fauna (Rashid et al. 2019) and flora (higher plants) was studies (Thapa et al. 1988; Pahlsson 1989).

3.1.1 Major Sources of Lead

The major sources of lead contamination incorporate metal purifying, toxic paints, lead arsenate, pesticides, vehicular emissions (Singh et al. 2018) and fertilizers containing phosphate (Lagerwerff and Armiger 1973; Goldsmith et al. 1976). Furthermore, use of antiknocking agents particularly lead alkyl in fuel makes the automobiles a major source of lead (Smith 1971) which gets accumulated on roads and near roadsides, and later on may be transported to far areas by different means. The soil and lushness occupying the wayside are regularly accumulating lead substance (Cannon and Bowlers 1962; Chow 1970; Motto et al. 1970; Page and Ganje 1970).

3.1.2 Concentration of Lead in the Earth

The lead concentration in the earth is dynamic and is expanding quickly because of the industrialization and urbanization amid the most recent couple of decades. Hussanin et al. (1993) announced in sandy soils an expanded aggregate and accessible lead in Egypt which had been flooded with water of household origin for up to 67 years. Moreover, during half a decade of irrigation metal concentration decrease with depth and, the total and available lead respectively, amplified from 0.01 gL^{-1}–0.04 gL^{-1} and 0.0007 gL^{-1}–0.003 gL^{-1}. Not with standing, Akhter and Madany (1993) reported 697.2 mgg^{-1} and 360.0 mgg^{-1} of lead and metals (lead, zinc, nickel, chromium) respectively, in road dust and house hold residue of Bahrain. A 0.003 mgL^{-1} lead concentration was reported in uncontaminated fresh water (Forstner and Wittman 1979), however as per, the worldwide scope of lead focus in aquifers differed from 0.001–0.06 mgL^{-1}. The amount of lead in water was observed to be as high as 0.0014 gL^{-1} that was utilized for cattle's and cleansing by countryside population (Chandra et al. 1993). An estimation by researchers of Chaudhary Charan Singh Agricultural University, Hissar (India) the amount of lead in Haryana was observed to vary from 0.01–0.02 g kg^{-1} and 0.007 to 0.022 g kg^{-1} in agriculture soils and in soils inundated with household water (Kuhad and Malik, 1989). Moreover, a lead concentration of 0.010–0.015 g kg^{-1} was also reported in soil (Kuhad and Malik 1989). Singh et al. (1997) while investigating the lead accumulation capability *Panicum maximum* reported 0.006 mg g^{-1} and 0.0008 mg g^{-1} lead in shoots when it was developed on the sewage and uncontaminated site, respectively.

Different water courses have additionally became polluted with abnormal state of lead. Amount of lead in representative water samples of Hussain Sagar Lake (Hyderabad, India) was accounted between 38,000 and 62,000 ngL^{-1}. Furthermore, it was also discovered that ground water gathered lead levels to a range of 200–1000 m, having 0.007–0.028 mg/l which diminished up to 0.001–0.009 mg/l water in the ground water from span of 1000–2000 meter of the water body (Srikanth et al. 1973). Waste deposition in the lake may be the possible reason. In India soils and water have turned out to be contaminated upto unsafe limits as there is greater lead collection in vegetation with no secondary lead sources in these belts (Kumar et al. 1993; Dabas 1992; Bharti and Singh 1993).

The metal persists to a great extent as a shallow store or topical airborne covering on flora surfaces (Schuck and Locke 1970). Also a 5–200 times more lead was reported in unwashed foliage surfaces (Smith 1971). Leaves with hairs accumulate more lead as compared to glabrous leaves perhaps because of the general wash-off of metal by precipitation (Zimdahl and Koeppe 1977).

3.1.3 Lead Uptake by Plants

Soil contains dissolvable and insoluble salts of lead that are steadfastly bound to particles of size 1–1000 nm (colloidal). Different life-forms may assimilate lead both from air as well as soil. The airborne part of lead, which comes out in the form of residue exhaust, and damp vapor, accumulate on the aboveground plant parts (Zimdahl and Koeppe 1977). Different vegetation types developing on the side of road are particularly enhanced with lead due to arrival of lead from vehicles which diminishes with distance from roadside (Wallace et al. 1974; Goldsmith et al. 1976; Wheeler and Rolfe 1979). The root biomass can remove a portion of this metal, however its movement to aboveground biomass is generally constrained. Different conditions of soil, for example, cation trade limit (Miller et al. 1975; Zimdahl and Faster 1976; John and Laerhoven 1976) and different particles (Rolfe and Reinbold 1977; Singh et al. 1994) change lead take-up from the soil arrangement. Welsh and Denny (1980) reported the accumulation of lead in aquatic plant cells was connected with tissue uronic acid. In maturing macrophyte tissue, an increased surface territory and hence lead binding sites which increment the take-up limit with respect to lead was reported (Odum and Drifmeyer 1978). Lead apparently progresses toward becoming complexed to anionic sites related with pectic substances with cell wall (Sharpe and Denny 1976; Guilizzoni 1991). It was discovered that in *Zea mays* lead may decrease cadmium take-up (Miller et al. 1977) and in *Sesame indicum* root and leaf biomass, it meddled with Cd (II), Cu (II) and Na (I) take-up (Singh et al. 1994). The presence of abundant measure of phosphate, insoluble accelerates of lead orthophosphate and lead pyrophosphate lessen lead assimilation, yet the presence of chelating agents, for example, Diethylenetriaminepentaacetic acid enhances lead take-up (Martin and Hammand 1966; Tandon and Crowdy 1971; Wallace et al. 1976). Diethylenetriaminepentaacetic corrosive likewise builds the take-up and movement of lead in *Hordeum vulgare* (Patel et al. 1977).

3.1.4 Accumulation and Localization of Lead in Aquatic Plant Parts

Pistia stratoites (water lettuce) is an aquatic plant that develops quickly and a high biomass crop with a broad root framework that has potential to improve the substantial metals evacuation. This plant displayed diverse pattern to lead removal, albeit, accumulated high concentrations mostly in the root framework. The sorption of weakened substantial metal particles, specifically Pb and Cd by dead *P. stratiotes* has appeared to be a proficient and easy choice to be considered in mechanical profluent treatment (Miretzky et al. 2005). *Eichhornia crassipes* (water hyacinth) has been recorded as most troublesome weed in aquatic system. It is a submerged aquatic plant, found bounteously throughout the year in very large amount and drainage channel system in and around the fields of irrigation.

Tiwari et al. (2007) clarified that overwhelming metals Pb, Zn, Mn demonstrate more noteworthy partiality towards bioaccumulation in their investigation. Nearness of higher grouping of substantial metals in plants implies the biomagnifications. *Eichhornia crassipes* has the extraordinary property to aggregate substantial metals Cd, Cu, Pb and Zn from the root tissues of the plant (Muramoto and Oki 1983).

A few investigations have been accounted on the utilization of dried plant material as a potential biosorbent in to remove Lead (II) in the wastewater., Liao and Chang (2004) argued that water hyacinth can retain and translocate the cadmium (Cd), lead (Pb), copper (Cu), zinc (Zn), and nickel (Ni) in the plant's tissue as a root or shoot. Water hyacinth plants had high bioconcentration with low convergences of the five components. This demonstrates water hyacinth can be a promising contender to eliminate the substantial metals. The generally low leaf substance, until the point when uncommon conditions are utilized, demonstrated the nearness of an anticipation instrument to repress Pb take-up (David et al. 2003). Wolverton (1988) and Brix (1993) clarified that explanation behind turbidity decrease i.e. the root hairs have electrical charges that draw in inverse charges of colloidal particles, for example, suspended solids and cause them to follow on the roots where they are gradually processed and acclimatized by the plant and microorganisms. Brix (1993) seen that *Eichhornia crassipes* has been utilized effectively in wastewater treatment framework to enhance the water quality by lessening the dimensions of natural and inorganic supplements. Consequently, the water hyacinth would most likely have high resilience and ought to be fit for removal of huge measures of lead (Sutcliffe 1962).

Duckweed is an assortment of aquatic plant free-gliding at the water surface. It is quickly developing and adjusts effectively to different amphibian conditions. Duckweed usually alludes to a gathering of gliding, blossoming plants of the family Lemnaceae. The distinctive species (*Lemna*, *Spirodela*, *Wolffia* and *Wolfiella*) are worldwide conveyed in wetlands, lakes and a few effluents tidal pond. The plants can develop at temperature running from 5 to 35 °C with ideal development between 20 °C and 31 °C and over an extensive variety of pH (3.5–10.5) (Cayuela et al. 2007). Wetlands and lakes are the most widely recognized destinations to discover duckweed.

The limit of aquatic plant, for example, duckweed (*Lemna* sp.) to expel lethal overwhelming metals from water are all around recorded and assumes an imperative job in extraction and aggregation of metals from wastewater. The normal seagoing plant *L. minor* can expel up to 90% of dissolvable Pb from water. *L. minor* can develop well in pH from 6 to 9 while the most reduced estimation of pH it can endure in between pH 5–6. Nonetheless, nitrate had couple of inhibitory on the development (Chong et al. 2003). Uysal and Taner (2009) inspected the capacity of the *L. minor* to expel dissolvable lead under various pH esteems (4.5–8.0) and temperature (15–35 °C) in nearness of various Pb concentrations 0.1–10.0 mgL^{-1} for 7 days. Their outcomes indicate Pb amassing was most noteworthy at pH 4.5 and after that it diminished to pH 6, yet it didn't change at pH 6–8 territory (Gallardo et al. 1999). *Hydrila verticillata* (Hydrilla) is a submerged aquatic weed that can develop up to the surface and frame thick tangles in all waterways. For expulsion of contaminants entire plant assumes imperative job. Showed that the dependence upon roots for substantial metal take-up was in rooted floating-leaved taxa with lesser dependence in submerged taxa. He likewise observed that inclination to utilize shoots as locales of overwhelming metal take-up rather than roots increments with movement towards submergence and effortlessness of shoot structure. Gallardo et al. (1999) discovered that following multi week of presentation to concentrated lead arrangement demonstrated greatest take-up (98%) of Pb by *Hydrilla*.

3.2 Cadmium (Cd)

Pollution of biosphere particularly the hydrosphere had been increased significantly, since the onset of industries (Nriagu 1990). The wastewater from industries containing hazardous pollutants particularly heavy metals, directly discharged into water bodies is one of the most important cause of contamination to living beings as they biomagnifies along the food chain due to their persistency and toxicity. In particular, Cadmium (Cd) is one of the common trace pollutants which are very harmful to organisms. Cadmium being non-essential heavy metal, is one of the "black list" substance in the Dangerous Substance Directory (Herrero et al. 2008). Apart from geochemical weathering of rocks, it is also dissipated into the ecosystems from power stations, metal working industries, Nickel-Cadmium batteries, fertilizers of phosphate origin and heating systems (Wagner 1993; Toppi and Gabbrielli 1999). The different methods that are employed to remove Cd from water are chemical precipitation, solvent extraction, adsorption, ion exchange, and electrophoresis and membrane separation. In addition to incomplete metal removal and generation of toxic sludge, expensiveness of these technologies had led to the exploration for low cost, low impact, visually benign and environmental friendly methods. Removal of heavy metals by plants generally called phytoremediation is a new cleanup method, that is environmental friendly and aesthetically pleasant (Chaney et al. 2005; Huang et al. 2004; Susarla et al. 2002). A number of laboratory experiments performed for Cd removal from aqueous media indicate that phytoremediation is a promising technique for metal removal. The most common species that are reported to have heavy metal removal properties are *Eichhornia crassipes, Alternanthera sessilis, Ceratophyllum demersum, Azolla pinnata, Chara coralline, Hygrorrhiza aristata, Hydrodictyon reticulatum, Hydrocotyle umbellate, Lemna minor, Salvinia, Pistia, Spirodela polyrhiza, Vallisneria spiralis* (Rai 2009).

Lu et al. (2004) investigated the phytoremediation potential of *Eichhornia crassipes* from tap water contaminated with different concentrations of Cd and Zn. They observed that metal removal was rapid in the first 4 days. The high values of bioconcentration factor for Cd and Zn suggested the heavy metal removal potential of water hyacinth and can be used for remediation purposes. Liao and Chang (2004) reported the removal of Cd and other heavy metals with *Eichhornia crassipes* Mart. Solms from the constructed wetlands of Taiwan. Similarly, *E. crassipes* can accumulate a substantial amount of Cd in shoot biomass (371 mg kg^{-1}) as well as root biomass (6103 mg kg^{-1}) (Zhu et al. 1999). They also observed that Cd along with other heavy metals results in the lesser accumulation of Cd in the aerial parts as compared to the shoots (Soltan and Rashed 2003). This property of heavy metal accumulation of *E. crassipes* makes it a favourable specie for Cd removal from water. Phytoremediation potential of *Eichhornia*, for the removal of cadmium (Cd), lead (Pb) and zinc (Zn) was also reported by Aisien et al. (2010). They observed that initially the metal removal from solution was rapid for first 2 weeks till saturation point was reached. Abhilash et al. (2009) reported 98% Cd removal with *Limnocharis flava* (L.) after 30 days from hydroponics experiment. Further, high bioconcentration factor (BCF) and translocation factor (TF), demonstrating that *L. flava* would be a suitable candidate for the phytofiltration. The translocation factor is calculated as the Cd concentration in shoots divided by the Cd concentration in roots, evaluates the capacity of plant to translocate heavy metals from root to shoot. Kashem et al. (2008) while investigating the Cd accumulation ability of *Colocasia antiquorum, Raphanus sativus* L. and *Ipomoea aquatica* reported that Cd accumulation intensified with rise in concentration. The Cd concentration in *Colocasia antiquorum* and *Ipomoea aquatica* was highest in roots in contrast to *Raphanus sativus* L. in which the concentration was maximum in aerial plant parts. The Cd concentration in dead leaves, normal leaves, stems, bulbs and roots increases from 158 to 1060, 37 to 280, 108 to 715, 42 to 290 and 1195 to 3840 mg kg^{-1}, respectively. These results demonstrated that *Colocasia antiquorum* had µgood Cd removal potential. An interesting absorption pattern was observed in H. verticillate. It was observed that maximum Cd absorption occurs during the daytime at the growth temperature (15–25 °C) (Dulay et al. 2010). *Azolla pinnata* which was considered to be more effective in comparison to *E. crassipes* has the BCF for Cd in roots as 24,000 which were quite high (Noraho and Gaur 1996). Similarly, Valderrama et al. (2013) evaluated the phytoremediation potential of *Azolla filiculoides* for Cd and Cu removal from the medium contaminated with 0.5–1.0 mg L^{-1} of Cd and 0.1–25 mg L^{-1} Cu. The results indicate that the concentration of Cd and Cu reached upto 1623.20 and 6013.1 µg/g, respectively. Wang et al. (2008) investigated that Cd phytoextraction of *Ipomoea aquatica* was correlated with the aqueous Cd ions in the free and complex form. It was also observed that high BCF of Cd in *Ipomoea aquatica* indicates its high potential to remediate Cd contaminated waste water.

Salvinia herzegoii in contrast to *Pistia stratiotes* accumulates a high level of Cd (Maine et al. 2001). Similarly, another species of *Salvinia, S. minima* was considered as Cd hyperaccumulator due to large surface area of roots with hydroxyl and carboxyl groups (Olguin et al. 2002). Phytoremediation capability was also reported other plants also such as, *Potamogeton natans, Myriophyllum aquaticum, Wolffia globosa*, and *Typha* (Cardwell et al. 2002; Fritioff and Greger 2006; Boonyapookana et al. 2002). Thus, Cd removal ability of macrophytes was studied very extensively and can be used to remediate Cd from contaminated water.

The primary target of Cd toxicity is unknown however, it was found to cause various phytotoxic indications *viz,* loss of chlorophyll, inhibition of growth, H_2O imbalance, phosphorus and nitrogen deficit etc. (Toppi and Gabbrielli 1999; Benavides et al. 2005). Cd can replace Zn, Ca and Fe from its compounds such as proteins, due to its chemical similarity. It can also react with the sulphur containing proteins and peptides and can cause oxidative stress (Benavides et al. 2005; Sandalio et al. 2001; Romero-Peurtas et al. 2004). Cd also interferes with mitochondrial electron transfer chain of plant cells, stimulates reactive oxygen species production which intern induces lipid peroxidation. The inhibition of water transport which causes proline accumulation was also reported due to Cd accumulation (Schat et al. 1997).

3.2.1 Cadmium Take-Up in Plants and Transportation

Cadmium is a one of the most dangerous heavy metals because of its high portability and the little fixation at which its impacts on plants start to appear (Barcelo and Poschenrieder 1992). Jarvis et al. (1976) found that the underlying foundations of lettuce discharged substantially more of their consumed Cd for translocation to the shoots than different products (ryegrass and orchard grass). The more prominent translocation is because of dynamic transport or to active transport or lack of metal absorption to fixed or soluble chelators in the root or perhaps due to exchange with the Ca, Mn and Zn traveling through the roots (John 1976). Moral et al. (1994) announced that Cd was effectively transported to ethereal parts of tomato and was not distinguished in natural products. Hinesly et al. (1984) detailed that the pH of the soils had extraordinary effect on cadmium transportation in corn (*Zea mays* L.). The highest grain Cd concentrations happened at soil pH at around 6.0. The take-up of cadmium by corn was less from the most acidic soil that additionally had the highest organic matter content (Street et al. 1977). Miragaya and Page (1976) found that the proportions of complexed to uncomplexed Cd were autonomous of Cd focus and somewhat influenced by pH over a scope of 6.0–8.5. Numerous components in the soils have been appeared to impact the take-up of substantial metals by plants. The cadmium take-up expanded with diminishing soil pH (Lagerwerff 1971; Miller et al. 1977) and diminished with expanding soil cation trade limit (Haghiri 1974). The cadmium seems, by all accounts, to be assimilated latently (Cutler and Rains 1974) and translocated openly (Jarvis et al. 1976). The chelators in supplement arrangements can help in cadmium take-up (Francis and Rush 1974). The articulated cooperation's amongst Zn and Cd happened in cadmium uptake and translocation. Evidently, some portion of Cd danger was an aftereffect of Cd obstruction in a Zn-dependent process (Falchuk et al. 1975).

The distinctions in the capacity of plants to collect overwhelming metals have been identified with contrasts in their root morphology (Hemphill 1972; Schierup and Larsen 1981). Plant with different thin roots would collect a bigger number of metals than one with couple of thick roots. Wahbeh (1984) contrasted with the ingestion and accumulation of Cd, Mn, Zn, Mg and Fe in three types of aquatic grasses. The two species *Halophila ovalis* and *Halophila uninervis* had higher photosynthetic and respiratory rates than *Halophila stipulacea* (Wahbeh 1984). Chukwuma (1993) reported the accumulation of cadmium, lead and zinc in developed and wild plant species in the abandoned lead-zinc mine. McKenna et al. (1993) reported that the associations between Zn and Cd in nutrient solution and their consequences for the aggregation of the two metals in plant roots and leaves. They detailed higher Cd focus is more in young leaves of lettuce and spinach. The potential accumulation of Cd in old leaves couldn't be exclusively because of the transpiration rate. The metal binding peptides are present in older leaves in higher amounts than in younger leaves in tobacco and cadmium was transported into the vacuoles as methods for detoxification (Lange and Wagner 1990). Cadmium concentrations were accounted higher in roots than in shoots (Cataldo et al. 1981; Rauser 1986). The suggestion set by for forage yields, this much of concentration is supposed to be less toxic for the animals feeding in these crops.

Rascio et al. (1993) contemplated development of the entire plant and the chlorophyll content, oxygen advancement, and chloroplast ultrastructure of leaf tissues in maize plants developed on a culture medium either without cadmium (Cd) or provided with expanding groupings of the metal. The plants treated with high Cd focuses demonstrated side effects of overwhelming metal danger, for example, length decrease of the two roots and shoots, leaf blanching, ultrastructural changes of chloroplasts and bringing down of photosynthetic action. A few manifestations showed up at 100 μM Cd, however, the lethal impacts of the metal were discovered just at 250 μM Cd. Quzounidou et al. (1997) contemplated the impacts of a 7-day exposure of 3-day-old wheat plants to expanding Cd concentration with exceptional consideration being given to chloroplast ultrastructural changes, chlorophyll fluorescence reactions, chlorophyll and supplement focus changes and development changes of the entire plant. Cadmium treatment was appeared to harm the structure of chloroplasts, as showed by exasperates shape and the expansion of the thylakoid films. These ultra structural changes recommend that Cd likely prompted untimely senescence.

Gratao et al. (2008) advocated that continuous increment in $CdCl_2$ concentration over some stretch of time seemed to adjust the plants to the harmful impacts of Cd, yet it additionally prompted a noteworthy higher accumulation of Cd in the organic products. Plants subjected to expanding convergences of Cd and different metals ought to give a superior comprehension of the instruments of detoxification, which may incorporate biochemical hereditary qualities with plant rearing to deliver pressure tolerant plants for detoxification or phytoremediation projects of polluted environments by heavy metals. Hsu and Kao (2007) obtained the detached rice leaves and intact leaves attached to rice seedlings, and found that Cd lethality in leaves of rice seedlings is because of H_2O_2 accumulation, which is fundamentally predictable with the aftereffects of Cho and Seo (2005) who detailed that a lower H_2O_2 collection gives Cd-resilience in *Arabidopsis* seedlings.

3.3 Mercury (Hg)

Lenka et al. (1990) investigated the bioconcentration and genotoxicity of aquatic mercury using *Eichhornia crassipes* by treating the plant to water supplemented with different concentrations of methyl mercury or phenyl mercuric acetate or mercury contaminated effluent for 4–96 h. The root samples collected at different exposure time intervals reveal that bioconcentration of mercury depends on concentration as well as duration of exposure. The results also indicate the potential of water hyacinth for mercury removal and hence, can be used for remediation. Dependence of rate of absorption on initial concentration was also reported by Gonzalez et al. (1994). Further, it was observed that mercury removal was rapid initially which decreases thereafter. Bioaccumulation (living plants) and bioabsorption (dry biomass) increases significantly as concentration of mercury ions increases in the solution in contrast to exposure time and pH values which show variations in bioaccumulation with different concentrations (Casagrande et al. 2018). Long before, similar type of results was reported in case of *Pistia stratiotes* L. (De et al. 1984). The absorption enhances as concentration of mercury (II) increases in the solution. It was also found that efficiency of as high as 90% can be achieved below 20 ppm mercury (II). Mercury removal capabilities of *Eichornia crassipes, Pistia stratiotes, Scirpus tabernaemontani* and *Colocasia esculenta* was also reported by Skinner et al. (2007). After exposing plants to different concentrations of mercury (0 mgL^{-1}, 0.5 mgL^{-1} or 2 mgL^{-1}) for 1 month, it was concluded that all these plant species can accumulate mercury. However, among all four species efficiency for metal removal of *Pistia stratiotes* and *Eichornia crassipes* was higher. Jana (1988), while studying the mercury and chromium accumulation in *Eichhornia crassipes* (Mart.) Solms., *Hydrilla verticillata* (L.f.) Royle and *Oedogonium areolatum* Lagarh, reported that mercury accumulation was highest in *Hydrilla* followed by *Oedogonium* and *Eichhornia*. The effect of phosphate concentration, light intensities and sediment: aqueous phase contamination ratios on mercury and methylmercury assimilation ability of *Eichhornia crassipes* was also evaluated (Chattopadhyay et al. 2012). It was observed that, in contrast to phosphate concentration which is required for mercury and methylmercury uptake, the light intensities increase the translocation of mercury and methylmercury which was preferentially concentrated in roots. The sediment: aqueous phase contamination ratios were also found to affects the bioavailability of mercury and methylmercury. Molisani et al. (2006) evaluated the role of *Elodea densa, Sagittaria montevidensis, Salvinia auriculata, Pistia stratiotes* and *Eichhornia crassipes* for mercury accumulation in two artificial reservoirs. The results showed that free floating and juveniles were more efficient for mercury removal. Among the plant organs roots accumulate more mercury. It was proposed that *Eichhornia crassipes* can remove a considerable amount of mercury from water. The mercury remediation ability of water hyacinth and pondweed was also reported by Romanova and Shuvaeva (2016). Gomes et al. (2014) established a pilot scale experimental design to evaluate the remediation potential of *Typha domingensis* for mercury. It was observed that metal removal potential varies with exposure time. In contrast to the other species it was also reported that *T. domingensis* accumulates higher mercury when the transfer coefficient was 7750.9864 ± 569.5468 L kg^{-1}. *Vallisneria spiralis,* an aquatic plant, when subjected to different concentration of mercury at different durations, disclosed a mercury concentration of 250 µmol/g and 1120 µmol/g in dehydrated weight of leaf and root, respectively (Gupta and Chandra 1998). Phytoremediation potential of *Limnocharis flava* for mine effluents containing mercury was investigated in a constructed wetland (Marrugo-Negrete et al. 2017). Metal removal potential was found to vary with exposure time. It was also observed that controlled rate of mercury removal was 9 times lower than continued rate.

Similar, types of mercury removal experiments by using different plants such as, *Elodea densa (*Maury et al. 1988*), Eriocaulon septangulare* (Coquery and Welbourn 1944), *Oryza sativa L.* (Heaton et al. 2003), *Azolla pinnata* and *Vallisneria,* (Rai and Tripathi 2009). *Salvinia natans* and *Lemna minor* (Sitarska et al. 2015) had been performed (Table 1).

4 Conclusion

The response of the aquatic plants to heavy metals *viz,* lead, cadmium and mercury can be traced from the literature. However, the level of accumulation in different plant parts varied with species, period of exposure, metal concentration and composition of soil/nutrient. Generally, a trade-off had been reported between nutrients and metal toxicity up to a certain concentration called threshold limit above which toxicity effect was reported. Toxic levels of lead, cadmium and mercury was found to affect both physiology and biochemistry of plants. Reaction of metals to functional groups of enzymes alter several important functions some of them are critical for photosynthesis and nitrogen assimilation. Although several aspects of metal tolerance were identified however, there is a lack of model specie in which the entire process is well defined. It seems that it is an intricate phenomenon with some genetic influence. An improved and good understanding of the information is obligatory to knob the critical problem of growing metal toxicity to the flora and fauna. There is an urgent need to address this growing problem. Apart from providing the necessary information for developing models that will accurately and precisely forecast the

Table 1 The different heavy metal and their presence in different parts of aquatic plants

Heavy metal	Plant species	Plant part	References
Lead	*Eichhornia crassipes*	Root	Miretzky et al. (2005)
	Eichhornia crassipes	Root tissue	Muramoto and Oki (1983) and Nor (1990)
	Water hyacinth	Root/ shoot	Sutcliffe (1962)
	Duckweed (*Lemna* sp.)		Chong et al. (2003)
	Hydrila verticillata	Roots	
Cadmium	*Eichhornia crassipes*	Shoot and root	Lu et al. (2004), Zhu et al. (1999), Soltan and Rashed (2003) and Aisien et al. (2010)
	Limnocharis flava (L.)	Shoot	Abhilash et al. (2009)
	Colocasia antiquorum, Raphanus sativus L. and *Ipomoea aquatica*	Root	Kashem et al. (2008)
	H. verticillate		Dulay et al. (2010)
	Azolla pinnata, E. crassipes	Root	Noraho and Gaur (1996)
	Salvinia herzegoii		Maine et al. (2001)
	S. minima	Root	Olguin et al. (2002)
	Potamogeton natans, Myriophyllum aquaticum, Wolffia globosa, and *Typha*		Cardwell et al. (2002), Fritioff and Greger (2006) and Boonyapookana et al. (2002)
Mercury	*Eichhornia crassipes*	Root	Lenka et al. (1990)
	Eichornia crassipes, Pistia stratiotes, Scirpus tabernaemontani and *Colocasia esculenta*	Root	Skinner et al. (2007) and Chattopadhyay et al. (2012)
	Hydrilla verticillata (L.f.) Royle and *Oedogonium areolatum*	Root	Jana (1988)
	Sagittaria montevidensis, Salvinia auriculata,	Root	Molisani et al. (2006)
	T. domingensis	Root	Gomes et al. (2014)
	Vallisneria spiralis	Root	Gupta and Chandra (1998)
	Elodea densa	Root	Maury et al. (1988)
	Eriocaulon septangulare	Root	Coquery and Welbourn (1944)
	Salvinia natans and *Lemna minor*	Root	Sitarska et al. (2015)
	Azolla pinnata and *Vallisneria,*	Root	Rai and Tripathi (2009)
	Oryza sativa L.	Root	Heaton et al. (2003)

influence of metals on the plant functions, it may also provide means to detoxify metal contaminated sites by the employing metal removing plant species. As such there is a need to study the detoxification pathways in detail.

References

Abhilash, P. C., Pandey, V. C., Srivastava, P., Rakesh, P. S., Chandran, S., Singh, N., & Thomas, A. P. (2009). Phytofiltration of cadmium from water by Limnocharis flava (L.) Buchenau grown in free-floating culture system. *Journal of Hazardous Materials, 170*, 791–797.

Ahmadpour, P., Ahmadpour, F., Mahmud, T. M. M., Abdu, A., Soleimani, M., & Tayefeh, F. H. (2012). Phytoremediation of heavy metals: A green technology. *African Journal of Biotechnology, 11*, 14036–14043.

Aisien, F. A., Faleye, O., & Aisien, E. T. (2010). Phytoremediation of heavy metals in aqueous solutions. *Leonardo Journal of Sciences, 17*, 37–46.

Akhter, M. S., & Madany, I. M. (1993). Heavy metals in street and house dust in Bahrain. *Water, Air and Soil Pollution, 66*, 111–119.

Barcelo, J., & Poschenrieder, C. H. (1992). Respuestas de las plantas ala contaminacion por metales pesados. In SEFV (Ed.), *Proceeding of the IV Simposium Nacional sobre Nutricion Mineral de las plantas, Alicante, Spain* (pp. 45–61).

Benavides, M. P., Gallego, S. M., & Tomaro, M. L. (2005). Cadmium toxicity in plants. *Brazilian Journal of Plant Physiology, 17*, 21–34.

Bhadra, R., Wayment, D. G., Hughes, J. B., & Shanks, J. V. (1999). Confirmation of conjugation processes during TNT metabolism by axenic plant roots. *Environmental Science & Technology, 33*, 446–452.

Bharti, N., & Singh, R. P. (1993). Growth and nitrate reduction by *Sesamum indicum* cv PB-I respond differentially to lead. *Phytochemistry, 33*, 531–534.

Bhat, R. A., Shafiq-ur-Rehman, Dervash, M. A., Mushtaq, N., Bhat, J. I. A., & Dar, G. H. (2017). Current status of nutrient load in Dal Lake of Kashmir Himalaya. *Journal of Pharmacognosy and Phytochemistry, 6*, 165–169.

Boonyapookana, B., Upatham, E. S., Kruatrachue, M., Pokethitiyook, P., & Singhakaew, S. (2002). Phytoaccumulation and phytotoxicity of cadmium and chromium in duckweed *Wolffia globose*. *International Journal of Phytoremediation, 4*, 87–100.

Brix, H. (1993). Wastewater treatment in constructed wetlands. System design, removal processes, and treatment performance. In G. A. Moshiri (Ed.), *Constructed wetlands for water quality improvement* (pp. 9–22). Boca Raton/Ann Arbor/London/Tokyo: Lewis.

Broyer, T. C., Johnson, C. M., & Paull, R. E. (1972). Some aspects of lead in plant nutrition. *Plant and Soil, 2*, 301–313.

Burken, J. G., & Schnoor, J. L. (1997). Uptake and metabolism of atrazine by poplar trees. *Environmental Science & Technology, 31*, 1399–1406.

Cannon, H. L., & Bowlers, J. M. (1962). Contamination of vegetation by tetraethyl lead. *Science, 137*, 765–766.

Cardwell, A. J., Hawker, D. W., & Greenway, M. (2002). Metal accumulation in aquatic macrophytes from southeast Queensland, Australia. *Chemosphere, 48*, 653–663.

Casagrande, G. C. R., Reis, C., Arruda, R., deAndrade, R. L. T., & Battirola, L. D. (2018). Bioaccumulation and biosorption of mercury by *Salvinia biloba* Raddi (Salviniaceae). *Water Air Soil Pollution, 229*, 166.

Cataldo, D. A., Garland, T. R., & Wildung, R. E. (1981). Cadmium distribution and chemical fate in soybean plants. *Plant Physiology, 68*, 835–839.

Cayuela, E. A., Munoz-Mayor, F., Vicente-Agullo, E., Moyano, J. O., Garcia-Abellan, M. T., Estan, M. C., & Bolarin. (2007). Drought pretreatment increases the salinity resistance of tomato plants. *Journal of Plant Nutrition and Soil Science, 170*, 479–484.

Chandra, P., Tripathi, R. D., Rai, U. N., Sinha, S., & Garg, P. (1993). Biomonitoring and amelioration of non-point source pollution on some aquatic bodies. *Water Science and Technology, 38*, 323–326.

Chaney, R. L., Angle, J. S., McIntosh, M. S., Reeves, R. D., Li, Y. M., Brewer, E. P., Chen, K. Y., Roseberg, R. J., Perner, H., Synkowski, E. C., Broadhurst, C. L., Wang, S., & Baker,

A. J. (2005). Using hyperaccumulator plants to phytoextract soil Ni and Cd. *Zeitschrift für Naturforschung, 60*, 190–198.
Chattopadhyay, S., Fimmen, R. L., Yates, B. J., Lal, V., & Randall, P. (2012). Phytoremediation of mercury and methyl mercury-contaminated sediments by water hyacinth (*Eichhornia Crassipes*). *International Journal of Phytoremediation, 14*, 142–161.
Cho, U. H., & Seo, N. H. (2005). Oxidative stress in *Arabidopsis thaliana* exposed to cadmium is due to hydrogen peroxide accumulation. *Plant Science, 168*, 113–120.
Chong, Y., Hu, H., & Qian, Y. (2003). Effects of inorganic nitrogen compounds and pH on the growth of duckweed. *Journal of Environmental Sciences, 24*, 35–40.
Chow, T. J. (1970). Lead accumulation in road side soil and grass. *Nature, 225*, 295–296.
Chukwuma, S. C. (1993). Comparison of the accumulation of Cd, Pb and Zn in cultivated and wild plant species in the derelict Enyigba lead-zinc mine. *Environmental Toxicology and Chemistry, 38*, 167–173.
Clark, R. E. (1993). IARC monographs on the evaluation of carcinogenic risks to humans: Chromium, nickel and welding. *Quarterly Review of Biology, 68*, 472–473.
Coquery, M., & Welbourn, P. M. (1944). Mercury uptake from contaminated water and sediment by the rooted and submerged aquatic macrophyte *Eriocaulon septangulare*. *Archives of Environmental Contamination and Toxicology, 26*, 335–341.
Cutler, J. M., & Rains, D. W. (1974). Characterization of cadmium uptake by plant tissue. *Plant Physiology, 54*, 67–71.
Dabas, S. (1992). Ph.D. Thesis, Department of Biosciences, M. D. India.
David, T. W., Myint, T. M., & Sein, T. (2003). Lead removal from industrial waters by water hyacinth. *AU Journal of Technology, 6*, 187–192.
Davis, L. C., Erickson, L. E., Narayanan, N., & Zhang, Q. (2003). Modeling and design of phytoremediation. In *Phytoremediation: Transformation and control of contaminants* (pp. 663–694). New York: Wiley.
De, A. K., Sen, A. K., Modak, D. P., & Jana, S. (1984). Studies of toxic effects of Hg(II) on *Pistia stratiotes*. *Water Air Soil Pollution, 24*, 351.
Dulay, J. A. L., Caldona, E. B., & Camacho, A. R. (2010). Phytoremediation of Cadmium contaminated water by Hydrilla (Hydrilla verticillate). *SLU Research Journal, 41*, 23–33.
Falchuk, K. H., Fawcett, D. W., & Vallee, B. L. (1975). Competitive antagonism of cadmium and zinc in the morphology and cell division of *Euglena gracilis*. *Journal of Submicroscopic Cytology, 7*, 139–152.
Forstner, U., & Wittman, G. T. (1979). *Metal pollution in aquatic environment*. Berlin/Heidelberg/New York: Springer.
Francis, C. W., & Rush, S. G. (1974). Factors affecting uptake and distribution of cadmium in plants. In D. D. Hemphill (Ed.), *Trace substances in environmental health* (Vol. 7, pp. 75–81). Jefferson City: Hemphill Columbia University.
Fritioff, A., & Greger, M. (2006). Uptake and distribution of Zn, Cu, Cd, and Pb in an aquatic plant *Potamogeton natans*. *Chemosphere, 63*, 220–227.
Gallardo, T., Benson, M., Robert, F., Martin, F., & Dean. (1999). Lead accumulation by three aquatic plants. Symposia papers presented before the division of Environmental Chemistry. *American Chemical Society, 39*, 46–47.
Gaur, A., & Adholeya, A. (2004). Prospects of arbuscular mycorrhizal fungi in phytoremediation of heavy metal contaminated soils. *Current Science, 86*, 528–534.
Goldsmith, C. D. J., Scanion, P. F., & Pirie, W. R. (1976). Lead concentrations in soil and vegetation associated with highways of different traffic densities. *Bulletin of Environmental Contamination and Toxicology, 16*, 66–70.
Gomes, M. V. T., de Souza, R. R., Teles, V. S., & Mendes, E. A. (2014). Phytoremediation of water contaminated with mercury using *Typha domingensis* in constructed wetland. *Chromrosphere, 103*, 228–233.
Gonzalez, H., Lodenius, M., & Martinez, L. (1994). Removal of mercury from polluted waters by the water hyacinth (*Eichhornia crassipes*). *Chemistry and Ecology, 9*, 7–12.

Gratao, P. L., Monteiro, C. C., Antunes, A. M., Peres, L. E. P., & Azevedo, R. A. (2008). Acquired tolerance of tomato (*Lycopersicon escultntum* cv. micro tom) plants to cadmium- induced stress. *Annals of Applied Biology, 153*, 321–333.

Guilizzoni, P. (1991). The role of heavy metals and toxic materials in the physiological ecology of submerged Macrophytes. *Aquatic Botany, 41*, 87–109.

Gupta, M., & Chandra, P. (1998). Bioaccumulation and toxicity of mercury in rooted-submerged macrophyte *Vallisneria spiralis*. *Environmental Pollution, 103*, 327–332.

Haghiri, F. (1974). Plant uptake of cadmium as influenced by cation exchanges capacity, organic matter, zinc and soil temperature. *Journal of Environmental Quality, 3*, 180–183.

Heaton, A. C. P., Rugh, C. L., Kim, T., Wang, N. J., & Meagher, R. B. (2003). Toward detoxifying mercury-polluted aquatic sediments with rice genetically engineered for mercury resistance. *Environmental Toxicology and Chemistry, 12*, 2940–2947.

Hemphill, D. D. (1972). Availability of trace elements to plants with respect to soil-plant interaction. *Annals of the New York Academy of Sciences, 199*, 46–61.

Herrero, R., Lodeiro, P., Rojo, R., Ciorba, A., Rodriguez, P., & Sastre de Vicente, M. E. (2008). The efficiency of the red alga *Mastocarpus stellatus* for remediation of cadmium pollution. *Bioresource Technology, 99*, 4138–4146.

Hinesly, T. D., Redberg, K. E., Pietz, R. I., & Ziegler, E. L. (1984). Cadmium and zinc uptake by corn *(Zea mays* L.) with repeated applications of sewage sludge. *Journal of Agricultural and Food Chemistry, 32*, 155–163.

Hsu, Y. T., & Kao, C. H. (2007). Toxicity in leaves of rice exposed to cadmium is due to hydrogen peroxide accumulation. *Plant and Soil, 298*, 231–241.

Huang, X. D., El-Alawi, Y., Penrose, D. M., Glick, B. R., & Greenberg, B. M. (2004). A multi-process phytoremediation system for removal of polycyclic aromatic hydrocarbons from contaminated soils. *Environmental Pollution, 130*, 465–476.

Hughes, J. B., Shanks, J., Vanderford, M., Lauritzen, J., & Bhadra, R. (1997). Transformation of TNT by aquatic plants and plant tissue cultures. *Environmental Science & Technology, 31*, 266–271.

Hussanin, E., Labib, A. S., & Dobal, A. T. (1993). Potential Pb, Cd, Zn and B contamination of sandy soils after different irrigation periods with sewage effluent. *Water, Air and Soil Pollution, 66*, 239–249.

Jadia, D. C., & Fulekar, M. H. (2009). Phytoremediation of heavy metals: Recent techniques. *African Journal of Biotechnology, 8*, 921–928.

Jana, S. (1988). Accumulation of Hg and Cr by three aquatic species and subsequent changes in several physiological and biochemical plant parameters. *Water, Air, and Soil Pollution, 38*, 105–109.

Jarvis, S. C., Jones, L. H. P., & Hopper, M. J. (1976). Cadmium uptake from solution by plants and its transport from roots to shoots. *Plant and Soil, 44*, 179–191.

John, M. K. (1976). Interrelationships between plant cadmium and uptake of some other elements from culture solutions by oats and lettuce. *Environmental Pollution, 11*, 85–95.

John, M. K., & Laerhoven, V. C. (1976). Revegetation of Metalliferous mine spoil contaminated by Pb & Zn. *Environmental Pollution, 100*, 163–173.

Kabata-Pendias, A. (2011). *Trace elements in soils and plant*. Boca Raton: CRC Press.

Kamran, M. A., Mufti, R., Mubariz, N., Syed, J. H., Bano, A., et al. (2014). The potential of the flora from different regions of Pakistan in phytoremediation: A review. *Environmental Science and Pollution Research, 21*, 801–812.

Kashem, M. A., Singh, B. R., Imamul Huq, S. M., & Kawa, S. (2008). Cadmium phytoextraction efficiency of arum (*Colocasia antiquorum*), radish (*Raphanus sativus* L.) and water spinach (*Ipomoea aquatica*) grown in hydroponics. *Water, Air, & Soil Pollution, 192*, 273.

Kinnersely, A. M. (1993). The role of phytochelatins in plant growth and productivity. *Plant Growth Regulation, 12*, 207–217.

Kuhad, M. S., & Malik, R. S. (1989). *Department of soil science*. Hissar: C.S.A. University.

Kumar, G., Singh, R. P., & Sushila. (1993). Nitrate assimilation and biomass production in *Sesamum indicum* L. seedlings in a lead enriched environment. *Water, Air and Soil Pollution, 66*, 163–171.

Lagerwerff, J. V. (1971). Uptake of cadmium, lead and zinc by radish from soil and air. *Soil Science, 111*, 129–133.

Lagerwerff, J. V., & Armiger, W. H. (1973). Uptake of lead by alfalfa and corn from soil and air. *Soil Science, 115*, 455–460.

Lange, V. R., & Wagner, G. J. (1990). Sub cellular localization of cadmium and cadmium-binding peptides in tobacco leaves. *Plant Physiology, 92*, 1086–1093.

Lenka, M., Panda, K. K., & Panda, B. B. (1990). Studies on the ability of water hyacinth (*Eichhornia crassipes*) to bioconcentrate and biomonitor aquatic mercury. *Environmental Pollution, 66*, 89–99.

Liao, S., & Chang, W. (2004). Heavy metal phytoremediation by water hyacinth at constructed wetlands in Taiwan. *Journal of Aquatic Plant Management, 42*, 60–68.

Lu, X., Kruatrachue, M., Pokethitiyook, P., & Homyok, K. (2004). Removal of Cadmium and Zinc by Water Hyacinth, *Eichhornia crassipes*. *Science Asia, 30*, 93–103.

Maine, M. A., Duarte, M. V., & Sune, N. L. (2001). Cadmium uptake by floating macrophytes. *Water Research, 35*, 2629–2634.

Marrugo-Negrete, J., Enamorado-Montes, G., Durango-Hernandez, J., Pinedo-Hernandez, J., & Díez, S. (2017). Removal of mercury from gold mine effluents using *Limnocharis flava* in constructed wetlands. *Chemosphere, 167*, 188–192.

Martin, G. C., & Hammand, P. B. (1966). Lead uptake by bromegrass from contaminated soil. *Agronomy Journal, 58*, 553–554.

Maury, R., Boudou, A., Ribeyre, F., & Engrand, P. (1988). Experimental study of mercury transfer between artificially contaminated sediment (CH_3HgC1) and macrophytes (*Elodea densa*). *Aquatic Toxicology, 12*, 213–228.

McKenna, I. M., Chaney, R. L., & Williams, F. M. (1993). The effects of cadmium and zinc interactions on the accumulation and tissue distribution of zinc and cadmium in lettuce and spinach. *Environmental Pollution, 79*, 113–120.

Mehmood, M. A., Qadri, H., Bhat, R. A., Rashid, A., Ganie, S. A., Dar, G. H., & Shafiq-ur-Rehman. (2019). Heavy metal contamination in two commercial fish species of a trans-Himalayan freshwater ecosystem. *Environmental Monitoring and Assessment, 191*, 104.

Miller, J. E., Hassat, J. J., & Koeppe, D. E. (1975). The effect of soil properties and extractable lead levels on lead uptake by soybean. *Communications in Soil Science and Plant Analysis, 6*, 339–347.

Miller, J. E., Hasset, J. J., & Koeppe, D. E. (1977). Uptake of cadmium by soybeans as influenced by soil cation exchange capacity, pH and available phosphorus. *Journal of Environmental Quality, 5*, 157–160.

Miragaya, G. J., & Page, A. L. (1976). Influence of ionic strength and inorganic complex formation on the sorption of traces amounts of Cd by montmorillonite. *Soil Science Society America Journal, 40*, 658–663.

Miretzky, P., Jardim, M. C. B. M. F., & Rocha, J. C. (2005). Factors affecting Hg(II) adsorption in soils from the Rio Negro Basin (Amazon). *Quimica Nova, 28*, 438–443.

Molisani, M. M., Rocha, R., Machadoachado, W., Barreto, R. C., & Lacerda, L. D. (2006). Mercury contents in aquatic macrophytes from two reservoirs in the Paraíba Do Sul: Guandú River System, Se Brazil. *Brazilian Journal of Biology, 66*, 101–107.

Moral, R., Palacios, G., Gomez, I., Navarro, P. J., & Mataix, J. (1994). Distribution and accumulation of heavy metals (Cd, Ni and Cr) in tomato plant. *Fresenius Environmental Bulletin, 3*, 395–399.

Motto, H. L., Daines, R. H., Chilko, D. M., & Motto, C. K. (1970). Lead in soil and plants; its relationship to traffic volume and proximity to highways. *Environmental Science Technology, 4*, 231–237.

Muramoto, S., & Oki, Y. (1983). Removal of some heavy metals from polluted water by water hyacinth (*Eichhornia crassipes*). *Bulletin Environmental Contamination and Toxicology, 30*, 170–177.

Newman, L. A., & Reynolds, C. M. (2004). Phytodegradation of organic compounds. *Current Option in Biotechnology, 15*, 225–230.

Noraho, N., & Gaur, J. R. (1996). Cadmium adsorption and intracellular uptake by two macrophytes, *Azolla pinnata* and *Spirodela polyrhiza*. *Fundamental and Applied Limnology, 136*, 135–144.

Nriagu, J. O. (1990). Global metal pollution. *Environment, 32*, 7–33.

Odum, W. E., & Drifmeyer, J. E. (1978). Sorption of pollutants by plant detritus. A review. *Environmental Health Perspective, 27*, 133–142.

Olguın, E. J., Hernandez, E., & Ramos, I. (2002). The effect of both different light conditions and the pH value on the capacity of *Salvinia minima* baker for removing cadmium, lead and chromium. *Acta Biotechnologica, 22*, 121–131.

Page, A. L., & Ganje, T. J. (1970). Accumulation of lead in soils for regions of high & low motor vehicles traffic density. *Environmental Science Technology, 4*, 140–142.

Pahlsson, A. B. (1989). Toxicity of heavy metal (Zn, Cu, Cd, Pb) to vascular plant. *Water, Air and Soil Pollution, 47*, 287–319.

Patel, M., Wallace, A., & Romney, H. M. (1977). Effect of chelating agents on phytotoxicity of lead and lead transport. *Communications in Soil Science and Plant Analysis, 8*, 733–740.

Pratas, J., Favas, P. J. C., Paulo, C., Rodrigues, N., & Prasad, M. N. V. (2012). Uranium accumulation by aquatic plants from uranium-contaminated water in Central Portugal. *International Journal of Phytoremediation, 14*, 221–234.

Quzounidou, G., Moustakes, M., & Eleftherious, E. P. (1997). Physiological and ultrastructural effect of cadmium on wheat. *Archives of Environmental Contamination and Toxicology, 32*, 154–160.

Rai, P. K. (2009). Heavy metal phytoremediation from aquatic ecosystems with special reference tomacrophytes. *Critical Reviews in Environmental Science and Technology, 39*, 697–753.

Rai, P. K., & Tripathi, B. D. (2009). Comparative assessment of *Azolla pinnata* and *Vallisneria spiralis* in Hg removal from G.B. Pant Sagar of Singrauli Industrial region, India. *Environmental Monitoring and Assessment, 148*, 75–84.

Rascio, N., Vecchia, N., Ferretti, F. D., Merio, M. L., & Ghisi, R. (1993). Some effect of cadmium on maize plants. *Archives of Environmental Contamination and Toxicology, 25*, 244–249.

Rashid, A., Bhat, R. A., Qadri, H., & Mehmood, M. A. (2019). Environmental and socioeconomic factors induced blood lead in children: An investigation from Kashmir, India. *Environmental Monitoring and Assessment, 191*(2), 76.

Rauser, W. E. (1986). The amount of cadmium associated with Cd-binding proteins in roots of Agrostis gigantea maize and tomato. *Plant Science, 43*, 85–91.

Rolfe, G. L., & Reinbold, K. A. (1977). *Environmental contamination by lead and other heavy metals* (p. 143). Urbana: Institute for Environmental Studies.

Romanova, T. E., & Shuvaeva, O. V. (2016). Fractionation of mercury in water hyacinth and pondweed from contaminated area of gold mine tailing. *Water Air Soil Pollution, 227*, 171.

Romero-Peurtas, M. C., Rodriguez-Serrano, M., Corpas, F. J., Gomez, M., Del-Rio, L. A., & Sandalio, L. M. (2004). Cadmium induced sub cellular accumulation of O_2- and H_2O_2 in pea leaves. *Plant Cell and Environment, 27*, 1122–1134.

Sandalio, L. M., Dalurzo, H. C., Gomez, M., Romero-Peurtas, M. C., & del-Rio, L. A. (2001). Cadmium induces changes in the growth and oxidative metabolism of pea plants. *Journal of Experimental Botany, 52*, 2115–2126.

Sarma, H. (2011). Metal hyperaccumulation in plants: A review focusing on phytoremediation technology. *Journal of Environmental Science and Technology, 4*, 118–138.

Schat, H., Sharma, S. S., & Vooijs, R. (1997). Heavy metal-induced accumulation of free proline in a metal-tolerant and a non tolerant ecotype of *Silene vulgaris*. *Physiologia Plantarum, 101*, 477–482.

Schierup, H., & Larsen, V. J. (1981). Macrophyte cycling of Zn, Cu, Pb and Cd in the littoral zone of a polluted and a non-polluted lake. I. Availability, uptake and translocation of heavy metals in *Phragmites australis* (Cav.). Trin. *Aquatic Botany, 11*, 179–210.

Schuck, E. A., & Locke, J. K. (1970). Relationship of automotive lead particulates to certain consumer crops. *Environmental Science & Technology, 4*, 324–330.

Sharpe, V., & Denny, P. (1976). Electron microscope studies on the absorption and localization of lead in the leaf tissue of *Potamogeton pectinatus*. *Journal of Experimental Botany, 27*, 1155–1162.

Singh, R. P., Maheshwari, R., & Sinha, S. K. (1994). Recovery of lead caused decreased in biomass accumulation of mung bean (*Vigna radiata* L.) seedlings of K_2HPO_4 and $Cacl_2$. *Indian Journal of Experimental Biology, 32*, 507–510.

Singh, R. P., Tripathi, R. D., Sinha, S. K., Maheshwari, R., & Srivastava, H. S. (1997). Response of higher plants to lead contaminated environment. *Chemosphere, 34*, 2467–2493.

Singh, D. V., Bhat, J. I. A., Bhat, R. A., Dervash, M. A., & Ganei, S. A. (2018). Vehicular stress a cause for heavy metal accumulation and change in physico-chemical characteristics of road side soils in Pahalgam. *Environmental Monitoring and Assessment, 190*(6), 353.

Sitarska, M., Traczewska, T., & Filyarovskaya, V. (2015). Removal of mercury (II) from the aquatic environment by phytoremediation. *Desalination and Water Treatment, 57*, 1–10.

Skinner, K., Wright, N., & Porter-Goff, E. (2007). Mercury uptake and accumulation by four species of aquatic plants. *Environmental Pollution, 145*, 234–237.

Smith, W. H. (1971). Lead contamination of road sides white pines. *Forest Science, 17*, 192–198.

Soltan, M. E., & Rashed, M. N. (2003). Laboratory study on the survival of water hyacinth under several conditions of heavy metal concentrations. *Advances in Environmental Research, 7*, 321–334.

Srikanth, R., Rao, A. M., Kumar, C. S., & Khanum, A. (1973). Lead, cadmium, nickel, and zinc contamination of ground water around Hussain Sagar lake, Hyderabad, India. *Bulletin Environmental Contamination and Toxicology, 50*, 138–143.

Street, J. J., Lindsay, W. L., & Sabey, B. R. (1977). Solubility and plant uptake of cadmium in soils mended with cadmium and sewage sludge. *Journal of Environmental Quality, 6*, 72–77.

Sun, H., Wang, Z., Gao, P., & Liu, P. (2013). Selection of aquatic plants for phytoremediation of heavy metal in electroplate wastewater. *Acta Physiologiae Plantarum, 35*, 355–364.

Susarla, S., Medina, V. F., & McCutcheon, S. C. (2002). Phytoremediation: An ecological solution to organic chemical contamination. *Ecological Engineering, 18*, 647–658.

Sutcliffe, J. F. (1962). *Mineral salts absorption in plants*. London: Pergamon Press.

Tandon, T. W., & Crowdy, S. H. (1971). The distribution of lead chelate in the transpiration stream of higher plants. *Pest Science, 2*, 211–213.

Thapa, D., Srivastava, H. S., & Ormrod, D. P. (1988). Physiological and biochemical effect of lead on higher plants. *Vegetas, 1*, 107–109.

Tiwari, S., Dixit, S., & Verma, N. (2007). An effective means of biofiltration of heavy metal contaminated water bodies using aquatic weed *Eichhornia crassipes*. *Environmental Monitoring and Assessment, 129*, 253–256.

Toppi, S. D. L., & Gabbrielli, R. (1999). Response to cadmium in higher plants. *Environmental and Experimental Botany, 41*, 105–130.

Uqab, B., Mudasir, S., Sheikh, A. Q., & Nazir, R. (2016). Bioremediation: A management tool. *Journal of Bioremediation & Biodegradation, 7*, 331.

Uysal, Y., & Taner, F. (2009). Effect of pH, temperature, and lead concentration on the bioremoval of lead from water using *Lemna minor*. *International Journal of Phytoremediation, 11*, 591–608.

Valderrama, A., Tapia, J., Peñailillo, P., & Carvajal, D. E. (2013). Water phytoremediation of cadmium and copper using *Azolla filiculoides* Lam. in a hydroponic system. *Water and Environment Journal, 27*, 293–300.

Wagner, R. C. (1993). Accumulation of cadmium in crop plants and its consequences to human health. *Advances in Agronomy, 51*, 173–212.

Wahbeh, M. I. (1984). Levels of Zn, Mn, Mg, Fe and Cd in three species of seagrass from Aqaba (Jordan). *Aquatic Botany, 20*, 179–183.

Wallace, A., Muller, R. T., Cha, J. W., & Alexander, G. V. (1974). Soil pH, excess lime and chelating agent on micronutrients in soybeans and bush beans. *Agronomic Journal, 66*, 698–700.

Wallace, A., Soufi, S. M., Alexander, C. V., & Cha, J. W. (1976). Comparison of the effect of high levels of DTPA and EDDHA on microelement uptake in bush bean. *Communication in Soil Plant Analysis, 7*, 111–116.

Wang, K. S., Huang, L. C., Lee, H. S., Chen, P. Y., & Chang, S. H. (2008). Phytoextraction of cadmium by *Ipomoea aquatica* (water spinach) in hydroponic solution: Effects of cadmium speciation. *Chemosphere, 72*, 666–672.

Welsh, R. P. H., & Denny, P. (1980). The uptake of lead and copper by submerged aquatic macrophytes in two English Lakes. *Journal of Ecology, 68*, 443–455.

Wheeler, G. L., & Rolfe, G. L. (1979). The relationship between daily traffic volume and the distribution of lead in roadside soil and vegetation. *Environmental Pollution, 18*, 265–274.

Wolverton, B. C. (1988). *Aquatic plant/microbial filters for treating septic tank effluent*. International Conference on Constructed Wetlands for Wastewater Treatment Chattanooga, TN.

Zhu, Y. L., Zayed, A. M., Qian, J. H., De Souza, M., & Terry, N. (1999). Phytoaccumulation of trace elements by wetland plants: II. Water hyacinth. *Journal of Environmental Quality, 28*, 39–344.

Zimdahl, R. L., & Faster, J. M. (1976). The influence of applied phosphorus, manure or lime or uptake of lead from soil. *Journal of Environmental Quality, 5*, 31–34.

Zimdahl, R. L., & Koeppe, D. E. (1977). Uptake by plants. In W. R. Boggerss & B. B. Wixon (Eds.), *Lead in the environment* (pp. 99–104). Washington, DC: National Science Foundation.

Impact of Climate Change on Freshwater Ecosystem and Its Sustainable Management

Birjees Hassan, Humaira Qadri, Md. Niamat Ali, Nissar Ahmad Khan, and Ali Mohd Yatoo

Abstract Freshwater ecosystems are vital for global biodiversity and ecosystem services. Freshwater ecosystems are susceptible to the impacts of environmental change, which may cause irreversible damage to these ecosystems upon which huge amount of biodiversity and ecosystem services are dependent. Within the next few decades the climate change will have considerable ecological impacts on most of the fresh water ecosystems as per the current climatic predictions. Different freshwater ecosystems will be affected differently by climate change. One of the most important and major impact to be caused by climate change will be on fresh water flow regime. The speed of climate change will be abrupt and uneven rather than slow and even. Impacts caused by climate change on freshwater ecosystems will be visible both physically and chemically. It is very hard and more complex to forecast the impact on freshwater recourses due to climate change. In most of the cases, climate change together with other man made pressures will cause much damage to freshwater ecosystems. It is very difficult to predict impact of climate change on freshwater ecosystems in the next few decades using current global climate models. Rather than focusing on impact assessment a risk–based approach should be adopted to assess and respond to climate change. A number of measures are required to protect freshwater ecosystems such as reducing extraction of water from ground and surface water, maintaining water flows, management of macrophytes, artificial

B. Hassan (✉) · A. M. Yatoo
Department of Environmental Science, University of Kashmir, Srinagar, Jammu and Kashmir, India

H. Qadri
Department of Environmental Sciences, School of Sciences, Sri Pratap College Campus, Cluster University of Srinagar, Srinagar, Jammu and Kashmir, India

M. N. Ali
Department of Environmental Science, University of Kashmir, Srinagar, Jammu and Kashmir, India

Cytogenetics and Molecular Biology Research Laboratory, Centre of Research for Development, University of Kashmir, Srinagar, Jammu and Kashmir, India

N. A. Khan
Department of Botany, Punjabi University, Patiala, Punjab, India

H. Qadri et al. (eds.), *Fresh Water Pollution Dynamics and Remediation*,
https://doi.org/10.1007/978-981-13-8277-2_7

oxygenation and mixing, sediment removal etc. so that fresh water ecosystems are not affected largely by small climate induced changes. When a diversity of healthy habitats of freshwater ecosystems can be maintained, the assimilative ability of freshwater ecosystems will be further strengthened. Incorporation of long lasting, observed study data with models and manipulative experiments will assist the progress of mechanistic, and hence predictive, perceptive of responses to future climate change.

Keywords Freshwater · Ecosystem · Biodiversity · Management · Macrophytes

1 Introduction

Climate change is considered as the global challenge in the twenty-first century. It has an impact on both natural and human systems by escalating their susceptibility at different scales and with undesirable intensity (Stocker et al. 2013; Hansen and Cramer 2015). Global warming has direct and widespread impact on the hydrological cycle and hence on the aquatic ecosystems (Huntington 2006; Oki and Kanae 2006). Human activities have directly led to an immense increase in green house gas emissions mainly carbon dioxide that contributes mainly in the warming of atmosphere. The concentration of carbon dioxide is expected to rise twice as high as those existing in pre-industrial period, within the next century. Freshwater ecosystems are highly exposed to stressors for eg eutrophication, species invasion, land-use change, and rising temperatures (e.g. Firth and Fisher 1992; Poff et al. 2002; Glen 2010; Boon and Raven 2012). Scientific studies also highlight a difficult situation recognized by managers and researchers such as: (1) ecological responses to change are most unpredictable and (2) the impacts of environmental change on freshwater resources is unsustainable with respect to gross generalization and prediction (e.g. Wilby et al. 2010). Stress associated with environmental change can cause un-predictable, speed transformations between various ecosystem states (i.e., regime shifts). However, some events depict an erosion of resilience by facilitating undesired regime shifts to freshwater ecosystems (Meerhoff et al. 2012) with uncertain outcomes regarding the provisioning of ecosystem services in the future. Although, there exists some warning indicators to detect and recognize regime shifts in various ecosystems (Carpenter et al. 2011; Seekell et al. 2012; Veraart et al. 2012) but the uncertainty and lack of generalization across ecosystems make this approach difficult to build up (Hughes et al. 2013) and put into practice (Biggs et al. 2009). (Covich et al. 2004; Gillson et al. 2013). Both synergistic or antagonistic ecological responses are caused by highly context dependent factors such as climatic and anthropogenic factors, that are difficult to interact (Covich et al. 2004; Gillson et al. 2013). Freshwater ecosystems will get affected due to change in temperature, quantity, and quality (Shuter and Meisner 1992), as well as through changes in the flow of timing and duration. Freshwaters are susceptible to climate change as most of the species within these uneven habitats have inadequate capability to scatter as the environment changes. Some species are hardly modified for other environmental conditions but can reside in some wetlands and can serve as essential elements of coastal and marine fish. Increased rainfall increases the velocity of water flow along the watershed, followed by displacement of sediments into the aquatic systems, causes the dislocation and loss of many fish (Pujolar et al. 2011).

In addition, high precipitation leads to a sharp rise in water level and consequently leads to the decline of protected areas for juvenile fish and other aquatic organisms, making them vulnerable to predators. For the reproduction and survival of wetland species, the magnitude and duration of flooding is a major factor (Pitchford et al. 2012). In addition, warmer temperatures are likely to result in higher ecosystem metabolism and productivity.

2 Physical Impact of Climate Change on Freshwater Ecosystems

2.1 Temperature

Rising temperatures lead to an enhancement in glacial melting, while in some areas increasing precipitation in winter season compensates for glacier melting (Arnell et al. 2001). The melting of glaciers will mainly depend on the rate of change in temperature; for example, Oerlemans et al. (1998) proposed that a rise of 0.4 °C per decade would eradicate nearly all of their study glaciers by 2100, while a rise of 0.1 °C per decade would only lead to a 10–20% loss of glacier volume, due to the absence of sufficient precipitation. Due to lack of seasonality in tropical temperatures, tropical glaciers may be particularly receptive to climate change as the glacial melting is significant the whole year (Kaser et al. 1996). The greenhouse effect will lead to a global rise in air temperature, with mean surface temperatures increasing 1.5–5.8 °C by the year 2100 (Houghton et al. 2001). In many regions, a diminution in the diurnal temperature range occurs because daily minimum air temperatures have increased more than daily maximum temperatures (Easterling et al. 1997). The variability in temperature i.e. 1 °C increase in the standard deviation of temperature will lead to a far greater frequency of extreme temperature events than a similar change in the mean temperature would (Meehl et al. 2000).

2.2 Precipitation

In mid and high latitude regions, surface precipitation has generally increased and in the tropics and subtropics, it is generally decreased (Easterling et al. 2000). Though, a little change in average precipitation could lead to extensive rise in the inconsistency of precipitation events; because the size of precipitation is not normally disseminated about the mean (Meehl et al. 2000). The soil distinctiveness and the extent of local precipitation changes could be determined through the soil moisture content and the volume of runoff could be determined through soil infiltration and water-holding capacity will in turn determine the volume of runoff. For instance, drier soil often shows decreased water infiltration, and severe freezing events can decrease water infiltration in limestone soils (Boix-Fayos et al. 1998). In addition, there will be increased water infiltration due to flooding.

2.3 Water Quantity and Flow Changes

Climate change also leads to significant changes in groundwater recharge. Smaller change in temperature has been observed on those freshwater ecosystems that receive input from groundwater than those dependent on precipitation. In tropical and arid regions, water flows depend primarily on precipitation. In tropical rivers, seasonal heavy rainfall events already surpass the natural infiltration rates of soil, leading to high sediment input and dangerous levels of pesticide runoff from agricultural lands (Pringle 2000). Temperature changes will affect water flow through changes in snowmelt and the form of falling precipitation in higher latitude regions. In large parts of Eastern Europe, European Russia, central Canada, and California, a major shift is observed in stream flow from spring to winter as high temperatures cause precipitation to fall as rain rather than snow (Dettinger and Cayan 1995; Westmacott and Burn 1997). Even in the absence of increased precipitation, the glacier fed rivers, lakes, and wetlands in tropical and temperate regions may experience increased flows due to glacial melting.

2.4 Effects on Physiology and Life History

Climate change will lead to change in the physiological processes and life history traits of animals. If higher ambient temperature will get increased, the metabolic demands of many animals could be changed for instance, even at sub-lethal temperatures, warming would lead to a several fold rise in the energy requirements of lake trout *(Salvelinus nomaycush)*. The growth metabolism of many organisms can be determined through the availability of food in that region.. In zooplankton with sufficient food supply, higher temperatures promote feeding, assimilation, growth, and reproductive rates (Schindler 1968). Increased temperatures can also lead to a rise in the frequency of toxic algal outbreaks and leads to their toxicity to other animals.

Temperature also affects the morphology of aquatic animals; increased rearing temperature causes a decrease in body size at a given developmental stage in over 90% of cold blooded, aquatic animals studied (Atkinson 1995). Temperature also determines the sex of offspring in the American alligator *(Alligator mississippiensis)* and several groups of turtles (Conover 1984). In one population of painted turtles

(Chrysemys picta), off spring sex was shown to be highly correlated with mean July air temperatures; statistical analyses indicate that a 2 °C rise in air temperatures would drastically skew sex ratios, and a 4 °C rise would virtually eliminate all males from the population. Moreover, warming temperatures lead to early breeding migrations and spawning dates in several species of amphibians (Beebee 1995).

2.5 Effect on Community Composition and Dynamics

The composition of many communities may get altered; due to differences in thermal tolerances and interactions between species. Climate change may affect motile vs. non-motile species differently, leading to change in species distribution. Atmospheric warming is expected to facilitate the spread and initiation of non-native species, especially in temperate and tropical systems (Stachowicz et al. 2002). Native species may be displaced by invaders with a competitive advantage by removing the heat barriers to invasion (Carpenter et al. 1992). In many riverine systems, reservoirs restrict flooding and facilitate the growth of exotic fish (Baron et al. 2003). Most of the animals that cannot adapt to increasing temperatures migrate from hot to cold regions. In view of the changing environmental conditions, most plants and animals display range shifts, rather than morphological change (Noss 2001). It is believed that in the next 100 years, the rate of warming is expected to increase almost ten times higher than warming after the last ice age (De Groot and Ketner 1994). It is very difficult to predict whether plants and animals are able to migrate quickly with respect to climate change (Malcolm and Markham 2000). Most of endemic fish goes extinct due to increased warming and the lack of northern migration (Matthews and Zimmerman 1990). Thus climate change directly or indirectly affects freshwater ecosystems and the communities that live within these ecosystems.

3 Effects of Climate Change on Lakes

3.1 Physical Effect

Increased mean surface temperatures likely leads to increased water temperatures and evaporation in many lakes, in both temperate and tropical areas (Schindler 2001; Zinyowera et al. 1998). If lake levels gets declined, the important spawning and rearing habitat (located at the edge of lake) would be lost (Tyedmers and Ward 2001), and there would be dramatic change on water outflow. Some lakes that currently supply outflow to downstream lake systems may become endorheic and saline (Schindler 2001). There is a 70% reduction in primary productivity since 1975 due to decrease in the amount of nutrients in the top layers of the lake which in turn increases the water transparency and hence light penetrates easily. Temperate lakes exhibit intense thermal gradients and shows larger seasonal alterations in water temperature than tropical lakes. Lakes at higher altitude experience thermal stratification seasonally, which are covered by ice in the winter and create a thermal gradient in the summer as upper layers of water warms up. Increased ambient temperatures have led to an earlier onset of thermal stratification, longer ice-free periods, and deeper thermoclines in many temperate lakes (Schindler et al. 1990). Drought and decreased groundwater flow may make some lakes more susceptible to acidification, as groundwater often contains acid-neutralizing chemicals important to lake buffering (Schindler 2001). However, the overall pH and chemical balance of lakes may be affected by temperature and precipitation changes in ways that are site-specific and difficult to predict. Finally, the physical effects of climate change on temperate lakes can be synergistic and complex.

3.2 *Biological Effect*

In tropical lakes, large scale decrease in the primary productivity have been observed and are likely to have a considerable impact on the rest of the food chain due to climate change. Climate warming cause changes in the physical and thermal stratification in temperate lakes and therefore effects the biotic communities. In spring and fall season, temperatures are most favourable and rate of growth are highest for cold water fish while the summer is optimal for cool and warm water. The upper layers of the lake becomes too hot during the summer and therefore coldwater fish migrate to cooler bottom layers. In Addition, climate warming leads to longer periods of thermal stratification, coldwater fish are restricted to cool bottom layers of the lake for longer duration of time, and form deeper thermoclines which in turn decrease the area of bottom layer of lake and enhance competition for food (Shuter and Meisner 1992). Rise in overall lake temperature leads to higher metabolic demands of biotic communities but coldwater fish generally shows reduced access to prey population. Winters become slightly more favorable through high temperatures, but not enough to compensate for losses in other seasons (Shuter and Meisner 1992). Overall, climate change decreases the growth rates and increases the heat mortality of almost all the cold water fish (Tyedmers and Ward 2001). Moreover, positive effects have also been observed on cold and warm water species because of climate warming and changes in thermal stratification. If poleward migration of species becomes possible, it then reduces winter kills, lengthens the growth season, and increases availability of habitat, locally as well as regionally (Shuter and Meisner 1992). Lakes that are not nutrient-limited, productivity is likely to increase and overall fish catch may increase, but the relative abundance and density of fish merely changes (Tyedmers and Ward 2001). Finally, although most studies examine the effects of climate change on only a few species of fish, negative effects on one species can have an impact on the entire community. For example, a summer kill of planktivorous herring in a Wisconsin (USA) lake, reduced predation on zooplankton by 50%, which led to an increase in large zooplankton and intensified zooplankton grazing, causing a substantial reduction in phytoplankton abundance (Kitchell 1992).

4 Effect of Climate Change on Rivers

The effects of climate change on rivers are likely to vary widely depending on latitude. Temperate rivers, like temperate lakes, will be affected primarily by temperature changes, while changes in precipitation timing and quantity could have dramatic effects on tropical rivers.

4.1 Physical Effects

Rise in atmospheric temperature will strongly influence water temperature in many rivers because of high surface to volume ratio (Tyedmers and Ward 2001). Warming of the atmosphere increases the temperature of water bodies throughout and as decreases the oxygen level. Temperate rivers experience seasonal thermal cycles, with uniform cold temperatures in winter and longitudinally stratified temperatures in summer, with lower temperatures at groundwater-fed headwaters and higher temperatures downstream (Shuter and Meisner 1992). High latitude rivers are already experiencing shorter periods of ice cover and earlier ice break-up (Magnuson et al. 2000), and many of the beneficial functions of ice jams may be compromised. Flow regime is a critical component of river ecosystems. Mean flow may increase or decrease depending on changes in average precipitation, evaporation, soil moisture, and groundwater recharge, but seasonal shifts in flow may be more significant to freshwater ecosystems (Carpenter et al. 1992. Spring snowmelts are likely to occur earlier due to warming, and winter flows are likely to increase in areas where winter precipitation falls as rain instead of snow. A shift in peak flows from spring/summer to winter will reduce the cooling effect of snowmelt on river temperature in summer (Tyedmers and Ward 2001). Where precipitation increases, stream flows may increase in volume and floods may become more frequent. Extreme flooding events and landslides could remove important woody debris from rivers and destabilize river channels (Carpenter et al. 1992). Where precipitation decreases, stream flow volume may also decrease, and reductions in runoff will lower the concentrations of DOC and organic matter in rivers. Increased evaporation could also lead to reduced stream flow, even in the absence of precipitation changes. Summer and ephemeral streams in arid regions are more vulnerable to drying up. A reduction in natural flooding events could eliminate many of the beneficial physical effects of seasonal flooding, such as creating floodplain habitat, displacing exotic plants, and determining river channel form.

4.2 Biological Effects

In tropical rivers, the rainy and dry seasons lead to large, expected seasonal variations in precipitation and yearly flooding of adjoining grasslands and forests, which supply plentiful feeding and breeding grounds for fish. Changes in water level have a strong affect on river fishes than by changes in temperature. As the floodplains dry up due to closer of rainy season, members of the "whitefish" guild (sensitive to reduced oxygen levels), move back to the main river channel. "Blackfishes", who are more tolerant of or adapted to low oxygen levels, remain in marginal floodplain habitats that become disconnected from the river and may even dry up completely (Welcomme 1979). Some of these species, such as the lungfishes, are able to aestivate and breathe air when their water supply evaporates. On the other hand, changes in floodplain dynamics will directly affect fish populations and fisheries yield, as growth rates and overall fish catch is correlated with the area of flooded land. Coldwater fish that are restricted to cool refugia at headwaters experience more competition, lower growth rate, and feasible range shifts during the summer (Shuter and Meisner 1992). At headwaters, warmer water and decreased oxygen shows negative effects on the eggs and larvae placed there (Carpenter et al. 1992). Diadromous stocks experience higher rates of pre-spawning mortality because of increased metabolic needs and disease outbreaks during the peak summer season (Tyedmers and Ward 2001). Even the stocks that do not perform migrations in summer, climate change results in a net specie decline due to fall in juvenile emergence, growth and survival (Henderson et al. 1992). Some invertebrates in northern rivers require a long duration of exposure to nearly 0 °C water, followed by warm spring for hatching. Moreover, the release of warm water in winter from dams shows massive decline of invertebrates for tens of kilometers downstream (Lehmkuhl 1974), and overall river warming would be expected to have a similar effect.

5 Impact of Climate Change on Wetlands

5.1 Physical Effects

Increase in atmospheric temperature will lead to drying of many wetlands, unless the rate of precipitation balances the rate of evaporation. Ephemeral wetlands with no channelized flow in or out, could be vanished completely, particularly if precipitation decreases and groundwater is withdrawn for domestic consumption. On the other hand, rise in precipitation could lead to flooding, extension and deepening of wetland habitat, and enlarged connectivity. Conversely, rise in precipitation may also lead to an increased contribution of sediment and pollutants, and could wipe out some wetlands if vegetation or other vital habitat features are entirely inundated. Furthermore, rising temperatures could lead dramatic changes in hydrological regime of arctic and subarctic bogs situated over permafrost. Peat lands underlain by permafrost could become net carbon sources rather than sinks. The drainage of tropical peat lands could lead to increased risk of fires due to rise in global climate and release large amounts of carbon dioxide into the atmosphere. Increase in carbon-dioxide concentrations in the atmosphere could lead to increase in net productivity in vegetation systems, causing carbon to accumulate in vegetation over time. The environments of tropical wetlands are considered highly vulnerable to climate change and might be affected in four different ways: by changes in the hydrological regime; changes in precipitation patterns; local changes in temperature/humidity and subsequently in patterns of evapo-transpiration, and increases in the frequency of extreme climate events. Added to these, the coastal wetlands, such as mangroves, might also be influenced by the rising of sea levels. Many coastal systems will experience increased levels of inundation and storm flooding, accelerated coastal erosion, seawater intrusion into fresh groundwater, encroachment of tidal waters into estuaries and river systems, and elevated sea surface temperatures. Coastal freshwater wetlands are particularly receptive to intense high tides resulting from an increase in storm frequency; these high tides can carry salts inland to salt resistant vegetation and soils, and could lead to the dislocation of freshwater flora and fauna by salt-resistant species (Michener et al. 1997). Destruction of coastal freshwater wetland communities will occurs due to rise in global warming as saline water invades, particularly if these communities cannot move inland due to development or dikes (Tyedmers and Ward 2001).

5.2 *Biological Impact*

Rare species may get lost if the ephemeral wetlands (especially in arid regions) dry up. For example, numerous endemic species of fairy shrimp in California that are critically threatened by habitat loss (Belk and Fugate 2000) could disappear if reduced rainfall and increased evaporation eliminates their shallow, vernal pool habitats. The loss of wetlands would decrease the number of ponds, size of available ponds and also enlarge inter pond distance (Gibbs 1993; Semlitsch and Brodie 1998), reduce the chances of amphibian re-colonization as adult frogs are generally only capable of traveling 200–300 m. Habitat connectivity on a regional scale would be decreased due to loss and drying of wetlands endangering migrating birds that rely upon a network of wetlands along their migration route. Wetlands in areas with higher rainfall undergo less negative effects, and may even benefit from increased wetland area and connectivity. Though, some rare species that are modified to drier, ephemeral wetlands may not be able to compete with invading species modified to wetter habitats and paddling birds may experience lower access to feeding areas (Butler and Vennesland 2000). Moist and stable wetlands would sustain more fish, which feed on vulnerable tadpoles and invertebrates that usually occupy seasonal wetlands with reduced predation pressure (Semlitsch and Brodie 1998).

6 Managing Freshwater Ecosystems to Withstand Climate Change

Climate change increases the air temperature due to which freshwater ecosystems gets warm up. Most of the lakes and streams have experienced decline in water level during summer droughts. These changes are expected to continue and accelerate in the future. In addition, these changes impact water quality and quantity, present harmful implications for freshwater ecosystems and the species that rely on these vulnerable habitats. Rivers, lakes, wetlands and their connecting ground waters are the "sinks" into which landscapes drain. Freshwater ecosystems are closely linked to the watersheds or catchments of which each is a part, far from being inaccessible conduits. The stream network itself is important to the continuum of river processes. Dynamic patterns of flow that are maintained within the natural range of variation will promote. The integrity and sustainability of freshwater aquatic systems are maintained within the natural range of variation through dynamic flow pattern.

6.1 Preserve Habitat Heterogeneity and Biodiversity

Climate change leads to increase in resistance and resilience to both species and habitat diversity, as diversity provides a greater range of stress tolerances and adaptive options. High biodiversity is often found in older or isolated aquatic habitats and in areas with high habitat heterogeneity especially dynamic habitats with seasonal changes in water level (river floodplains of seasonal wetlands; Abell et al. 2002). Most of these areas also harbor rare species that have evolved in and remain restricted to a particular habitat. High biodiversity sites may be protected by protecting rare or vulnerable species. Protecting rare species, can help in drawing public attention and funding to conservation efforts, but those policies aimed exclusively at conserving single species and probably may restrict from the more desirable goals of protecting entire ecosystem function (Junk 2002) and increasing resistance and resilience to climate change. Areas where natural physical barriers separate biota and transition zones between different habitats or ecosystems may also harbor high biodiversity (Abell et al. 2002). Protecting transitional zones has the added benefit of accommodating possible range shifts due to climate change and can help preserve diverse habitat types. Protecting a variety of potential habitats may help increase resistance and resilience in vulnerable species; for example, protecting an array of natural ponds with a wide range of sizes and hydro periods will help ensure that amphibians have access to suitable breeding sites regardless of climatic variation (Semlitsch 2002).

6.2 Protect Physical Features Rather Than Individual Species

Ecosystem function is determined by basic physical features such as water flow, channel morphology, and nutrient balance, rather than by species assemblages (Moss 2000). Protecting flow patterns, water quality and quantity is very important for protecting biodiversity in freshwater habitats (Abell et al. 2002), whereas conservation efforts that focus exclusively on preserving particular species or groups of species without bearing in mind wider physical features of the system may be destined to failure. In many cases, the function of a particular species in a freshwater ecosystem is actually more important than its identity; for example, plants are essential components of some aquatic habitats (floodplain vegetation and aquatic plants in shallow lakes), but the exact species of plant may be less important than the physical features it provides (Moss 2000).

Because of global warming and precipitation variability many physical features of freshwater ecosystems such as rivers, lakes and wetlands are expected to undergo a number of changes. Removing barriers to water flow, maintaining healthy river basins, and decreasing input of nutrients and toxic substances will increase the probability that freshwater ecosystems will be able to adopt to climate change. For example, removing levees and other barriers to the lateral expansion of rivers could prevent the loss of critical edge habitats and the species that depend on them.

Connectivity is a necessary feature of many freshwater ecosystems, as it can help to maintain flow regimes, encourage ecological integrity, and allow migrating animals to move between different habitats at various life history stages. Connectivity is important not only between different freshwater habitats but also between freshwater habitats and subterranean systems or groundwater sources (Abell et al. 2002). Although some species may be able to adapt to climate change in their current habitats, warmer waters will force other species to move into cool, thermal refugia, where temperatures remain below their thermal tolerance limits and metabolic demands are lower. Many species (i.e. coldwater fish) already rely on thermal refugia at certain times of the year, and these species are likely to become even more dependent on these refuges for year round survival.

6.3 Protect Reserves from Human Pressures and Non Native Species

Stressed ecosystems display reduced resistance and resilience to change, particularly those that tend to reduce diversity are likely to become more important as local climates become more variable (Noss 2001). Human stresses, such as overexploitation and poor land use practices, should be reduced as much as possible. It is important to boost efforts to avoid access of invading species and to eliminate harmful exotic species present in these reserves. However, as thermal barriers that previously excluded invaders will be removed, many ecosystems are likely to become more vulnerable to invasions and the communities gets invaded by warmer-adapted species due to climate change (Carpenter et al. 1992; Schindler 2001). Unfortunately, in some ecosystems, preventing access to motile, invasive species may interfere with the aim of maintaining connectivity and to allow climate-induced migrations. Separating vulnerable habitats from other freshwater ecosystems may be responsible in some cases, but making barriers that restrict flow by preventing access of non native species may do more harm. In cases where the migration is important to native species and there are less chances of invasion, it is better to maintain current levels of connectivity while enacting careful monitoring of ecosystems.

7 Watershed Management

Watershed management has pivotal role in adequate protecting of aquatic patches. The rising population led to increased deforestation and industrial development inside watersheds. Notably, the pressure raised by habitat modification has put tremendous stress and enhanced the effects caused by climate change. Unwanted cutting of trees around freshwater environs has warm impacts on concerned water bodies. The vegetation envelop near flowing water bodies decreases the chances of woody debris intrusion in it. Besides, loss of shield to the direct sunlight enhances the water temperature, which is a cause of global warming.

7.1 Drainage Basins Restoration

The majority of drainage basins experienced huge stress of deprivation due to varied natural and anthropogenic activities in immediate land setting and amplified quantity of water removal (McCarty and Zedler 2002). Reinstatement of despoiled sites holds enormous pledge for freshwater environs, both by means of improving the conservation reliability of spoiled and providing apparatus for safe level improvement of freshwater environs from further harm with climate change. Unbeaten re-establishment techniques include neutralization of more acidic aquatic environs by the application of lime (Schindler 1997; Venema et al. 1997), and restoring the hydrology of lakes by cleaning obstacles to the normal flow route (Gilbert and Anderson 1998). The above cited problems in aquatic environs could be the sole potential of climate change. Therefore, awareness from existing reinstatement schemes can give us an idea about outlook comeback from climate change.

8 Management Policies

It is based upon the identification that ambiguity is inborn in all the natural activities, and the belief that executive management processes will modify with time (Parma et al. 1998). Passive and delaying management policies involves adjusting management practices based on what is learned from the past practices, but learning about the fundamental system is not an open target, whileas, energetic management practices is rather like performing a well-designed trial. The underlying principle behind this loom is that we cannot hope to recognize the complex systems (Parma et al. 1998). However, an effective management practices towards the global climate change are only beginning to be felt in many areas, may provide enough time to learn about the underlying processes governing how a particular system responds to change, and provide an understanding of how to best manage the system in the face of global climate change.

8.1 Monitoring of Species Change and Characterization

Before selecting an adaptation strategy for climate change one of the important steps is to differentiate life cycle, species, community and quality of a water bodies that are mainly inclined to changes in average climate or extreme climatic events (Solomon 1994). Study and monitoring on these physical features can help establish the range of management practices. However, change in rainfall, runoff and intensity of flow from deforested, agricultural lands will dramatically add to the sediment load and nutrient content, along with large-scale changes in species composition, distribution and abundance. By calculate which species are likely to be most susceptible to climate change before large effects will be observed, a careful monitoring before effects are noticeable will provide a baseline against which future changes can be compared (Herman and Scott 1994; Noss 2001).

9 Freshwater Conservation Strategies

One of the unique potential threats to freshwater environs is the conventional increase in human water requirements, due to population growth and development. Climate change together with pressure caused by excessive water extraction will almost certainly work together, thereby increasing the effects change in climate on aquatic biota. The IPCC recommends using Integrated Water Resource Management (IWRM) to adapt to increasing water resource demands (Arnell et al. 2001). An immense development more than this water management tactics in terms of maintaining health of the aquatic environs is ecologically sound water management, and is sustainable in nature.

10 Conclusion

Climate change nowadays is a concern not only to the scientific society but also to the people from all regions of the world. Increasing human population, industrialization, unscientific and agricultural practices has put tremendous pressure to the climate related entities. Burning of fossil fuels, vehicular exhausts, construction of artificial dams and generation of MSW directly or indirectly increasing the green house gases (GHGs) into the atmosphere. The increasing concentration of these gases into the atmosphere enhances the green house effect, which has numerous impacts on freshwater environs. The climate change not only raises the temperature on aquatic environs but also has many core negative impacts on freshwater quality. The intrusion of new plant and animal species, change in the season, duration and location of precipitation and unexpected rise of microbial load are various prominent impacts of climate change on aquatic environs. There is now no doubt about that increased unorthodox anthropogenic activities somehow are responsible for climate change, which has direct and indirect impacts on freshwater environs. Therefore, priority should be given to reduce the releasing of more and more GHGs into the atmosphere, which somehow will reduce the unexpected change in meteorological factors.

References

Abell, R., Thieme, M., Dinerstein, E., & Olson, D. (2002). *A sourcebook for conducting biological assessments and developing biodiversity visions for eco-region conservation* (Volume II: Freshwater eco-regions) (201 pp). Washington, DC: World Wildlife Fund.

Arnell, N., Liu, C., et al. (2001). Chapter 4: Hydrology and water resources. In J. Mc-Carthy, O. Canziana, N. Leary, D. Dokken, & K. White (Eds.), *Climate change (2001): Impacts, adaptation, and vulnerability* (pp. 191–233). Contribution of Working Group II to the Third Assessment Report of the Intergovernmental Panel on Climate Change. Cambridge: Cambridge University Press.

Atkinson, D. (1995). Effects of temperature on the size of aquatic ectotherms: Exceptions to the general rule. *Journal of Thermal Biology, 20*(1/2), 61–74.

Baron, J. S., Poff, N. L., Angermeier, P. L., Dahm, C. N., Gleick, P. H., Hairston, N. G., Jr., Jackson, R. B., Johnston, C. A., Richter, B. D., & Steinman, A. D. (2003). Sustaining healthy freshwater ecosystems. *Issues in Ecology, 10*, 1–16.

Beebee, T. J. C. (1995). Amphibian breeding and climate change. *Nature, 374*, 219–220.

Belk, D., & Fugate, M. (2000). Two new Branchinecta (Crustacea: Anostraca) from the southwestern United States. *The Southwestern Naturalist, 45*(2), 111–117.

Biggs, R., Carpenter, R., & Brock, W. R. (2009). Turning back from the brink: Detecting an impending regime shift in time to avert it. *Proceedings of the National Academy of Sciences USA, 106*, 826–831.

Boix-Fayos, C., Calvo-Cases, A., Imeson, A. C., Soriano Soto, M. D., & Tiemessen, I. R. (1998). Spatial and short-term temporal variations in runoff, soil aggregation and other soil properties along a Mediterranean climatological gradient. *Catena, 33*, 123–138.

Boon, P. J., & Raven, P. J. (2012). *River conservation and management*. London: Wiley-Blackwell.

Butler, R. W., & Vennesland, R. G. (2000). Integrating climate change and predation risk with wading bird conservation research in North America. *Waterbirds, 23*(3), 535–540.
Carpenter, S. R., Fisher, S. G., Grimm, N. B., & Kitchell, J. F. (1992). Global change and freshwater ecosystems. *Annual Review of Ecology and Systematics, 23*, 119–139.
Carpenter, S. R., Cole, J. J., Pace, M. L., Batt, R., Brock, W. A., Cline, A., Coloso, J., & Hodgson, H. R. (2011). Early warnings of regime shifts: A whole-ecosystem experiment. *Science, 332*, 1079–1082.
Conover, D. O. (1984). Adaptive significance of temperature-dependent sex determination in a fish. *American Naturalist, 123*(3), 297–313.
Covich, A. P., Austen, M. C., Bärlocher, F., Chauvet, E., Cadrinale, B. J., Biles, C. L., Inchausti, P., & Dangels, O. (2004). The role of biodiversity in the functioning of freshwater and marine benthic ecosystems. *Bioscience, 54*, 767–775.
De Groot, R. S., & Ketner, P. (1994). Sensitivity of NW European species and ecosystems to climate change and some implications for nature conservation and management. In J. Pernetta, R. Leemans, D. Elder, & S. Humphrey (Eds.), *Impacts of climate change on ecosystems and species: Implications for protected areas* (pp. 28–53). Gland: The World Conservation Union (IUCN), Gland.
Dettinger, M. D., & Cayan, D. R. (1995). Large-scale forcing of recent trends toward early snowmelt runoff in California. *Journal of Climate, 8*, 606–623.
Easterling, D. R., Horton, B., Jones, P. D., Peterson, T. C., Karl, T. R., Parker, D. E., Salinger, M. J., Razuvayev, V., Plummer, N., Jamason, P., & Follard, C. K. (1997). Maximum and minimum temperature trends for the globe. *Science, 277*, 364–367.
Easterling, D. R., Meehl, G. A., Parmesan, C., Changnon, S. A., Karl, T. R., & Mearns, L. O. (2000). Climate extremes: Observations, modeling, and impacts. *Science, 289*, 2068–2074.
Firth, P., & Fisher, S. G. (1992). *Global change and freshwater ecosystems*. New York: Springer.
Gibbs, J. P. (1993). Importance of small wetlands for the persistence of local populations of wetland-associated animals. *Wetlands, 13*, 25–31.
Gilbert, O. L., & Anderson, P. (1998). *Habitat creation and repair* (288 pp). New York: Oxford University Press.
Gillson, L., Dawson, T. P., Jack, S., & McGeoch, M. A. (2013). Accommodating global change contingencies in conservation strategy. *Trends in Ecology and Evolution, 28*, 135–142.
Glen, G. (2010). *The impact of global change on European lakes*. New York: Springer.
Hansen, G., & Cramer, W. (2015). Global distribution of observed climate change impacts. *Nature Climate Change, 5*, 182–185.
Henderson, M. A., Levy, D. A., & Stockner, J. S. (1992). Probably consequences of climate change on freshwater production of Adams River sockeye salmon. *Geo Journal, 28*(1), 51–59.
Herman, T. B., & Scott, F. W. (1994). Protected areas and global climate change: assessing the regional or local vulnerability of vertebrate species. In J. Pernetta, R. Leemans, D. Elder, & S. Humphrey (Eds.), *Impacts of climate change on ecosystems and species: Implications for protected Areas* (pp. 13–27). Gland: The World Conservation Union (IUCN).
Houghton, J. T., Ding, Y., Griggs, D. J., Noguer, M., van der Linden, P. J., & Xiaosu, V. (Eds.) (2001). *Climate change: The scientific basis*. Intergovernmental Panel on Climate Change: Working Group I. Cambridge: Cambridge University Press. 881 pp.
Hughes, T. P., Linares, C., Dakos, V., van de Leemput, I. A., & van Nes, E. H. (2013). Living dangerously on borrowed time during slow, unrecognized regime shifts. *Trends in Ecology & Evolution, 28*, 149–155.
Huntington, T. G. (2006). Evidence for intensification of the global water cycle: Review and synthesis. *Journal of Hydrology, 319*, 83–95.
Junk, W. J. (2002). Long-term environmental trends and the future of tropical wetlands. *Environmental Conservation, 29*(4), 414–435.
Kaser, G., Hastenrath, S., & Ames, A. (1996). Mass balance profiles on tropical glaciers. *Zeitschrift für Gletscherkundeund Glazialgeologie, 32*, 75–81.

Kitchell, J. F. (Ed.). (1992). *Food web management: A case study of Lake Mendota* (553 pp). New York: Springer.

Lehmkuhl, D. M. (1974). Thermal regime alterations and vital environmental physiological signals in aquatic systems. In J. W. Gibbons & R. R. Sharitz (Eds.), *Thermal ecology.* Atomic energy commission symposium series, CONF-730505, Augusta, GA, USA (pp. 216–222).

Magnuson, J. J., Robertson, D. M., Benson, B. J., Wynne, R. H., Livingstone, D. M., Arai, T., Assel, R. A., Barry, R. G., Card, V., Kuusisto, E., Granin, N. G., Prowse, T. D., Stewart, K. M., & Vuglinski, V. S. (2000). Historical trends in lake and river cover in the Northern Hemisphere. *Science, 289*, 1743–1746.

Malcolm, J. R., & Markham, A. (2000). *Global warming and terrestrial biodiversity decline. A report prepared for the WWF*. http://www.panda.org/downloads/climate_change/speedkills_c6s8.pdf

Matthews, W. J., & Zimmerman, E. G. (1990). Potential effects of global warming on native fishes of the southern Great Plains and the southwest. *Fisheries, 15*(6), 26–32.

McCarty, J. P., & Zedler, J. B. (2002). Restoration, ecosystem. In H. A. Mooney & J. G. Canadell (Eds.), *The earth system: Biological and ecological dimensions of global environmental change* (Vol. 2, pp. 532–539). Chichester: Wiley.

Meehl, G. A., Karl, T., Easterling, D. R., Changnon, S., Pielke, J. R., Changnon, D., Evans, J., Groisman, P. Y., Knutson, T. R., Kunkel, K. E., Mearns, L. O., Parmesan, C., Pulwarty, R., Root, T., Sylves, R. T., Whetton, P., & Zwiers, F. (2000). An introduction to trends in extreme weather and climate events: Observations, socioeconomic impacts, terrestrial ecological impacts, and model projections. *Bulletin of the American Meteorological Society, 81*(3), 413–416.

Meerhoff, et al. (2012). Environmental warming in shallow lakes: A review of potential changes in community structure as evidenced from space-for-time substitution approaches. *Advances in Ecological Research, 46*, 259–394.

Michener, W. K., Blood, E. R., Bildstein, K. L., Brinson, M. M., & Gardner, L. R. (1997). Climate change, hurricanes and tropical storms, and rising sea level in coastal wetlands. *Ecological Applications, 7*, 770–801.

Moss, B. (2000). Biodiversity in fresh waters—An issue of species preservation or system functioning? *Environmental Conservation, 27*(1), 1–4.

Noss, R. F. (2001). Beyond Kyoto: Forest management in a time of rapid climate change. *Conservation Biology, 15*(3), 578–590.

Oerlemans, J., Anderson, B., Hubbard, A., Huybrechts, P., Johannesson, T., Krap, W. H., Schmeits, M., Stroeven, A. P., van der Wal, R. S. W., Wallinga, J., & Zuo, Z. (1998). Modeling the response of glaciers to climate warming. *Climate Dynamics, 14*, 267–274.

Oki, T., & Kanae, S. (2006). Global hydrological cycles and world water resources. *Science, 313*, 1068–1072.

Parma, A. M., et al. (1998). What can adaptive management do for our fish, forests, food, and biodiversity? *Integrative Biology, 1*(1), 16–26.

Pitchford, J. L., Wu, C., Lin, L. S., Petty, J. T., Thomas, R., Veselka, I. V. W. E., Welsch, D., Zegre, N., & Anderson, J. T. (2012). Climate change effects on hydrology and ecology of wetlands in the mid-Atlantic highlands. *Wetlands, 32*(1), 21–33. https://doi.org/10.1007/s13157-011-02593.

Pringle, C. M. (2000). River conservation in tropical versus temperate latitudes. In P. J. Boon, B. R. Davies, & G. E. Petts (Eds.), *Global perspectives on river conservation: Science, policy and practice* (pp. 371–384). West Sussex: Wiley.

Poff, N. L., Brinson, M., & Day, J. B. (2002). *Freshwater and coastal ecosystems and global climate change: A review of projected impacts for the United States* (44 p). Virginia: Pew Center on Global Climate Change.

Pujolar, J. M., Vincenz, I. S., Zane, L., Jesensek, D., De Leo, G. A., & Crivelli, A. J. (2011). The effect of recurrent floods on genetic composition of marble trout populations. *Plos One*. https://doi.org/10.1371/journal.pone.0023822. PMid: 21931617.

Schindler, D. W. (1968). Feeding, assimilation and respiration rates of Daphnia magna under various environ- mental conditions and their relation to production estimates. *Journal of Animal Ecology, 37*, 369–385.

Schindler, D. W. (1997). Liming to restore acidified lakes and streams: A typical approach to restoring damaged ecosystems? *Restoration Ecology, 5*, 1–6.

Schindler, D. W. (2001). The cumulative effects of climate warming and other human stresses on Canadian freshwaters in the new millennium. *Canadian Journal of Fisheries and Aquatic Sciences, 58*(18), 29.

Schindler, D. W., Beaty, K. G., Fee, E. J., Cruikshank, D. R., DeBruyn, E. R., Findlay, D. L., Linsey, G. A., Shearer, J. A., Stainton, M. P., & Turner, M. A. (1990). Effects of climate warming on lakes of the central boreal forest. *Science, 250*, 967–970.

Seekell, D. A., Carpenter, S. R., Cline, T. J., & Pace, M. L. (2012). Conditional heteroskedasticity forecasts regime shift in a whole ecosystem experiment. *Ecosystems, 15*, 741–747.

Semlitsch, R. D. (2002). Principles for management of aquatic-breeding amphibians. *Journal of Wildlife Management, 64*, 615–631.

Semlitsch, R. D., & Brodie, J. R. (1998). Are small isolated wetlands inexpendable? *Conservation Biology, 12*, 1129–1133.

Shuter, B. J., & Meisner, J. D. (1992). Tools for assessing the impact of climate change on freshwater fish populations. *Geo Journal, 28*(1), 7–20.

Solomon, A. M. (1994). Management and planning of terrestrial parks and reserves during climate change. In J. Pernetta, R. Leemans, D. Elder, & S. Humphrey (Eds.), *Impacts of climate change on ecosystems and species: Implications for protected areas* (pp. 1–12). Gland: The World Conservation Union (IUCN).

Stachowicz, J. J., Terwin, J. R., Whitlatch, R. B., & Osman, R. W. (2002). Linking climate change and biological invasions: Ocean warming facilitates non-indigenous species invasions. *Proceedings of the National Academy of Sciences, 99*(24), 15497–15500.

Stocker, T. F., Qin, D., Plattner, G. K., Tignor, M., Allen, S. K., Boschung, J., Nauels, A., Xia, Y., et al. (2013). *Climate change 2013: The physical science basis. Contribution of working group I to the fifth assessment report of the Intergovernmental Panel on Climate Change*. Cambridge: Cambridge University Press.

Tyedmers, P. C., & Ward, B. (2001). A review of the impacts of climate change on BC's freshwater fish resources and possible management responses. *Fisheries Centre Research Reports, 9*(7), 1–12.

Venema, H. D., Schiller, E. J., Adamowski, K., & Thizy, J. M. (1997). A water resources planning response to climate change in the Senegal River basin. *Journal of Environmental Management, 49*, 125–155.

Veraart, A. J., Faassen, E. J., Dakos, V., van Nes, E. H., Lurling, M., & Scheffer, M. (2012). Recovery rates reflect distance to a tipping point in a living system. *Nature, 481*, 357–359.

Welcomme, R. L. (1979). *Fisheries ecology of flood plain rivers* (317 pp). London: Longman.

Westmacott, J. R., & Burn, D. H. (1997). Climate change effects on the hydrologic regime within the Churchill-Nelson River Basin. *Journal of Hydrology, 202*, 263–279.

Wilby, R. L., Orr, H., Watts, G., Battarbee, R. W., Berry, P. M., Chadd, R., Dugdale, S. J., Dunbar, M. J., et al. (2010). Evidence needed to manage freshwater ecosystems in a changing climate: Turning adaptation principles into practice. *Science of the Total Environment, 408*, 4150–4164.

Zinyowera, M. C., Jallow, B. P., Maya, R. S., Okoth-Ogendo, H. W. O., et al. (1998). Africa. In R. T. Watson, M. C. Zinyowera, R. H. Moss, & D. J. Dokken (Eds.), *The regional impacts of climate change; An assessment of vulnerability, A special report of IPCC Working Group II* (pp. 30–84). Cambridge: Cambridge University Press.

Role of Biotools in Restoration of Freshwater Ecosystems

Irfan-ur-Rauf Tak

Abstract Climate change, rapidly increasing population and depleting water resources have resulted in prolonged floods and droughts that have resulted in drinking water becoming a cut-throat resource. The ability of toxins to accumulate in the aquatic systems is a vital concern for environmental safety. In this connection, the newest approaches in biotechnology have been employed which include biomineralization, biosorption, phytostabilization, hyperaccumulation, biostimulation, mycoremediation, cyanoremediation and genoremediation. The ample renovation of the environment requires incorporation, assimilation and assistance of these advances along with conventional methods so as to ascertain the mystery of nature. Besides, the need of water industry is to ensure economical and constant supply of fresh water in adequate amounts. The present book chapter will provide better understanding of the problems associated with the toxicity of freshwater ecosystems as well as the feasible and eco-friendly technologies required for cleaning up of the water resources. However, the challenges involved in adopting the new initiatives for cleaning the polluted freshwater ecosystems from both greener and natural point of view must not be ignored.

Keywords Biomarker · Bioremediation · Biotransformation · Freshwater · Toxic metals · Genoremediation

1 Introduction

There has been a considerable increase in the level of environmental pollution in the last decade which is mainly due to human activities (UNEP 2012). Industries, agricultural resources and man related activities mainly in the urban areas cause environmental pollution. The nature of the pollutants is biological, chemical or may be physical (Lee et al. 2006; Govarthanan et al. 2013). Sulphur dioxide, nitrogen oxide, pesticides, toxic metals, herbicides and carcinogens are mainly included in the

I. Tak (✉)
Department of Zoology, SP College, Cluster University Srinagar, Srinagar, India

H. Qadri et al. (eds.), *Fresh Water Pollution Dynamics and Remediation*,
https://doi.org/10.1007/978-981-13-8277-2_8

chemical pollutants (Deng et al. 2007; Kumar and Mani 2010; Mapanda et al. 2005; Robinson et al. 2011). Pathogenic organisms, some poisonous and dangerous biological products are mainly included in the biological pollutants while as the heat, sound, radiation and radioactive substances are included in the physical quantities (Nasr and Ismail 2015). Unprecedented health hazards due to pollution have recently come into light resulting in an increasing need for additional legislature (Schwarzenbach et al. 2006).

Preservation of environment and biodiversity besides controlling the environmental pollution are the prime focus of countries around the world and it is in this context the importance of biotechnological approaches and their implications need to be properly evaluated. There has however been a serious concern regarding the use of biotechnological products and their impact assessment due to their interaction with the environmental factors (Anouti 2014). The effluents from the biotechnological companies are also a cause of concern and demand proper strategies that need to be employed with regard to the safety of their use. Recently economists are aiming to develop methods of true economic valuation and in this direction, multicriteria analysis is a method aimed to take into account both the quantitative as well as qualitative data including the non-monetary variables (Barbier et al. 1997).

Environmental biotechnology basically refers to the use of microorganisms to improve the environmental quality (Chen et al. 2005). Biotechnology includes a set of techniques that make use of the living organisms or their parts to make or to modify the products which may include plants or animals (Kastenhofer 2007). It involves the development of specific organisms that are used for specific application or purposes and includes the use of technologies such as the recombinant DNA technology, cell fusion and various other new bioprocesses. It is basically the use of biological processes, living organisms or their derivatives to make or modify products or processes. It finds its application in number of areas that include health care, crop production and agriculture, wastewater treatment, waste degradation and drug industries. Biotechnological tools include those processes of biological interest that use the chemistry of living organisms through cell manipulation in order to develop new and alternative methods that are aimed at more cleaner and effective ways of producing traditional products and also at the same time help in maintaining the natural and aesthetic beauty of the environment. As opposed to the conventional methods of synthesis of products, biotechnology is the latest trend in production processes around the world, the reason being the eco-friendly nature of biotechnological methods whereas the later methods add pollutants and waste into our environment. A number of problems associated with the traditional treatment methods like incineration or landfills have generated the need for alternative, economical and more reliable methods for treatment of pollution.

2 Bioremediation: A Sustainable Approach for Cleaning the Contaminated Freshwater Ecosystem

Bioremediation is a waste management technique that uses the organisms to remove pollutants from the contaminated site. As per EPA, it is a treatment that uses naturally occurring organisms to break down the hazardous substance into very less toxic substances. Technologies can be grouped into either in situ or ex situ. In situ bioremediation involves the contaminated material to be treated at site where as ex situ involves the removal of contaminated material to be treated elsewhere. Bioremediation may occur on its own or may also occur through the addition of fertilizers etc. that help in encouraging the growth of pollution eating microbes within the medium (O'Loughlin et al. 2000; Prasad 2004; Meagher 2000). Microorganisms used for bioremediation purpose are called as bioremediators. However, it has been seen that not all contaminants are easily treated by bioremediation using microorganisms. Heavy metals such as Cd and Pb are not easily absorbed by microorganisms and the elimination of a wide range of pollutants as well as wastes from the environment requires the need of increasing our understanding of the importance of different pathways and also regulatory networks (Brim et al. 2000).

Compounds that are contaminated are transformed by living organisms with the help of reactions that take place as part of their metabolic processes. Biodegradation of a particular compound is as a result of action of multiple organisms. Microorganisms when imported to a contaminated site help in enhancing degradation by a process that is known as bio augmentation (Tong et al. 2011). For the process of bio augmentation to be effective, microorganisms must attack the pollutant enzymatically and then convert them to harmless products. A bioremediation can only be effective where the environmental conditions allow microbial growth and degradation to proceed at a very fast rate. Like other technologies, bioremediation has a number of limitations (Segura et al. 2009). Some of the contaminants such as highly aromatic hydrocarbons have been found to be resistant to microbial attack as they are degraded very slowly or they are not degraded at all and hence it is not easy to predict the rates of clean up for a remediation. Bioremediation techniques are more economical than traditional methods as some of the pollutants can be treated at site and therefore it helps in reducing the exposure risks for clean up personal. Since the phenomenon of bioremediation is based on the concept of natural attenuation, the public considers it as more accepting. Most of the remediation techniques are done under aerobic conditions and running a system under aerobic conditions may therefore permit microbial organisms to degrade (Siegrist et al. 2004). The bioremediation process can therefore be broadly categorized into two groups: in situ and ex situ bioremediation (Kensa 2011).

2.1 In Situ Bioremediation

It is a kind of bioremediation which involves the process of supplying oxygen and nutrients by the help of circulating aqueous solutions through the contaminated soils so as to stimulate naturally occurring bacteria so that they can degrade the contaminants. It is a very cheap method which uses harmless microbial consortium so as to degrade the pollutants that are especially useful for saturated soil as well as groundwater remediation. The technique involves conditions such as infiltration of water containing nutrients as well as oxygen as electron acceptors (Vidali 2001; Chauhan and Jain 2010; Rayu et al. 2012). Besides this, in situ bioremediation is classified as intrinsic bioremediation and engineered bioremediation (Hazen 2010). The first approach is concerned with the stimulation of the indigenously occurring microbial population by giving them nutrients and oxygen so as to increase their metabolic activities. The second type of approach involves any type of stimulated biological remediation of an environment (Hazen 2010). The introduction of microorganisms to the contaminated site helps in accelerating the degradation process by generating conductive physiochemical conditions (Kumar et al. 2011). When the site conditions are not feasible, engineered bioremediation is used to the particular site especially using the genetically engineered bacteria (Singh et al. 2011). The major advantage of in situ bioremediation is its cost effectiveness, besides having no excavation, minimal site disruption and also the possibility of simultaneous treatments of soil and groundwater. However its major drawback is that it is time consuming, besides the seasonal variation in the microbial activity and also the problematic applications of treatment additives in the natural environment (Rayu et al. 2012).

2.2 Ex Situ Bioremediation

It is a kind of bioremediation which involves the removal of contaminated soil or water from the ground. The method is classified as solid phase system that includes land treatment and soil piles and slurry phase system that includes solid liquid suspension in case of bioreactors. Solid phase treatment includes the organic wastes as well as problematic wastes. The treatment process includes soil biopiles, composting, hydroponics and land farming (Ramos et al. 2011; Kumar et al. 2011; Rayu et al. 2012; Li et al. 2004). Under slurry phase bioremediation, contaminated soil is combined with water and other additives in a large tank called as bioreactor are added and then mixed to keep the microorganisms in contact with the contaminants that are present in the soil so as to create the optimum environment for the microorganisms to degrade the contaminants. By using proper sampling techniques and also maintaining controlled conditions with collected core sample, the effective ex situ bioremediation can be achieved (Paliwal et al. 2012; Cheng et al. 2009; Duong et al. 2013; Norstrom et al. 2004). Also hydroponics is a method of growing plants using mineral nutrient solutions in water and it has now become a common method for the characterization of plant response to the metal stress. For effective bioremediation with the help of hydroponics, this method is integrated with other remedial techniques. It has been seen that the rate and extent of biodegradation are greater in a bioreactor system than in situ because the contaminated environment is more manageable and hence more predictable. The major drawback that is associated with this system is that the contaminant can be stripped from the soil with the help of soil washing or physical extraction before being finally placed in a bioreactor. Besides this, some other bioremediation methods are also discussed below.

2.3 *Bioventing*

It is most important and most common in situ treatment and it involves supplying nutrients and air through wells to contaminated soil so as to stimulate the indigenous aerobic bacteria and is also an example of sub-surface bioremediation. It makes use of low air flow rates and provides only the amount of oxygen that is necessary for biodegradation while also minimizing release of contaminants to the atmosphere. Pollutants are mostly biodegraded in aerobic conditions with the help of indigenous heterotrophic microorganisms that are naturally occurring in the soil. But in order to promote the microbial degradation, poor oxygen is delivered to anaerobic and permeable polluted soil profiles and that too at a very low flow rate so that the oxygen supply rate meets the demand by the microbes and therefore minimizes volatilization of pollutants (USEPA 2004). Besides, subsurface bioremediation remediates the shallow aquifers with the help of geochemical reactions and which then ultimately leads to remediating the soils from heavy metals and thereby providing safe groundwater for the purpose of drinking and irrigation (Robinson et al. 2011).

2.4 *Biosparging*

It involves the insertion of air under pressure below the water table so as to increase groundwater oxygen concentration and therefore enhance the rate of biological degradation of contaminants by naturally occurring bacteria (Adams and Reddy 2003). It increases the mixing that occurs in the saturated zone and therefore increases the contact between soil and the groundwater. The low cost of installing the small diameter air injection points allows considerable flexibility in the overall design and construction of the system. Biosparging can also be used to lower the concentration of petroleum constituents that are dissolved in the groundwater. It has proved to be very effective in reducing the petroleum products at underground storage tank sites (USEPA 2004). The remediation of large scale petroleum contamination of soil and the groundwater has provided very important information about biosparging efficiency in the sandstone sedimentary bedrock (Machackova et al. 2012; Kumar and Mani 2012).

2.5 Bioaugmentation

It is the addition of pre-grown microbial cultures to the sites that are contaminated so as to enhance the degradation of unwanted compounds (Tyagi et al. 2011). Exogenous culture in very less number of instances competes with an indigenous population in order to develop and sustain useful levels of population (USEPA 2004; Kumar et al. 2011). In general, like most of the other bioremediation processes, bioaugmentation may not stone all alone on its own. The combined biostimulation as well as bioaugmentation along with the use of degrading bacteria, biosurfactants have been found to produce better results (Cheng et al. 2009).

2.6 Biodegradation

It is a generic term that is used to describe the methodologies that are affecting the cleanup of environmental pollutants. It has become as an improved substitute for expensive physiological remediation methods but however because of the lack of the information about the growth and metabolism of microorganisms in the polluted environment that often limits its implementation. Recent advances that have been made in the understanding of biogeochemical processes as well as genomics have opened up new perspectives towards the new opportunities of pollution abatement (Chauhan and Jain 2010; Jeffries et al. 2012; Rayu et al. 2012; Tyagi et al. 2011). In order for aerobic biodegradation to occur, enough amount of dissolved oxygen must exist within the surface that will serve as an electron acceptor (Adams and Reddy 2003). The drawbacks that are associated with current bioremediation techniques have made it necessary to seek more and more eco-friendly and cost-effective techniques for sites that are contaminated with heavy metals. In order to provide alternative ways to solve these problems, microbial induced calcite precipitation that is MICP has proven to be very effective. These MICP products are able to strongly absorb heavy metals onto their surfaces and during precipitation of calcite, heavy metal ions may be incorporated into the calcite crystal by substitution reaction (Pan 2009).

2.7 Mycoremediation

It is a form of bioremediation that uses fungi to degrade or sequester contaminants that are present in the environment and also to repair the weakened immune system of environment. Mycofiltration is a similar process that uses fungal mycelia in order to filter toxic waste and microorganisms from water into the soil with the help of stimulation of microbial and enzyme activity. Saprophytic, endophytic and mycorrhizal fungi are capable of recovering the soil water ecosystems and therefore balancing the biological population. The mycelium secretes extracellular enzymes and enzymes that help in breaking down lignin and cellulose that are the two main building blocks of plant fiber. The key to mycoremediation is to determine the right fungal species that will target a specific pollutant (Stamets 2005; Dudhane et al. 2012). Fungal species such as *A. niger*, *A. pullulans*, *C. resinae*, *F. trogii* and various other fungal species of capable of recovering the heavy metals from the polluted environment (Loukidou et al. 2003; Say et al. 2003; Tastan et al. 2010; Ramasamy et al. 2011). The recent advances that have been made regarding the mycoremediation have been highlighted under Table 1.

2.8 *Cyanoremediation*

The rate at which the heavy metals enter into the atmosphere exceeds the rate of their elimination through natural processes and this leads to the accumulation of the heavy metals in the aquatic ecosystem (Shirdam et al. 2006). Several living and non-living organisms have been found suitable for the treatment of the contaminated aquatic ecosystems. Recently, it has been seen that there has been an increasing awareness about the cyanoremediation as bioremediation as well as pollution control agents (Norstrom et al. 2004; Deng et al. 2007; Singhal et al. 2004; Tripathi et al. 2008; Yin et al. 2012). The ability of blue algae for As accumulation has been found to serve as cyanoremediation that will efficiently remove arsenic from aquatic environments. Although the role of cyanobacteria has already been established for the remediation of wetland ecosystems (Fiset et al. 2008) and also of agricultural rice fields (Tripathi et al. 2008) for metal recovery, yet the beneficial application of cyanobacteria in the process remediation of contaminated natural aquatic environments or in case of industrial effluents has still not been properly defined (Fiset et al. 2008). Deng et al. (2007) observed that green marine algae *Cladophora fascicularis* can be used as an efficient biosorbent material for removal of Pb from wastewater. Further, Dubey et al. (2011) evaluated the potential of cyanobacterial species which were found suitable for bioremediation, especially in biodegradation and biosorption of contaminants either as individuals or mixtures. Recently, Saunders et al. (2012) cultured three species of algae (*Hydrodictylon*, *Oedogonium* and *Rhizoclonium* species) to test its metal uptake and bioremediation potential in wastewater contaminated with heavy metals derived from coal-fired power generation. It was seen that all these species achieved high concentration of heavy metals. Thus algae have been found to be efficient biological vector that is having a beneficial role in the practical application of wastewater bioremediation. The recent advances that have been made regarding cyanoremediation have been highlighted in the Table 1.

Table 1 Microbes having bioremediation (cyanoremediation, biostimulation, mycoremediation) potential and heavy metals they can remediate (Mani and Kumar 2014)

Class	Microorganism	Metals	References
Algae	*Chlorella pyrendoidosa*	U	Singhal et al. (2004)
	Aspergillus niger, Ascophyllum nodosum, Bacillus firmus, Chlorella fusca, Oscillatoria anguistissima	Pb, Zn, Cd, Cr, Cu, Ni	Ahluwalia and Goyal (2007)
	Cladophora fascicularis	Pb	Deng et al. (2007)
	Saccharomyces cerevisiae	Cr, Ni, Cu, Zn	Machado et al. (2010)
	Spirogyra sp. and *Cladophora* sp.	Pb, Cu	Lee and Chang (2011)
	Spirogyra sp. and *Spirullina* sp.	Cr, Cu, Fe, Mn, Zn	Mane and Bhosle (2012)
	Hydrodictylon, Oedogonium and *Rhizoclonium species*	V, As	Saunders et al. (2012)
	Spirogyra sp. and *Spirulina* sp.	Cr, Cu, Fe, Mn, Se, Zn	Mane and Bhosle (2012)
Bacteria	*Pseudomonas veronii*	Cd, Zn, Cu	Vullo et al. (2008)
	Burkholderia species	Cd, Pb	Jiang et al. (2008)
	Bacillus and Pseudomonas	U	Kumar et al. (2008)
	Bradyrhizobium sp. and *Rhizobacteria* sp.	Cd, Pb, Cu	Dary et al. (2010)
	Bacillus sp.	Cd, Pb, Cu	Guo et al. (2010)
	Kocuria flava	Cu	Achal et al. (2011)
	Serratia marcescens	U	Kumar et al. (2011a, b)
	Pseudomonas aeruginosa	U	Choudhary and Sar (2011)
	Bacillus cereus	Cd, Zn	Hrynkiewicz et al. (2012)
	Bacillus cereus	Cr	Kanmani et al. (2012)
	Halomonas sp.	Sr	Achal et al. (2012a)
	Sporosarcina ginsengisoli	As	Achal et al. (2012b)
	Species of *Bacillus, Streptococci, Salmonella, Pseudonomnas, Micrococcus* and *E. coli*	Cd, Cu, Fe	Fulekar et al. (2012)
	Bacillus cereus strain XMCr-6	Cr	Dong et al. (2013)
Fungi	*Penicillium canescens*	Cr	Say et al. (2003)
	Ganoderma lucidum, Penicillium sp.	Ar	Loukidou et al. (2003)
	Aspergillus versicolor	Cr, Ni, Cu	Tastan et al. (2010)
	Aspergillus fumigatus	Pb	Ramasamy et al. (2011)
	Cladonia rangiformis (lichen)	Pb	Ekmekyapar et al. (2012)
	Species of *Aspergillus, Mucor, Penicillium* and *Rhizopus*	Cd, Cu, Fe	Fulekar et al. (2012)

2.9 Biostimulation

Microorganisms like bacteria and fungi are nature's original recyclers and their capacity to transform natural and synthetic chemicals into energy sources (Tang et al. 2007) and raw materials for their growth suggests the fact that expensive chemical and physical remediation processes should get replaced by the biological processes that are low in cost and are more environmental friendly. Introduction of nutrients and other supplementary components to the microbial population in order to induce propagation at a hastened rate is one of the most common approaches that is used for in situ bioremediation of accidental spills and other chronically contaminated sites worldwide (Cheng et al. 2009; Tyagi et al. 2011). Biostimulation activity is stimulated with the help of supplementing nutrients through microbes and also by introducing microorganisms with catalytic capabilities (Ma et al. 2007; Baldwin et al. 2008; Kanmani et al. 2012). With the introduction of molecular engineering, it is now possible to derive strains that show improved performance even under stressful field conditions and in this direction, although significant progress has been made, much more still needs to be done.

2.10 Biomineralization

Mining activities are a responsible for heavy metal contamination in the ecosystem. Several studies have showed that elevated levels of metals are present around metalliferous mines as well as industrial areas and the resulting contamination of agricultural soils is among the major environmental concerns. It has been seen that metalliferous soils provide very limited habitat for plants due to their phytotoxicity which results in severe selection pressures. Species comprising of heavy metal plant communities are actually genetically altered ecotypes that have specific tolerance to heavy metals and that is adapted through micro evolutionary processes. Evolution of this metal tolerance takes place at each specific site (Ernst 2006). A very high degree of metal tolerance that is shown mainly depends on the bioavailable fractions of metalloids that are present in the soil and also on the type of mineralization. The synthesis of materials that resemble complex morphology of the various natural biominerals is one of the key fields in today's biomimetic science. Biomineralization is basically a natural process with the help of which we produce complicated structured inorganic materials that possess vital functions in biological systems. The various morphologies of biominerals have helped scientists to copy these materials through the underlying chemistry of biomineralization which has then enabled the replication of outstanding optical as well as mechanical properties of biominerals with their biological functions such as navigation, storage etc. Chen et al. (2013) prepared MSPs by using hyaluronic acid as a reaction site for deposition of calcium phosphate minerals. Microbial process that have been shown to bind metals and then form minerals represent a very fundamental part of important biogeochemical cycles which can help in the formation of minerals by the process of mineralization.

This process offers an competent way to impound heavy metals within the relatively stable solid phases.

2.11 Biosensors

Biosensors are biophysical devices that are able to detect as well as measure the quantities of specific substances in a variety of environments. They include enzymes, antibodies as well as microorganisms and these all can be used for clinical, immunological as well as genetic research purposes. Biosensor probes are used in the detection as well as monitoring of pollutants in the environment. Biosensors are non-destructive in nature and they can make use of whole cells as biomimetic for detection purposes. There other advantages include rapid analysis, specificity as well as accurate reproducibility. They can be created by linking one gene with the other. The biosensor cell when used in a polluted site can signal by emitting light and which therefore suggests that low levels of inorganic mercury or toluene are present at the polluted site. This can be measured further by making use of fibre optic flourimeters. They can also be created by making use of the enzymes, nucleic acids and antibodies that are attached to synthetic membranes as molecular detectors. Another application of the biosensors is biomonitoring which is defined as the measurement and assessment of toxic chemicals in a tissue or any other related combination. It involves the uptake, biotransformation, accumulation and then removal of toxic chemicals and this then helps minimizing the risk to the industrial workers that are directly exposed to toxic chemicals.

3 Biodegradation of Xenobiotic Compounds

Xenobiotics are man-made compounds of recent origin and include dyestuffs, solvents, nitrotoluenes, explosive oils and surfactants. As these are unnatural substances, the microbes that are present in the environment do not have a proper mechanism for their degradation. Hence they tend to remain in the ecosystem for many years. The degradation of xenobiotic compounds depends upon the stability, size and also the environment in which the molecule exists. Biotechnology tools can therefore be used to understand their molecular properties and thus help in designing suitable mechanisms so as to attack these compounds.

4 Biotechnological Tools in Cleanup of Fresh Water Ecosystem

Biotechnological control of fresh water pollution includes following processes:

4.1 Activated Sludge Treatment Process

It is a type of wastewater treatment process for treating sewage or industrial wastewaters using aeration and biological floc that is composed of bacteria and protozoa. In sewage treatment plant, the activated sludge is a biological process that is used for one or several purposes like oxidising carbonaceons biological matter, oxidising nitrogenous matter mainly ammonium and nitrogen in biological matter, removing nutrients (N and P).The process takes advantage of aerobic microorganisms that can digest organic matter in sewage and then dump together. It thereby produces a liquid that is relatively free from suspended solids and organic matter and flocculated particles that will readily settle out and can be then be removed (Fig. 1).

4.2 Trickling Filter

It consists of a fixed bed of rocks, lava, coke, gravel, polyurethane foam, sphagnum peat moss or plastic media over which sewage or waste water flows downwards and causes a layer of microbial slime to grow. Removal of pollutants from waste water involves both adsorption and absorption of organic compounds and some inorganic species such as nitrite and nitrate ions by the layer of microbial bio film (Fig. 2).

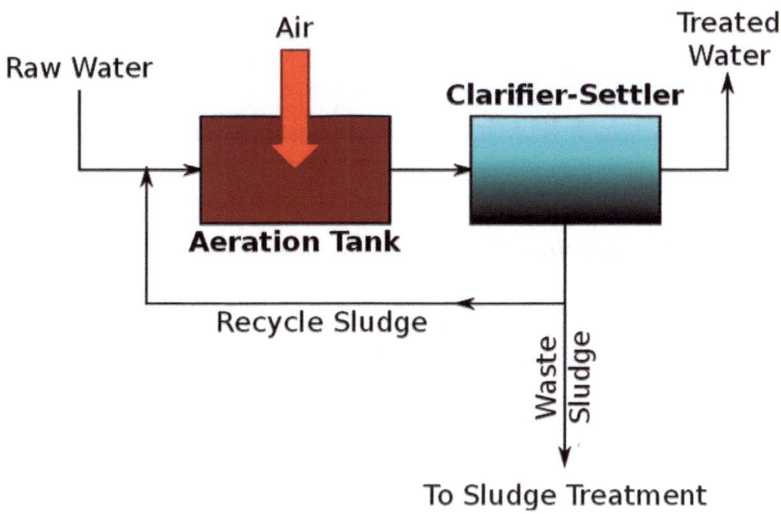

Fig. 1 Sludge treatment process

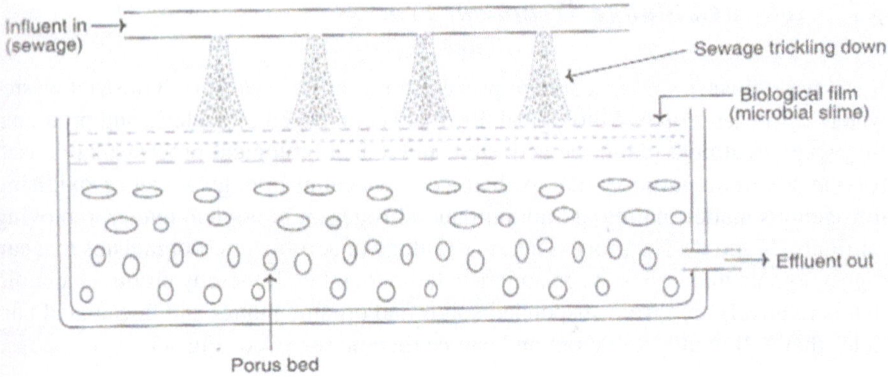

Fig. 2 Trickling filter

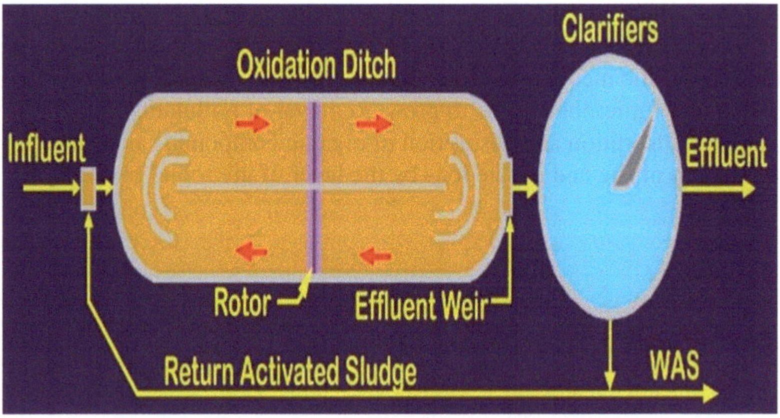

Fig. 3 Oxidation ditch

4.3 Oxidation Ditches Treatment Process

The oxidation ditch is a modified activated sludge biological treatment process that utilises long solids retention times (SRTs) to remove biodegradable organics (Fig. 3).

4.4 Oxidation Pond Treatment

They are large, shallow ponds designed to treat waste water through interaction of sunlight, bacteria and algae. Algae grow within pond and utilise sunlight to produce O_2 during photosynthesis. This O_2 is used by aerobic bacteria in the oxidation pond

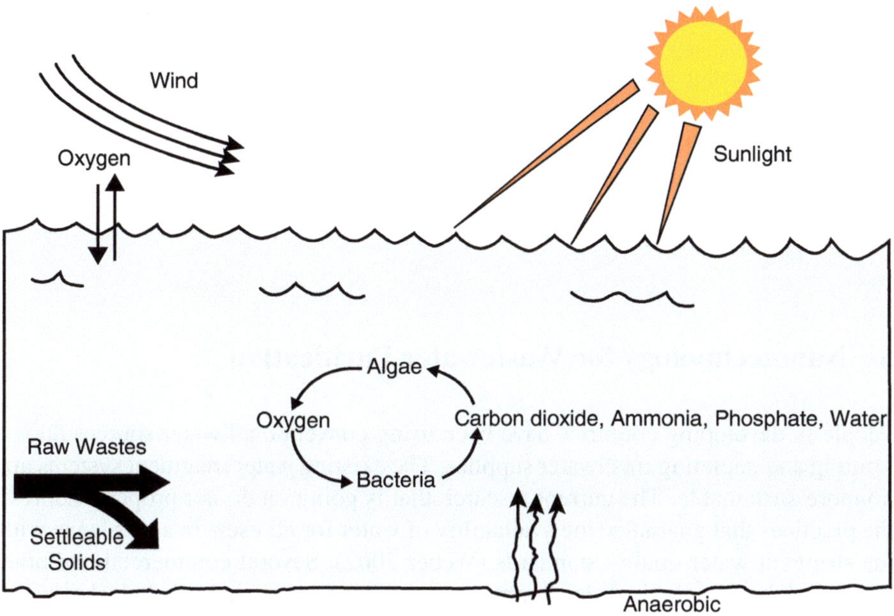

Fig. 4 Aerobic oxidation pond

to breakdown. The broken down solids settle down in the ponds, remitting in effluent that is relatively well treated (Fig. 4).

5 Values of Bioremediation

Bioremediation is a natural process and is therefore seen as an acceptable process that helps in treatment of contaminated materials such as soil. Microbes which are able to degrade the contaminants increase in numbers when the contaminant is present and when the contaminant is degraded the biodegradative population declines. The residues from the treatment are usually harmless products that include carbon dioxide, water and cell biomass. Theoretically it has been seen that bioremediation is useful for the complete destruction of a wide variety of contaminants and many compounds that are normally considered to be hazardous can be transformed to harmless products. This eliminates the chances of future liability that is associated with the treatment and disposal of contaminated material. Instead of transferring the contaminants from one environment to another, for example from land to water or air, the complete destruction of target pollutant is therefore possible. It can be most of the times carried out on site and often without causing a major disruption of normal activities. This also eliminates the need that is required to transport the quantities of waste off site and the consequent potential threats to human health and the environment that can arise during the process of transportation. It can also prove less expensive than other technologies that are used in cleaning up of hazardous wastes (Salt et al. 1998).

6 Nanotechnology for Wastewater Purification

People in developing countries have been using conventional water sources due to limiting and depleting freshwater supplies. The existing water treatment systems are no more sustainable. The current research that is going on do not properly address the practices that guarantee the availability of water for all users in accordance with the stringent water quality standards (Weber 2002). Several commercial and non-commercial technological developments have been employed on daily basis but nanotechnology has proved to be the most advanced method for waste water treatment. Developments that have taken place in the nanoscale research have made it possible to invent economically feasible and environmentally stable treatment technologies that effectively treat waste water and thereby helping in meeting the ever increasing water quality standards. It is suggested that nanotechnology can address many of the water quality issues with the help of different types of nanoparticles and nanofibers (Savage and Diallo 2005). Nanotechnology uses materials that have sizes smaller than 100 nm and that too in one dimension that is at the levels of atoms and molecules as compared to other disciplines such as chemistry and material sciences (Masciangioli and Zhang 2003; Eijkel and den Berg 2005; Rickerby and Morrison 2007; Vaseashta et al. 2007). The unique properties that are associated with nanomaterials such as high reactivity and strong sorption are explored for using it in waste water treatment depending upon their functions in unit operations as highlighted in Table 2 (Qu et al. 2013).

Table 2 Potential applications of nanotechnology in wastewater treatment

Technique	Nanomaterials	Innovative properties
Adsorption	CNTs and nanofibers	Huge surface area and high density of active sites
Disinfection	Ag/TiO_2 and CNTs	Powerful antimicrobial property, least toxicity, cheap and stability
Photocatalysis	Nano-TiO_2 and Fullerene derivatives	Photocatalytic activity in solar spectrum, low human toxicity, high stability and selectivity
Membranes	Nano-Ag/TiO_2/Zeolites/Magnetite and CNTs	Powerful antimicrobial property, least toxic to humans and mechanical stability

Nanoparticles have the capability of penetrating deeper and can treat wastewater efficiently than conventional technologies (Riu et al. 2006; Theron et al. 2008; Gautam et al. 2013). In the context of remediation and treatment, nanotechnology has been seen to have the potential of both providing better water quality as well as quantity in the long run. Nanotechnology has tremendously contributed to the development of more efficient and cost effective water filtration processes and in this context membrane technology is considered as one of the most advanced waste water treatment processes (Bhattacharyya et al. 1998; Ritchie et al. 1999, 2001; DeFriend et al. 2003; Hollman and Bhattacharyya, 2004).

7 Biotechnology and Aquatic Resource Profiling, Human Health and Ecosystem Health

This investigation idea deals with the activities that are linked to the thoughtful genetic makeup of freshwater assets. Biotechnology and genomics in this quarter includes the process of studying the genome of aquatic species. In this context, aquatic resource profiling supports the concept of sustainable fisheries, aquaculture and the overall protection of biodiversity. Employing biotechnological tool will manage and protect the aquatic health and also will meet the standards of water quality. Useful management and fortification of this exceptionally imperative valuable source continues to linger a challenge with so much still to learn about the living organisms in aquatic environs Healthy ecosystems are the very fundamental basis for biodiversity, healthy communities as well as development.

8 Conclusion

Bioremediation is a cheap technology than the present day available water treatment technologies. It has a wide scope as an innovative technology to cleanup wide range of contaminates effectively without disturbing the non target components of the polluted sites. Furthermore, it is an alternative option to wipe out contaminants using potential biota. In context of remediation, nanotechnology has emerged as a potential technology for ensuring better water quality as well as quantity in the long run. There is an imperative need to implement the modern biotechnological advances for maintaining the health of the aquatic ecosystems.

References

Achal, V., Pan, X., & Zhang, D. (2011). Remediation of copper-contaminated soil by Kocuria flava CR1, based on microbially induced calcite precipitation. *Ecological Engineering, 37*, 1601–1605.

Achal, V., Pan, X., & Zhang, D. (2012a). Bioremediation of strontium (Sr) contaminated aquifer quartz sand based on carbonate precipitation induced by Sr resistant Halomonas sp. *Chemosphere, 89*, 764–768.

Achal, V., Pan, X., Fu, Q., & Zhang, D. (2012b). Biomineralization based remediation of As (III) contaminated soil by Sporosarcina ginsengisoli. *Journal of Hazardous Materials, 202*, 178–184.

Adams, J. A., & Reddy, K. R. (2003). Extent of benzene biodegradation in saturated soil column during air sparging. *Groundwater Monitoring and Remediation, 23*(3), 85–94.

Ahluwalia, S. S., & Goyal, D. (2007). Microbial and plant derived biomass for removal of heavy metals from wastewater. *Bioresource Technology, 98*(12), 2243–2257.

Anouti, F. A. (2014). Concerns regarding food biotechnology: An ongoing debate. *Journal of Biodiversity, Bioprospecting and Development, 1*, 106.

Baldwin, B. R., Peacock, A. D., Park, M., & Ogles, D. M. (2008). Multilevel samplers as microcosms to assess microbial response to biostimulation. *Ground Water, 46*, 295–304.

Barbier, E. B., Acreman, M., & Knowler, D. (1997). *Economic valuation of wetlands: A guide for policy makers and planners* (127 pp). Gland: Ramsar Convention Bureau.

Bhattacharyya, D., Hestekin, J. A., Brushaber, P., Cullen, L., Bachas, L. G., & Sikdar, S. K. (1998). Novel poly-glutamic acid functionalized microfiltration membranes for sorption of heavy metals at high capacity. *Journal of Membrane Science, 141*, 121–135.

Brim, H., McFarlan, S. C., Fredrickson, J. K., Minton, K. W., & Zhai, M. (2000). Engineering Deinococcus radiodurans for metal remediation in radioactive mixed waste environments. *Nature Biotechnology, 18*, 85–90.

Chauhan, A., & Jain, R. K. (2010). Biodegradation: Gaining insight through proteomics. *Biodegradation, 21*, 861–879.

Chen, W., Mulchandani, A., & Deshusses, M. A. (2005). Environmental biotechnology: Challenges and opportunities for chemical engineers. *AICHE Journal, 51*, 690–695.

Chen, Z., Li, Z., Lin, Y., Yin, M., Ren, J., & Qu, X. (2013). Biomineralization inspired surface engineering of nanocarriers for pH-responsive, targeted drug delivery. *Biomaterials, 34*, 1364–1371.

Cheng, S. S., Hsieh, T. L., Pan, P. T., Gaop, C. H., Chang, L. H., Whang, L. M., & Chang, T. C. (2009). Study on biomonitoring of aged TPHcontaminated soil with bioaugmentation and biostimulation (Conference paper). In *10th international in situ and on-site bioremediation symposium*, Baltimore MD, May 5–8.

Choudhary, S., & Sar, P. (2011). Uranium biomineralization by a metal resistant Pseudomonas aeruginosa strain isolated from contaminated mine waste. *Journal of Hazardous Materials, 186*, 336–343.

Dary, M., Chamber-Perez, M. A., Palomares, A. J., & Pajuelo, E. (2010). "In situ" phytostabilisation of heavy metal polluted soils using Lupinus luteus inoculated with metal resistant plant-growth promoting rhizobacteria. *Journal of Hazardous Materials, 177*(1–3), 323–330.

DeFriend, K. A., Wiesner, M. R., & Barron, A. R. (2003). Alumina and aluminate ultrafiltration membranes derived from alumina nanoparticles. *Journal of Membrane Science, 224*, 11–28.

Deng, L., Su, Y., Su, H., Wang, X., & Zhu, X. (2007). Sorption and desorption of lead (II) from wastewater by green algae *Cladophora fascicularis*. *Journal of Hazardous Materials, 143*(1–2), 220–225.

Dong, G., Wang, Y., Gong, L., Wang, M., Wang, H., He, N., Zheng, Y., & Li, Q. (2013). Formation of soluble Cr(III) end-products and nanoparticles during Cr(VI) reduction by Bacillus cereus strain XMCr-6. *Biochemical Engineering Journal, 70*, 166–172.

Dubey, S. K., Dubey, J., Mehra, S., Tiwari, P., & Bishwas, A. J. (2011). Potential use of cyanobacterial species in bioremediation of industrial effluents. *African Journal of Biotechnology, 10*(7), 1125–1132.

Dudhane, M., Borde, M., & Jite, P. K. (2012). Effect of aluminium toxicity on growth responses and antioxidant activities in Gmelina arborea Roxb inoculated with AM Fungi. *International Journal of Phytoremediation, 14*(7), 643–655.

Duong, T. T. T., Verma, S. L., Penfold, C., & Marschner, P. (2013). Nutrient release from composts into the surrounding soil. *Geoderma, 195–196*, 42–47.

Eijkel, J. C. T., & van den Berg, A. (2005). Nanofluidics: What is it and what canwe expect from it. *Microfluidics and Nanofluidics, 1*(3), 249–267.

Ekmekyapar, F., Aslan, A., Bayhan, Y. K., & Cakici, A. (2012). Biosorption of Pb(II) by nonliving lichen biomass of Cladonia rangiformis Hoffm. *International Journal of Environmental Research, 6*(2), 417–424.

Ernst, W. H. O. (2006). Evolution of metal tolerance in higher plants. *Forest Snow and Landscape Research, 80*, 251–274.

Fiset, J. F., Blais, J. F., & Riverso, P. A. (2008). Review on the removal of metal ions from effluents using seaweeds, alginate derivatives and other sorbents. *Revue des Sciences de l'Eau, 21*(3), 283–308.

Fulekar, M. H., Sharma, J., & Tendulkar, A. (2012). Bioremediation of heavy metals using biostimulation in laboratory bioreactor. *Environmental Monitoring and Assessment, 184*(12), 7299–7307.

Gautam, P., Madathil, D., & Nair, A. N. B. (2013). Nanotechnology in waste water treatment: A review. *International Journal of ChemTech Research, 5*, 2303–2308.

Govarthanan, M., Lee, K. J., Cho, M., Kim, J. S., Kamala-Kannan, S., & Oh, B. T. (2013). Significance of autochthonous Bacillus sp. KK1 on biomineralization of lead in mine tailings. *Chemosphere, 8*, 2267–2272.

Guo, H., Luo, S., Chen, L., Xiao, X., Xi, Q., Wei, W., Zeng, G., Liu, C., Wan, Y., Chen, J., & He, Y. (2010). Bioremediation of heavy metals by growing hyperaccumulator endophytic bacterium Bacillus sp. L14. *Bioresource Technology, 101*(22), 8599–8605.

Hazen, T. C. (2010). In situ: Groundwater bioremediation. In K. N. Timmis (Ed.), *Handbook of hydrocarbon and lipid microbiology* (pp. 2583–2594). Berlin: Springer.

Hollman, A. M., & Bhattacharyya, D. (2004). Pore assembled multilayers of charged polypeptides in microporous membranes for ion separation. *Langmuir, 20*, 5418–5424.

Hrynkiewicz, K., Dabrowska, G., Baum, C., Niedojadlo, K., & Leinweber, P. (2012). Interactive and single effects of ectomycorrhiza formation and Bacillus cereus on metallothionein MT1 expression and phytoextraction of Cd and Zn by Willows. *Water, Air, and Soil Pollution, 223*, 957–968.

Jeffries, T. C., Seymour, J. R., Newton, K., Smith, R. J., Seuront, L., & Mitchell, J. G. (2012). Increases in the abundance of microbial genes encoding halotolerance and photosynthesis along a sediment salinity gradient. *Biogeosciences, 9*(2), 815–825.

Jiang, C. Y., Sheng, X. F., Qian, M., & Wang, Q. Y. (2008). Isolation and characterization of heavy metal resistant Burkholderia species from heavy metal contaminated paddy field soil and its potential in promoting plant growth and heavy metal accumulation in metal polluted soil. *Chemosphere, 72*, 157–164.

Kanmani, P., Aravind, J., & Preston, D. (2012). Remediation of chromium contaminants using bacteria. *International Journal of Environmental Science and Technology, 9*, 183–193.

Kastenhofer, K. (2007). Converging epistemic cultures? *Innovations, 20*(4), 359–373.

Kensa, M. V. (2011). Bioremediation: An overview. *Journal of Industrial Pollution Control, 27*(2), 161–168.

Kumar, C., & Mani, D. (2010). *Enrichment and management of heavy metals in sewage irrigated soil*. Saarbrucken: LAP LAMBERT Academic Publishing AG & KG.

Kumar, C., & Mani, D. (2012). *Advances in bioremediation of heavy metals: A tool for environmental restoration*. Saarbrucken: LAP LAMBERT Academic Publishing AG & KG.

Kumar, R., Joshi, S. R., & Acharya, C. (2008). Metal tolerant Bacillus and Pseudomonas from uranium rich soils of Meghalaya. *Research Journal of BioTechnology* (Special Issue), 345–350.
Kumar, R., Acharya, C., & Joshi, S. R. (2011). Isolation and analyses of uranium tolerant Serratia marcescens strains and their utilization for aerobic uranium U(VI) bioadsorption. *Journal of Microbiology, 49*(4), 568–574.
Kumar, A., Bisht, B. S., Joshi, V. D., & Dhewa, T. (2011a). Review on bioremediation of polluted environment: A management tool. *International Journal of Environmental Sciences, 1*(6), 1079–1093.
Lee, Y. C., & Chang, S. P. (2011). The biosorption of heavy metals from aqueous solution by Spirogyra and Cladophora filamentous macroalgae. *Bioresource Technology, 102*(9), 5297–5304.
Lee, C. S., Li, X. D., Shi, W. Z., Cheung, S. C., & Thornton, I. (2006). Metal contamination in urban, suburban and country park soils of Hong Kong: A study based on GIS and multivariate statistics. *Science of the Total Environment, 356*, 45–61.
Li, L., Cunningham, C. J., Pas, V., Philp, J. C., Barry, D. A., & Anderson, P. (2004). Field trial of a new aeration system for enhancing biodegradation. *Waste Management, 24*, 127–137.
Loukidou, M. X., Matis, K. A., Zouboulis, A. I., & Liakopoulou-Kyriakidou, M. (2003). Removal of As(V) from wastewaters by chemically modified fungal biomass. *Water Research, 37*(18), 4544–4552.
Ma, X., Nonvak, P. J., Ferguson, J., Sadowsky, M., et al. (2007). The impact of H2 addition or dechlorinating microbial communities. *Bioremediation Journal, 11*, 45–55.
Machackova, J., Wittlingerova, Z., Vlk, K., & Zima, J. (2012). Major factors affecting in situ biodegradation rates of jet-fuel during largescale biosparging project in sedimentary bedrock. *Journal of Environmental Science and Health, Part A, 47*(8), 1152–1165.
Machado, M. D., Soares, E. V., & Soares, H. M. V. M. (2010). Removal of heavy metals using a brewer's yeast strain of Saccharomyces cerevisiae: Chemical speciation as a tool in the prediction and improving of treatment efficiency of real electroplating effluents. *Journal of Hazardous Materials, 180*(1–3), 347–353.
Mane, P. C., & Bhosle, A. B. (2012). Bioremoval of some metals by living Algae Spirogyra sp. and Spirullina sp. from aqueous solution. *International Journal of Environmental Research, 6*(2), 571–576.
Mani, D., & Kumar, C. (2014). Biotechnological advances in bioremediation of heavy metals contaminated ecosystems: An overview with special reference to phytoremediation. *International Journal of Environmental Science and Technology, 11*, 843–872.
Mapanda, F., Mangwayana, E. N., Nyanangara, J., & Giller, K. E. (2005). The effect of long-term irrigation using wastewater on the heavy metal contents of soils under vegetables in Harare, Zimbabwe. *Agriculture, Ecosystems & Environment, 107*, 151–165.
Masciangioli, T., & Zhang, W. X. (2003). Peer reviewed: Environmental technologies at the nanoscale. *Environmental Science & Technology, 37*, 102A–108A.
Meagher, R. B. (2000). Phytoremediation of toxic elemental and organic pollutants. *Current Opinion in Plant Biology, 3*, 153–162.
Nasr, M., & Ismail, S. (2015). Performance evaluation of sedimentation followed by constructed wetlands for drainage water treatment. *Sustainable Environment Research, 25*, 141–150.
Norstrom, A., Larsdotter, K., Gumaelius, L., Jansen, J. L. C., & Dalhammar, G. (2004). A small scale hydroponics wastewater treatment system under Swedish conditions. *Water Science and Technology, 48*(11–12), 161–167.
O'Loughlin, E. J., Traina, S. J., & Sims, G. K. (2000). Effects of sorption on the biodegradation of 2-methylpyridine. *Environmental Toxicology and Chemistry, 19*, 2168–2174.
Paliwal, V., Puranik, S., & Purohit, H. J. (2012). Integrated perspective of effective bioremediation. *Applied Biochemistry and Biotechnology, 166*, 903–924.
Pan, X. L. (2009). Microbially induced carbonate precipitation as a promising way to in situ immobilize heavy metals in groundwater and sediment. *Research Journal of Chemistry and Environment, 13*, 3–4.

Prasad, M. N. V. (2004). *Heavy metal stress in plants: From biomolecules to ecosystems* (2nd ed.). Heidelberg: Springer.
Qu, X., Alvarez, P. J. J., & Li, Q. (2013). Applications of nanotechnology in water and wastewater treatment. *Water Research, 47*, 3931–3946.
Ramasamy, R. K., Congeevaram, S., & Thamaraiselvi, K. (2011). Evaluation of isolated fungal strain from e-waste recycling facility for effective sorption of toxic heavy metals Pb(II) ions and fungal protein molecular characterization-a mycoremediation approach. *Asian Journal of Biological Sciences, 2*(2), 342–347.
Ramos, J. L., Marques, S., Dillewijn, P. V., Espinosa-Urgel, M., Segura, A., & Duque, E. (2011). Laboratory research aimed at closing the gaps in microbial bioremediation. *Trends in Biotechnology, 29*(12), 641–647.
Rayu, S., Karpouzas, D. G., & Singh, B. K. (2012). Emerging technologies in bioremediation: Constraints and opportunities. *Biodegradation, 23*, 917–926.
Rickerby, D. G., & Morrison, M. (2007). Nanotechnology and the environment: A European perspective. *Science and Technology of Advanced Materials, 8*, 19–24.
Ritchie, S. M. C., Bachas, L. G., Olin, T., Sikdar, S. K., & Bhattacharyya, D. (1999). Surface modification of silica- and cellulosebased microfiltration membranes with functional polyamino acids for heavy metal sorption. *Langmuir, 15*, 6346–6357.
Ritchie, S. M. C., Kissick, K. E., Bachas, L. G., Sikdar, S. K., Parikh, C., & Bhattacharyya, D. (2001). Polycysteine and other polyamino acid functionalized microfiltration membranes for heavy metal capture. *Environmental Science and Technology, 35*, 3252–3258.
Riu, J., Maroto, A., & FX, X. R. (2006). Nanosensors in environmental analysis. *Talanta, 69*, 288–301.
Robinson, C., Bromssen, M. V., Bhattacharya, P., Haller, S., Biven, A., Hossain, M., Jacks, G., Ahmed, K. M., Hasan, M. A., & Thunvik, R. (2011). Dynamics of arsenic adsorption in the targeted arsenic-safe aquifers in Matlab, south-eastern Bangladesh: Insight from experimental studies. *Applied Geochemistry, 26*, 624–635.
Salt, D. E., Smith, R. D., & Raskin, I. (1998). Phytoremediation. *Annual Review of Plant Physiology and Plant Molecular Biology, 49*, 643–668.
Saunders, R. J., Paul, N. A., Hu, Y., & de Nys, R. (2012). Sustainable sources of biomass for bioremediation of heavy metals in wastewater derived from coal-fired power generation. *PLoS One, 7*(5), e36470. https://doi.org/10.1371/journal.pone.0036470.
Savage, N., & Diallo, M. S. (2005). Nanomaterials and water purification: Opportunities and challenges. *Journal of Nanoparticle Research, 7*, 331–342.
Say, R., Yimaz, N., & Denizli, A. (2003). Removal of heavy metal ions using the fungus Penicillium canescens. *Adsorption Science and Technology, 21*, 643–650.
Schwarzenbach, R. P., Escher, B. I., Fenner, K., Hofstetter, T. B., & Johnson, C. A. (2006). The challenge of micropollutants in aquatic systems. *Science, 313*, 1072–1077.
Segura, P. A., Francois, M., Gagnon, C., & Sauve, S. (2009). Review of the occurrence of anti-infectives in contaminated wastewaters and natural and drinking waters. *Environmental Health Perspectives, 117*, 675–684.
Shirdam, R., Khanafari, A., & Tabatabaee, A. (2006). Cadmium, nickel and vanadium accumulation by three strains of marine bacteria. *Iranian Journal of Biotechnology, 4*(3), 180–187.
Siegrist, H., Ternes, T. A., & Joss, A. (2004). Scrutinizing pharmaceuticals and personal care products in wastewater treatment. *Journal of Environmental Science and Technology, 38*, 392A–399A.
Singh, J. S., Abhilash, P. C., Singh, H. B., Singh, R. P., & Singh, D. P. (2011). Genetically engineered bacteria: An emerging tool for environmental remediation and future research perspectives. *Gene, 480*, 1–9.
Singhal, R. K., Joshi, S., Tirumalesh, K., & Gurg, R. P. (2004). Reduction of uranium concentration in well water by Chlorella (Chlorella pyrenoidosa) a fresh water algae immobilized in calcium alginate. *The Journal of Radioanalytical and Nuclear Chemistry, 261*, 73–78.

Stamets, P. (2005). *Mycelium running: How mushroom can help save the world.* New York: Ten Speed Press, Crown Publishing Group.

Tang, C. Y., Criddle, Q. S., Fu, C. S., & Leckie, J. O. (2007). Effect of flux (Transmembrane pressure) and membranes properties on fouling and rejection of reverse osmosis and nanofiltration membranes treating perfluorooctane sulfonate containing wastewater. *Environmental Science & Technology, 41*, 2008–2014.

Tastan, B. E., Ertugrul, S., & Donmez, G. (2010). Effective bioremoval of reactive dye and heavy metals by Aspergillus versicolor. *Bioresource Technology, 101*(3), 870–876.

Theron, J., Walker, J. A., & Cloete, T. E. (2008). Nanotechnology and water treatment: Applications and emerging opportunities. *Critical Reviews in Microbiology, 34*, 43–69.

Tong, A., Peake, B., & Braund, R. (2011). Disposal practices for unused medications around the world. *Environment International, 37*, 292–298.

Tripathi, R. D., Dwivedi, S., Shukla, M. K., Mishra, S., Srivastava, S., Singh, R., Rai, U. N., & Gupta, D. K. (2008). Role of blue green algae biofertilizer in ameliorating the nitrogen demand and fly-ash stress to the growth and yield of rice (Oryza sativa L.) plants. *Chemosphere, 70*, 1919–1929.

Tyagi, M., Fonseca, M. M. R. D., & Carvalho, C. C. C. R. D. (2011). Bioaugmentation and biostimulation strategies to improve the effectiveness of bioremediation processes. *Biodegradation, 22*, 231–241.

UNEP. (2012). *Global environment outlook 5—Environment for the future we want.* Nairobi: United Nations Environment Programme.

USEPA. (2004). *Cleaning up the nation's waste sites: Markets and technology trends.* Washington: US Environmental Protection Agency.

Vaseashta, A., Vaclavikova, M., Vaseashta, S., Gallios, G., Roy, P., & Pummakarnchana, O. (2007). Nanostructures in environmental pollution detection, monitoring, and remediation. *Science and Technology of Advanced Materials, 8*, 47–59.

Vidali, M. (2001). Bioremediation. An overview. *Pure and Applied Chemistry, 73*(7), 1163–1172.

Vullo, D. L., Ceretti, H. M., Hughes, E. A., Ramyrez, S., & Zalts, A. (2008). Cadmium, zinc and copper biosorption mediated by Pseudomonas veronii 2E. *Bioresource Technology, 99*, 5574–5581.

Weber, W. J. (2002). Distributed optimal technology networks: A concept and strategy for potable water sustainability. *Water Science and Technology, 46*, 241–246.

Yin, X. X., Wang, L. H., Bai, R., Huang, H., & Sun, G. X. (2012). Accumulation and transformation of arsenic in the blue-green alga Synechocysis sp. PCC6803. *Water, Air, and Soil Pollution, 223*(3), 1183–1190.

Bioremediation: A Sustainable and Emerging Tool for Restoration of Polluted Aquatic Ecosystem

Bhat Mohd Skinder, Baba Uqab, and Bashir Ahmad Ganai

Abstract The most important and visible factors like the population explosion, urbanization and economic growth are accountable for ecological degradation and contamination. Ecological detoxification is a riddle that needs to be solved through ecological concepts and techniques. Thus, the application of advanced science and technology helps us to apply diverse biota for pollution abatement. Diverse and potential biota has efficiency to reinstate the polluted environment effectively, but dearth of knowledge about the factors *viz.*, pH, moisture content, temperature, redox potential, soil type and oxygen controlling the growth and metabolism of microorganism in polluted environments often limits its implementation. The enhancements in bioremediation have been realized through the help of the various areas of microbiology, biochemistry, molecular biology, analytical chemistry, chemical and environmental engineering. The techniques involved in the process of bioremediation are Ex-Situ and In-Situ, depends on the type and site of contamination. In the present context it has been revealed bioremediation plays an important role in the restoration of polluted ecosystem through environmental friendly mechanisms.

Keywords Bioremediation · Phytoremediation · Aquatic ecosystem · Pollutants · Contamination · Heavy metals · Bioaugmentation · Biostimulation

1 Introduction

Bioremediation process: The biological restoration and rehabilitation of contaminated sites and cleanup of the contaminated areas as a result of the manufacture, storage, transport, and use of inorganic and organic chemicals (Hamer 1993; Baker and Herson 1994). The process offers the possibility of immobilizing, removing, degrading, altering or otherwise decontaminating various chemicals from the environs through the action of bacteria (Gadd 2001; Morel et al. 2002), plants and fungi

B. M. Skinder (✉) · B. Uqab · B. A. Ganai
Department of Environmental Science/Centre of Research for Development, University of Kashmir, Srinagar, India

H. Qadri et al. (eds.), *Fresh Water Pollution Dynamics and Remediation*,
https://doi.org/10.1007/978-981-13-8277-2_9

(Kvesitadze et al. 2006). In this process, biological agents mainly bacteria, fungi or yeast are used to clean polluted environment (Strony and Burgess 2008). The bioremediation technology endorses growth of microbes, native to degraded sites and performs ideal activities (Agarwal 1998). The growth of microorganisms can be achieved in several ways for example, through the addition of nutrients, by terminal electron receptor, by controlling temperature and moisture conditions (Hess et al. 1997). The main requirements for microbes are energy and carbon source (Vadali 2001). The energy source or nutrients needed by microbes for their body metabolism are provided by contaminants that are present in degraded environment (Tang et al. 2007).

The advancement in agriculture and industries has led to the production of different pollutants being added into our environment, thus make shortage of clean waters and soil which led to the less production of crop yield (Kamaludeen et al. 2003). The food demand has direct proportionality with the increase in population so farmers are forced to go for intensive agriculture and excessive use of pesticides. The use of pesticides degrades the quality of soil as well as aquatic system. The population explosion increases the pressure on natural resources, makes it impossible to maintain quality of environs where we inhabit. Biotechnology offers an appropriate answer for managing degraded environments. Many contaminations like chlorinated solvents, hydrocarbons, polycyclic aromatic hydrocarbons (PAHs), heavy metals etc. have been expertise by environmental biotechnology investigators'. Bioremediation process is not a mystic formula, but it is a natural process alternative to incineration, catalytic destruction, or the use of absorbents and is cost effective (Blaylock et al. 1997).

Pollution of aquatic system is an issue of great concern at global level, and broadly divided into three main categories, i.e. contamination by organic and inorganic compounds, heavy metals and microorganisms. The various sources and concentrations of heavy metals are shown in the Table 1. These heavy metals find its way in the aquatic system through different pathways. Several different physicochemical and biological processes are commonly employed to remove heavy metals from industrial wastewaters. Conventional physicochemical methods are not cost-effective and some of them are not environmentally friendly like electrochemical treatment, ion exchange, osmosis, precipitation, evaporation and sorption. Alternatively, bioremediation processes are eco-compatible and economically feasible option and show promising results for the removal of metals, even present in very low concentrations, where physicochemical removal approaches fail to operate (Ojuederie and Babalola 2017). The high metal binding capacity of biological agents, which can remove heavy metals from contaminated sites with high efficiency, is basis of bioremediation strategy. Microbes can be considered as a biological tool for metal removal because of being used to concentrate, remove, and recover heavy metals from contaminated aquatic systems. They (microbes) are very useful due to the action on pollutants even present in very dilute solutions, and can also

Table 1 Different sources and concentrations of heavy metals in soil annually in the world (1000 t a^{-1})

Source	As	Cd	Cr	Cu	Hg	Ni	Pb	Zn
Agriculture and food waste	0–0.60	0–0.30	4.5–90	3–38	0–1.50	6–45	1.5–27	12–150
Commodity impurities	36–41	0.78–1.6	305–610	395–790	0.55–0.82	6.5–32	195–390	310–620
Logging and timber Industry wastes	0–3.30	0–2.20	2.2–18	3.3–52	0–2.20	2.2–23	6.6–8.2	13–65
Municipal wastes	0.09–0.70	0.88–7.50	6.6–33	13–40	0–0.26	2.2–10	18–62	22–97
Municipal sludge	0.01–0.24	0.02–0.34	1.4–11	4.9–21	0.01–0.8	5.0–22	2.8–9.70	18–57
Farmyard manure	1.2–4.4	0.2–1.20	10–60	14–80	0–0.20	3–36	3.2–20	150–320
Coal ash	6.7–37	1.5–13	149–446	93–335	0.37–4.8	56–279	45–242	112–484
Organic wastes	0–0.25	0–0.01	0.1–0.48	0.04–0.61	–	0.17–3.2	0.02–1.6	0.13–2.1
Marl	0.04–0.5	0–0.11	0.04–0.19	0.15–2.0	0–0.02	0.22–3.5	0.45–2.6	0.15–3.5
Atmospheric deposition	8.4–18	2.2–8.4	5.1–38	14–36	0.63–4.3	11–37	202–263	49–135
Fertilizer	0–0.02	0.03–0.25	0.03–0.38	0.05–0.58	–	0.20–3.5	0.42–2.3	0.25–1.1
Metal processing solid wastes	0.01–0.21	0–0.08	0.65–2.4	0.95–7.6	0–0.08	0.84–2.5	4.1–11	2.7–19
Total	52–112	5.6–38	484–1309	541–1367	1.6–15	106–544	479–1113	689–2054

adapt to extreme conditions. The mechanisms associated with metal biosorption by microbe are still not well understood, but studies revealed that they play an important role in the uptake of metals and such action involves accumulation or resistance. In the marine ecosystem, microorganisms are advantageous in the elimination of petroleum hydrocarbons, demonstrate the eco-sustainable bioremediation attained in sensitive marine ecosystem and may the only approach for biodiversity rich and fragile ecosystem (Paniagua-Michel and Rosales 2015). The hierarchy of complexity of bioremediation, limitations and scope in modern day science has been shown in the Fig. 1. Bioremediation has potential to restore contaminated ecosystem inexpressively get effectively (Ayangbenro and Babalola 2017), but lack of information about the factors controlling the growth and metabolism of microorganism in polluted environments often limits its implementation.

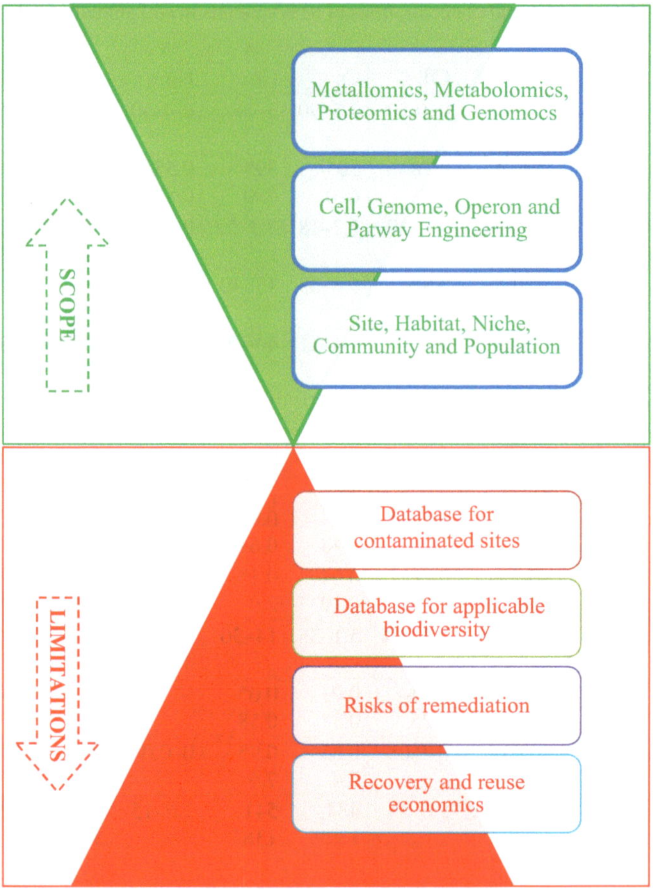

Fig. 1 Bioremediation: Hierarchy of complexity (Scope and limitations of bioremediation application)

2 Principles of Bioremediation

Bioremediation is the use of biological interventions of biodiversity for mitigation (and if possible, complete elimination) of the toxic effects caused by environmental pollutants at a given site. Bioremediation process involves use of microorganisms to biodegrade the contaminants in the contaminated environment (Sharma 2012; Azubuike et al. 2016). The various types of microorganisms like bacteria, fungi and yeasts are used for breakdown the hazardous substance into less toxic or non-toxic substances. The contaminants can be used as nutrients or energy sources by micro-organisms (Tang et al. 2007; Mbhele 2007). In this process microorganism degrade and metabolize chemical substances and restore environment quality (Dave and Ghaly 2011). It operates through the principles of biogeochemical cycling (Figs. 2 and 3).

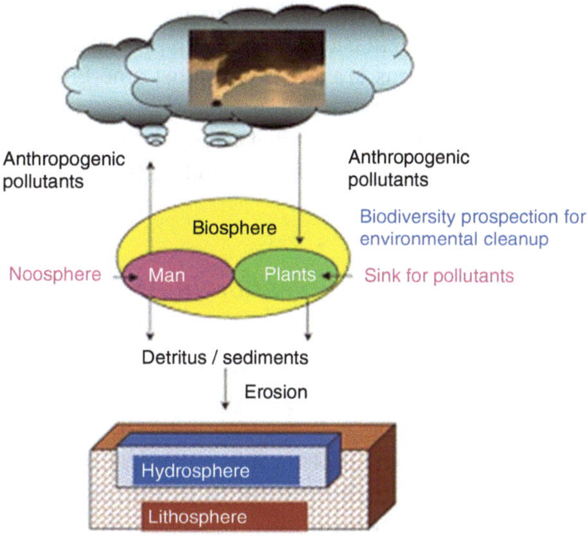

Fig. 2 Relationship between biogeochemical cycle and bioremediation. (Prasad 2004a, b)

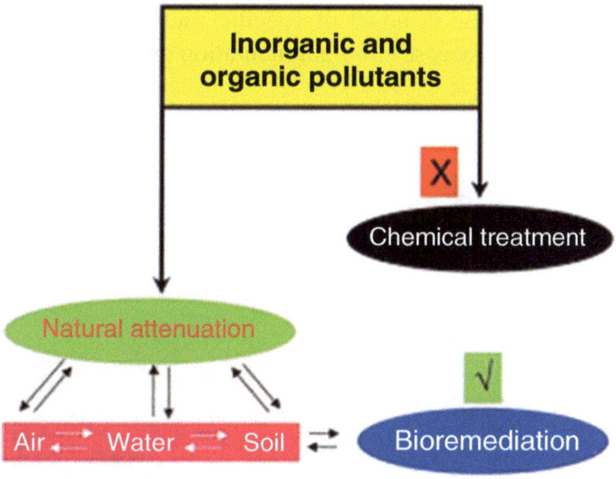

Fig. 3 Natural attenuation and bioremediation: Environmental cleanup

2.1 *Mechanism of Bioremediation*

In general, the mechanism of bioremediation is not a single step process but multiple processes are interrelated and dependent upon sun energy driven plant physiological processes, rhizospheric processes and other available precursors to perform the cleanup process in environmental friendly manner. The mechanisms involved are logical approach as the sequence of how contaminants come into contact with the plant system, rhizosphere and transportation processes. In bioremediation process, several mechanisms are involved subject to the designed application. Bioremediation, an integral part of all Environmental Biotechnology Program

(EBP), discover the uses of biological mechanisms to transform or immobilize and destroy environmental contaminants to protect potential sensitive receptors. The use of living organisms are one of the most emerging and useful alternative technologies for removing contaminants, restoring contaminated sites and preventing further pollution of the environment (Dave and Ghaly 2011). The roles of bioremediation for cleanup of toxic substances including miscellaneous uses are described in Fig. 4.

2.2 *Microbial Intervention in Aquaculture System*

In polluted aquatic system or aquaculture, there are range of microbes having natural affinity to pollutants (hydrocarbons) through a primary mechanism of control known as competitive exclusion, limit the presence of pathogenic microbes in aquatic environment. Others like biological nutrient removal (excess of nitrogen, ammonia and phosphorous) reduces the overload of organic matter in water. Bioremediation through microorganism in polluted aquatic ecosystem works because of the competitive exclusion of the indigenous microflora and the antagonisticcharacteristics against potential pathogens. The mechanisms that likely relate directly to the removal of specific metals, oxyanions and organic contaminants entail two processes: bio-augmentation and bio-stimulation.

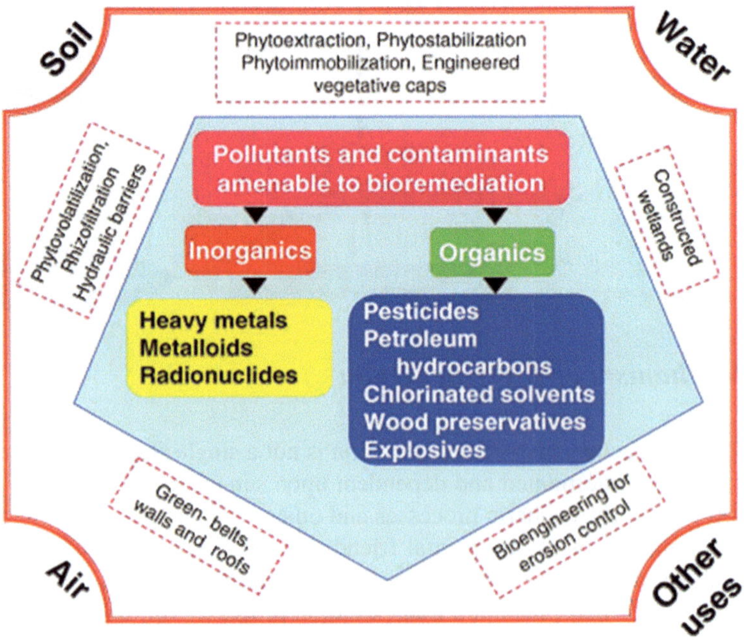

Fig. 4 Multiple general mechanisms involved in bioremediation

2.2.1 Bioaugmentation

It is the direct addition of pre-grown microorganisms that can break down contaminants and accelerate their destruction as pre-grown microbial cultures enhance microbial populations at a site to improve contaminant clean up and reduce clean up time and cost (Tyagi et al. 2011; Azubuike et al. 2016). If biodegrading microbial populations are not present in soil because of contaminant toxicity, specific microorganisms can be added as "introduced organisms" to boost existing populations. For example, during phytoremediation of metal-contaminated estuaries, bioaugmentation with endogenous rhizobacteria with *Spartinamaritima*, resulted in increased plant subsurface biomass, metal accumulation and enhanced metal removal (Mesa et al. 2015). The process is known as bioaugmentation. It has been reported native microbes are usually present in very less in quantities and may not have potential to prevent the extent of the contaminant, or they have not ability to degrade a particular contaminant at a particular polluted site. Thus, bioaugmentation proposal in a way to provide specific microbes in bulk mass's to complete the biodegradation (National Academy Press 1993).

2.2.2 Biostimulation

It is enriching the environment by adding nutrients such as phosphorus (P) and nitrogen (N), boost the competency of naturally occurring microbes to break down toxic substances or chemicals (Sharma 2012). For example, during oil spill the increase in carbon concentration stimulates the growth of already oil degrading microbes and the addition of supplemental nutrients in the proper concentrations, increases the degradation of hydrocarbon by microbes (Fig. 5). This happened because the microbes achieved maximum growth rate and therefore, the maximum power of pollutant uptake (Boufadel et al. 2006; Zahed et al. 2010). The important factor for achieving the maximum biostimulation is ideal concentration of nutrients for the maximum growth of microorganisms and keeps that concentration as long as possible (Lee et al. 2007). One of the best advantages of biostimulation is biodegradation occurs due to already present indigenous microbes, are well suited to the environment and well distributed within the subsurface. However, the main disadvantage of biostimulation is local geology of the subsurface that determines

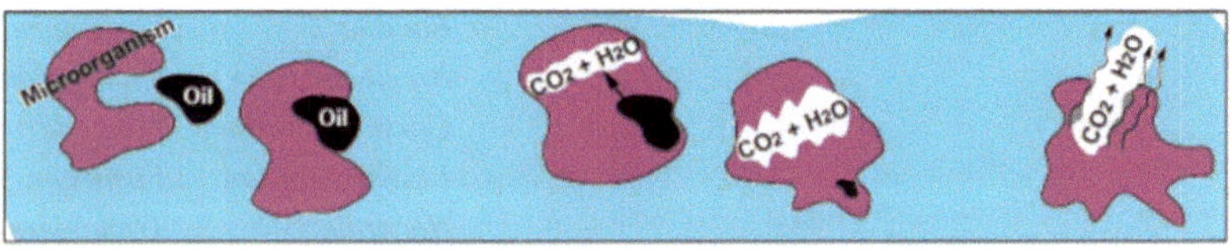

Fig. 5 Biodegradation of hydrocarbons (**a**) Microorganism eats oil or other organic contaminant. (**b**) Microorganism digests oil and converts it into CO_2 and H_2O. (**c**) Microorganism gives off carbon dioxide and water into the environment

delivery and availability of additives to microorganisms. Thus, Biostimulation is case specific, depends on the nature of nutrient, chemical properties and the characteristic of the contaminated sites or environment. Optimal nutrient level is one of the keys for the success of Biostimulation (if oxygen is not a limiting factor).

2.3 How Microbes Destroy Contaminants?

At present bioremediation is given most preference to cleaning a limited range of contaminants from the polluted aquatic ecosystem, mostly hydrocarbons found in gasoline. Microorganisms have the capability to biodegrade almost all organic contaminants and many inorganic contaminants at commercial level (National Research Council 1993), for example, in case of oil spillage, various microbes (bacteria, fungi and yeast) are used for degradation of petroleum hydrocarbons as shown in the Table 2. When hydrocarbons are released in the marine ecosystem, several processes took place, which contributes to the biodegradation and bioremediation (Das and Chandaran 2011) (Fig. 6). The contact between bacteria and contaminants is the basic condition for degradation. The uneven spread of microbes doesn't maintain this association in the soil. However, some bacteria's show chemo tactic response (sensing the contamination) and moves towards it. As there are various types of pollutants present in the degraded environment hence, diversity of microbes are needed to tackle the situation (Table 3) (Watanabe et al. 2000).

The organic contaminants get transformed into less toxic substances because the contaminant becomes the food for the growth and reproduction of microorganisms (Mbhele 2007). Microbes breakdown chemical bonds of contaminant and the released electron get transferred to the electron acceptor, such as oxygen, during this process, the microbes get energy. The carbon that is one of the basic building blocks of new cell constituents to produce more cells is being provided by the organic contaminants (Fig. 7) (National Research Council 1993).

Table 2 List of microorganisms degrading Hydrocarbons

Bacteria	Yeast and fungi
Achromobacter	*Aspergillus*
Acinetobacter	*Candida*
Alcaligenes	*Cladosporium*
Arthrobacter	*Pencillium*
Bacillus	*Rhodotorula*
Brevibacterium	*Sporobolomyces*
Corynebacterium	*Trichoderma*
Flavobacterium	*Fusarium*
Nocardia	*Trichoderma*
Pseudomonas	
Vibrio	

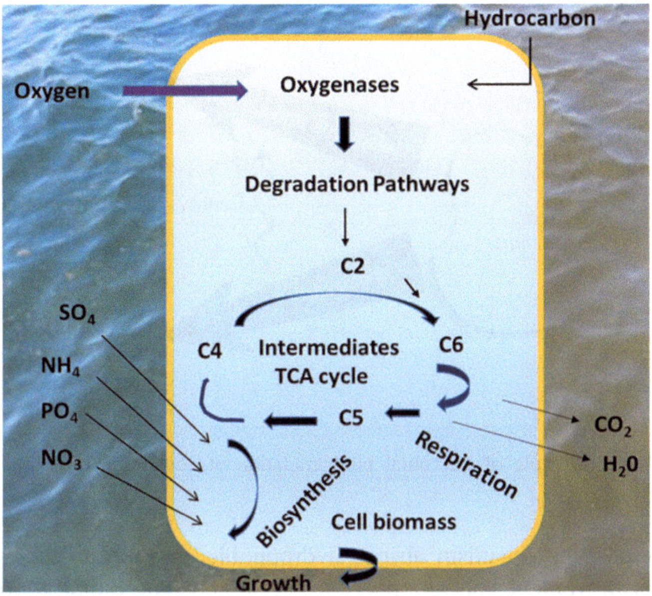

Fig. 6 Biodegradation of petroleum hydrocarbons in aerobic conditions in the Marine ecosystem. (Das and Chandaran 2011)

Table 3 Biodegradation potential of microorganisms for xenobiotic

Organisms	Toxic chemicals	References
Pseudomonas spp.	Benzene, PCBs, Anthracene,	
Alcaligens spp.	Aromatics, PCBs	Lal and Khanna (1996)
Arthrobacter spp.	Benzene, Polycyclic aromatic, long chain alkanes, Phenols	Jogdand (1995) and Deam-Ross et al. (2002)
Bacillus spp.	Halogenated hydrocarbons	Cybulski et al. (2003)
Azotobacter spp.	Benzene, cyloparaffins	Deam-Ross et al. (2002)
Rhodococcus spp.	Aromatics	Park et al. (1998)
Mycobacterium spp.	Hydrocarbons, Polycyclic hydrocarbons	Jogdand (1995)
Methosinu ssp.	PCBs, formaldehyde	Ijah (2002)
Xanthomonas	PCBs, biphenyls	Jogdand (1995)

2.4 *How Microbes Demobilize Contaminants?*

Microbes not only converting the contaminant into less toxic substances but also makes mobile contaminants to be demobilized, a useful strategy for holding hazardous substances. There are three basic ways through which microbes are used to demobilize the contaminants

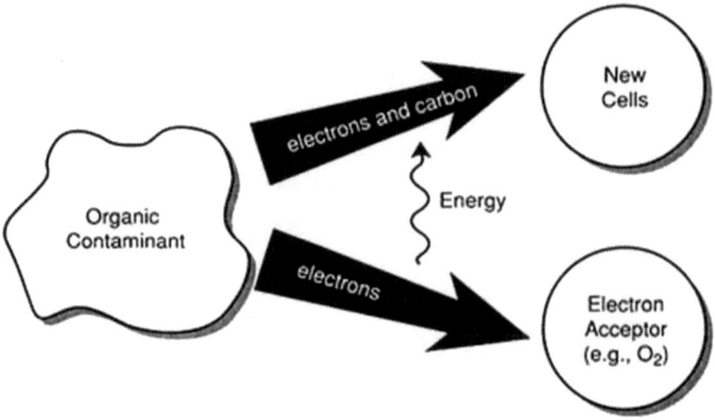

Fig. 7 The basic principle of microbial biodegradation of contaminants. (National Research Council 1993)

I. Biomes of microorganism absorb hydrophobic organic molecules. The contaminant movement gets stopped due to the adequate biomass growth in the path of contaminant migration. Sometimes this concept is known as biocurtain.
II. Organic substances are degraded by microbes and binds with metals and keep the metals in solution. An unbound metal often gets immobilized and precipitate.
III. Microbes produce reduced or oxidized species, which cause metals to precipitate. For instances: Oxidation of Fe^{2+} to Fe^{3+}, gets precipitates as $FeOH_{3(s)}$; reduction of SO_4^{2-} to sulfide (S^{2-}), that precipitates with Fe^{2+} as pyrite ($FeS_{(s)}$) or with Hg^{2+} as $HgS_{(s)}$; reduction of hexavalent chromium (Cr^{6+}) to trivalent chromium (Cr^{3+}), which can precipitate as chromium oxides, sulfides, or phosphates; and, reduction of soluble uranium to insoluble U^{4+}, which precipitates as uraninite (UO_2) (National Research Council 1993).

There occur some changes in the environment when microbial activity took place while degrading the contaminants (National Research Council 1993). Some of the changes are:

- **Chemical change**

During the process of bioremediation the ground water chemistry gets altered. Specific chemical reactants and products are determined by reactions catalyzed by microbes in chemical equations. For instance, the best familiar biochemical equation for the degradation of toluene (C_7H_8) is: $C_7H_8 + 9O_2 \rightarrow 7CO_2 + 4H_2O$. Thus, during the bioremediation process the concentration of inorganic carbon (represented by CO_2) must increase as the concentration of toluene and oxygen decrease. Similarly, dechlorination of $C_2H_3Cl_3$, (trichloroethaneor TCA) to $C_2H_4Cl_2$, (dichloroethaneor DCA) by hydrogen-oxidizing anaerobic bacteria: $C_2H_3Cl_3 + H_2 \rightarrow C_2H_4Cl_2 + H^+ + Cl^-$.

Here, $C_2H_3Cl_3$and H_2 decrease as $C_2H_4Cl_2$, hydrogen ion (H^+), and chloride ion (Cl^-) increase. Thus, the formation of hydrogen ion may cause the pH to decrease, depending on the ground water chemistry. So, in general, it is expected there is fall in concentration of oxygen in aerobic conditions when microorganisms are active and in the anaerobic conditions electron acceptors (NO_3^-, SO_4^{2-}, Fe^{3+}, Mn^{4+}) will get decline, with the parallel increase in the reduced species of these compounds (N_2, H_2S, Fe^{2+}, and Mn^{2+}) respectively. When organic carbon gets oxidized, inorganic carbon concentration must increase under both conditions. The inorganic carbon may take the form of bicarbonate ion (HCO_3^-), dissolved carbon dioxide or gaseous form (CO_2)

- **Native Organisms get adapted:** Bioremediation also alter the metabolic capabilities of indigenous microbes. Most often, microbes do not breakdown contaminants upon initial contact, but they may develop the capability after prolonged exposure to degrade the contaminant. Adaptation not only occurs in a single microbial community but among distinct microbial communities. Although the proper mechanism of adaptation yet to be verified, but adaptation is important because it is a critical principle in ensuring the existence of microorganisms that can exterminate the innumerable newly produced chemicals that humans have created/creating and introduced into the environment.

2.5 *Microbial Populations for Bioremediation*

Microorganisms used in bioremediation are known as bioremediators. Bioremediation is a complex system of many factors (Table 4) (Vidali 2001). Microorganisms can inhabit in aerobic as well as anaerobic conditions and can grow and adapt subzero temperatures. Microorganisms present in nature are indigenous or extraneous and are needed for the process of bioremediation (Prescott et al. 2002). They can be isolated from any sources from the range of optimal to extreme conditions, but most of them have shown optimal growth over a narrow range, so it is important to maintain optimal conditions. The diverse adaptability and biological systems presented by microbes make them perfect to be utilized in remediation of environmental hazards. The use of these microbes depends upon on the chemical nature of the polluting agents and selection is to be very careful as they survive within a limited range

Table 4 Different factors and conditions for Bioremediation (Vidali 2001)

Factor	Conditions required
Microbes	Aerobic or Anaerobic
Biological processes	Catabolism and Anabolism
Environmental factors	O_2, pH, Temperature, Electron acceptor/donor
Nutrients	C, N, O_2, etc.
Soil moisture	Water holding capacity 25–28%
Type of soil	Clay or silt content (Low)

of chemical contamination (Prescott et al. 2002; Dubey 2004). In 1991, it has been reported that more than 70 microbial agents were discovered to degrade petroleum compounds (US Congress 1991) and in successive decades equal number of microbes has been added to the list (Glazer and Nikaido 2007).

2.6 Environmental Constrains

The growth and activity of microbes are affected by moisture content, pH and temperature (Verma and Jaiswal 2006) (Table 5) (Shanahan 2004).

2.6.1 Temperature

The metabolism of microbial system is substantially affected by temperature (Rike et al. 2008). The range of 10–38 °C is considered most suitable for microorganisms. Biochemical reaction rates are affected by temperature and the rates double for every rise of 10 °C in temperature. But the cells die above a certain temperature. Bioremediation within the subsurface (down to 100 m) where temperature remains within 1–2 °C (annual mean temperature) would occur more quickly in temperate climates (Freeze and Cherry 1979).

2.6.2 pH

The pH range nearly 5.5–8.5 most suitable for bioremediation processes and is considered optimum range for many heterotrophic bacteria major microorganisms in most bioremediation technologies. The pH range is site specific and is influenced by a complex relationship of chemistry of contaminants and organisms, physiochemical properties of the local environs. There is dissolution or precipitation of metals in

Table 5 Environmental constrains (Shanahan 2004)

Environmental factor	Optimum conditions	Conditions required for microbial activity
Available soil moisture	25–28% water holding capacity	25–28% water holding capacity
Oxygen	DO >0.2 mg/L, >10% air filled pore space for aerobic degradation	Aerobic, minimum air filled pore space of 10%
Redox potential	Eh >50 mV	
Nutrients	C/N/P ratio = 120/10/1	Nitrogen and Phosphorus
pH	6.5–8.0	5.5–8.5
Temperature	20–30 °C	15–45 °C
Contaminants	Hydrocarbon 5–10% of dry weight of soil	Not too toxic
Heavy metals	700 ppm	Total content 2000 ppm.

soil due to change in pH and thus, may increase the mobility of hazardous metals. Therefore evaluation of soil buffering capacity is necessary prior to application of amendments (Pichtel 2007).

2.6.3 Moisture Content

Moisture content plays an important role in soils where bioremediation taking place as it changes the availability of pollutants, gases transfer, toxicity level of pollutants, the growth stage and movement of microbes, and species distribution. Availability of water is defined in terms of a parameter called water activity. It is the ratio of the system's vapor pressure to that of pure water (at the same temperature) (Suthersan et al. 2016. Too much moister content stops the penetration of oxygen into the soil and it becomes a limiting factor for the efficiency of bioremediation. It has been reported about 20–80% water content is adequate but in some cases like in surface contamination 20% moisture content is enough. However if continuous oxygen supply is being provided to deeper contamination, 80% moisture content would be adequate.

2.6.4 Redox Potential

The redox potential of the soil (oxidation-reduction potential, *Eh*) is directly related to the concentration of Oxygen (O_2) in the gas and liquid phases. Aerated soils have an *Eh* of about 0.8–0.4 V; moderately reduced soils have 0.4–0.1 V; reduced soils 0.1 to −0.1 V and highly reduced soils have about 0.1 to −0.3 V. In the process of respiration oxygen gets depleted, leads to lowering of redox potential and producing anaerobic (i.e., reducing) conditions. Such conditions restrict aerobic reactions and encourage anaerobic processes such as fermentation, denitrification, and sulfate reduction. Redox potentials are difficult to measure in the soil or groundwater and are not widely used in the field (Pichtel 2007).

2.6.5 Mass Transfer Characteristics

These are used to determine potential rates of liquids or gases movement through soil and include soil texture, moisture content, unsaturated hydraulic conductivity and dispersivity, vs. soil moisture tension, porosity, bulk density,, hydraulic conductivity and infiltration rate (Hillel 1998; Sara 2003). Hydro geologic characteristics and factors taking into consideration include aquifer type, hydraulic conductivity, hydro geologic gradient, permeability, recharge capability, groundwater depth, moisture content or field capacity, thickness of the saturated zone, homogeneity, plume stability, depth and extent of contamination. These are some parameters that should be considered while framing the design of any bioremediation system (Hillel 1998; Suthersan et al. 2016; Sara 2003).

2.7 *Strategies of Bioremediation*

The two major bioremediation techniques are *Ex-situ* and *In-situ* (Fig. 8).

2.7.1 Ex-Situ Bioremediation Techniques

It is a done by relocation of a contaminated material to another site to hasten biocatalysis. These techniques are usually considered on: the depth of pollution, degree of pollution, types of pollutant, the cost of treatment, geology and geographical location of the polluted site. Performance criteria are also being considered in these techniques (Philp and Atlas 2005).

Biopile

It is a type of *ex-situ* bioremediation techniques. It involves above-ground piling of quarried polluted soil, followed by aeration and nutrient amendment to enhance bioremediation through increasing microbial activities. The components of this procedure are: treatment bed aeration, irrigation, nutrient and leachate collection systems. Because of its constructive, effective biodegradation and cost effectiveness features this technique is increasingly being considered (Whelan et al. 2015). It can also be used effectively in extreme environments (very cold regions) (Dias et al. 2015; Gomez and Sartaj 2014; Whelan et al. 2015).

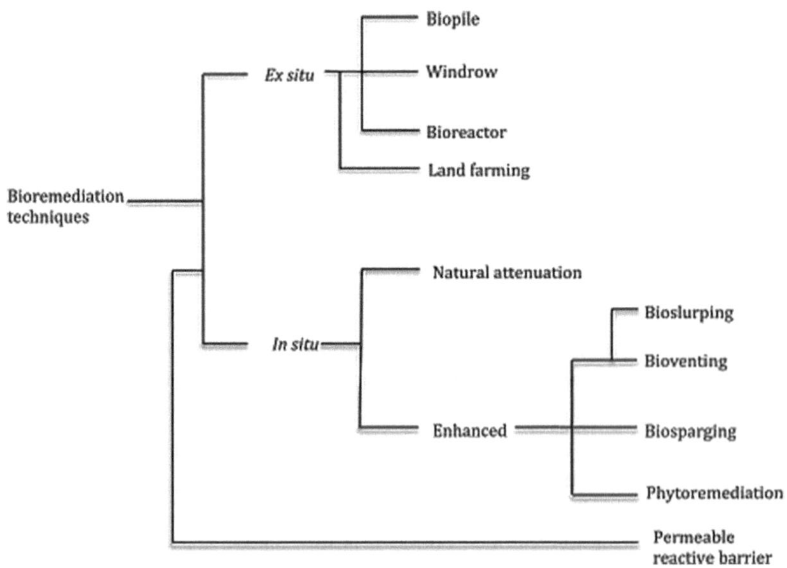

Fig. 8 Schematic representation of major bioremediation techniques

Windrows

This technique involves periodic turning of piled polluted soil to improve bioremediation by increasing degradation activities of native and/or transient hydrocarbonoclasticbacteria found in contaminated soil. The periodic turnings of piled contaminated soil, with the addition of water brings about increase in uniform distribution of pollutants, aeration, nutrients and microbial degradation activities, consequently speedy up the level of bioremediation, through biotransformation, mineralization and assimilation (Barr 2002). However, it is not best suited for remediating soils polluted with toxic volatiles (Azubuike et al. 2016)

Bioreactor

In this technique raw materials are converted to a particular product(s) following the series of biological reactions in a container. There are different operating modes: batch, sequencing batches, fed-batch, continuous and multistage. Market economy and capital expenditure determines the choice of operating mode of bioreactor (Azubuike et al. 2016). This technique has several advantages as compared to Biopile and Windrows. Excellent control of bioprocess parameters like pH, temperature, substrate and inoculum concentrations, agitation, and aeration rates are one of the major advantages of this technique. There is maximum biological degradation and minimum abiotic losses (Mohan et al. 2004).

Land Farming

Land farming is regarded as both *Ex-situ* and *In-situ* bioremediation technique. The site of treatment determines the type of bioremediation. When polluted soil is treated on-site, it can be regarded as *In-situ*; else, it is *Ex-situ*. Besides the site of treatment, pollutant depth also plays an important role in determining type of bioremediation. It has been revealed that when a pollutant depth is less than 1 m, bioremediation might proceed without digging, if pollutant lying >1.7 m depth, needs to be excavated and relocating to the ground surface for effective enhanced bioremediation (Nikolopoulou et al. 2013). The autochthonous microorganisms perform aerobic biodegradation of pollutants (Philp and Atlas 2005; Paudyn et al. 2008; Volpe et al. 2012; Silva-Castro et al. 2015). It is cost-effective and less equipment's are required for operation.

2.7.2 In-Situ Bioremediation

This involves treatment of pollutants at the site of contamination. There is no need of excavation or relocation of pollutants. These techniques were successfully used to treat heavy metals, chlorinated solvents, dyes and hydrocarbons at contaminated sites (Folch et al. 2013; Kim et al. 2014; Frascari et al. 2015; Roy et al. 2015). Some of the *In-situ* bioremediation techniques are enhanced (bioventing, biosparging and phytoremediation), others proceed without any form of enhancement (intrinsic bioremediation or natural attenuation).

Bioventing

This involves controlled stimulation of airflow (oxygen) to an unsaturated (vadose) zone in order to enhance bioremediation process, due to increasing activities of native microorganisms. In bioventing, amendments are made by supplying moisture content and nutrients to increase the microbial transformation of pollutants into harmless product (Philp and Atlas 2005).

Bioslurping

The technique works on the combination of bioventing, vacuum-enhanced pumping and soil vapour extraction to attain the groundwater remediation and soil remediationby indirect delivery of oxygen and stimulation of biodegradation of contaminants (Gidarakos and Aivalioti 2007).

Biosparging

The technique is very similar to bioventing however, unlike bioventing; air is injected at the saturated zone, which can cause upward movement of volatile organic compounds to the unsaturated zone to promote biodegradation. The two major factors that determine the effectiveness of biosparging are soil permeability and pollutant biodegradability (Philp and Atlas 2005).

2.8 Phytoremediation

This technique is an emerging technology which involves use of plants to decontaminate the polluted soil and water (Bhadra et al. 1999). Some of the types of phytoremediation processes are highlighted in the given Table 7. In this process plants and their linked microorganisms (microbial rhizosphere) are used for environmental cleanup of variety of organic and inorganic pollutants (Raskin et al. 1994; Salt et al. 1998). Organics are degraded in the root zone depending on the properties of plants or taken up, followed by sequestration, degradation or volatilization. Different internal reactions are catalyzed by enzymes produced by plants with various activities and functions (Table 6). Some of the enzymes like oxygenases in

Table 6 Role of enzymes in bioremediation (Husain et al. 2009)

Enzyme	Target pollutant
Aromatic dehalogenase	DDT, PCBs etc. (Chlorinated aromatics)
Cytochrome P450	Xenobiotics (PCBs)
Dehalogenase	Chlorinated solvents and Ethylene
Laccase	Oxidative step in degradation of explosives
Nitrilase	Herbicides
Nitroreductase	RDX and TNT
O-demethylase	Metalachor and Alachlor
Peroxdase	Phenols
Phosphatase	Organophosphates
Carboxyl esterases, Glutathione s-transferase, N-malonyltransferases, N-glucosyltransferases, O-glucosyltransferases, O-malonyltransferases, Peroxidases and Peroxygenases	Xenobiotics

plants are able to address hydrocarbons aliphatic and aromatic compounds. Likewise, nitroreductases can reduce and breakdown energetic compounds such as explosives TNT(trinitrotoluene), 1,3,5-trinitroperhydro-1,3,5-triazine (RDX) and 1,3,5,7-tetranitro-1,3,5,7-tetrazocine (HMX, High melting explosive) (Hughes et al. 1997; Anonymous 2009). Some other pollutants are PAHs the fuel additive MTBE (Davis et al. 2003), and polychlorinated biphenyls (PCBs). Metabolism of trichloroethylene (TCE) in plant tissues completes in three phases.

Phase I: The activation/transformation of TCE to trichloroethanol.
Phase II: Conjugation with a plant molecule.
Phase III: Sequestration of the conjugate into the cell wall or within the vacuole (Figs. 9 and 10) (Table 7).

2.9 *Permeable Reactive Barrier (PRB)*

The technique due to its design and mechanism of pollutant removalperceived as a physical method for remediation of polluted groundwater. But biological reaction is one of the several mechanisms (Sorption, precipitation and degradation) of contaminant removal in PRB technique (Thiruvenkatachari et al. 2008; Obiri-Nyarko et al. 2014). Some of the alternative proposed terms such as biological PRB, bio-enhanced PRBand passive bio-reactive barrier, to put up the bioremediation feature of the technique, microorganisms play an important role in enhancement of bioremediation as compared to independent biotechnology (Philp and Atlas 2005).

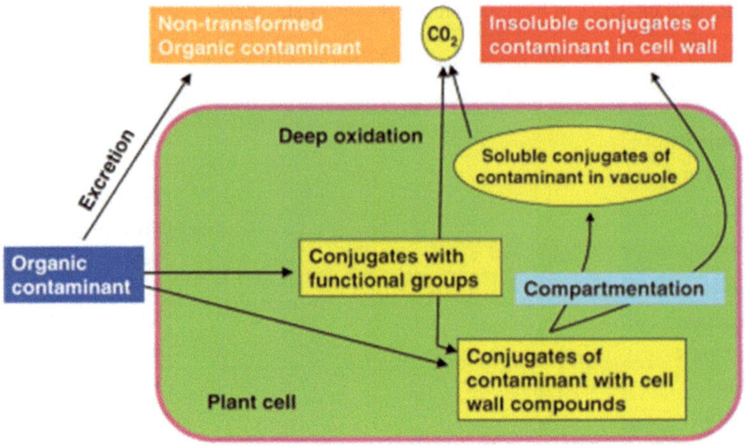

Fig. 9 Detoxification of xenobiotics. (Reichenauer and Germida 2008; Van Aken 2009)

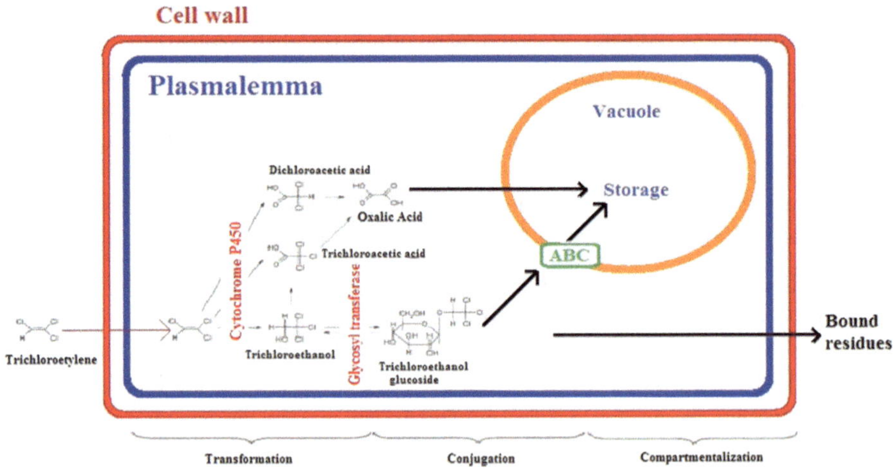

Fig. 10 Schematic representation of the metabolism of TCE in plant tissues. (Reichenauer and Germida 2008; Van Aken 2009)

3 Conclusion

Bioremediation technology offers a great role in detoxifying the contaminated sites with high efficiency, low cost and environmental friendly. The disadvantages are outshined by its advantages. Thus, this technology is in increasing demand and has been recognized as sustainable management tool for almost every type of contamination.

Table 7 Types of phytoremediation, functions, plant species which remove the pollutants

Process	Function	Pollution	Medium	Plants	References
Phytoextraction	Removal of heavy metals and organics from soil	Cadmium, lead, Zinc, oil spills and radionuclide's	Soil and groundwater	*Viola baoshanensis*, *Sedum alferedii*, *Rumex crispus*	Macek et al. (2000)
Phytotransformation	Sequestration and degradation of organic compounds	Xenobiotic substances	Soil	*Cannas*	
Phytodegradation	Microorganisms associated with plants degrade organic pollutants	DDT, Explosives and nitrates	Groundwater	*Elodea Canadensis*, Pueraria	Newman and Reynolds (2004)
Rhozofiltration	Absorbs mainlymetals from polluted aquatic system	Cadmium, Arsenic, lead and Zinc	Ground water	*Brassica juncea*	Verma et al. (2006)
Phytostabilization	Plants are used to reduce the bioavailability of pollutants in the environment	Copper, Cadmium, Chromium, Nickel, lead and Zinc	Soil	*Anthyllisvulneraria*, *Festucaarvernensis*	Vazquez et al. (2006)

References

Agarwal, S. K. (1998). *Environmental biotechnology* (1st ed., p. 267289). New Delhi: APH Publishing Corporation.

Anonymous. (2009). ITRC (Interstate Technology and Regulatory Council). Phytotechnology technical and regulatory guidance and decision trees, revised. PHYTO-3. Washington DC.

Ayangbenro, A. S., & Babalola, O. O. (2017). A new strategy for heavy metal polluted environments: A review of microbial biosorbents. *International Journal of Environmental Research and Public Health, 14*, 94.

Azubuike, C. C., Chikere, C. B., & Okpokwasili, G. C. (2016). Bioremediation techniques-classification based on site of application: Principles, advantages, limitations and prospects. *World Journal of Microbiology and Biotechnology, 32*, 180.

Baker, K. H., & Herson, D. S. (1994). Introduction and overview of bioremediation. In K. H. Baker & D. S. Herson (Eds.), *Bioremediation*. New York: McGraw-Hill.

Barr, D. (2002). *Biological methods for assessment and remediation of contaminated land: Case studies*. London: Construction Industry Research and Information Association.

Bhadra, R., Wayment, D. G., Hughes, J. B., & Shanks, J. V. (1999). Confirmation of conjugation processes during TNT metabolism by axenic plant roots. *Environmental Science & Technology, 33*, 446–452.

Blaylock, M. J., Salt, D. E., Dushenkov, S., Zakharova, O., Gussman, C., Kapulnik, Y., Ensley, B. D., & Raskin, I. (1997). Enhanced accumulation of Pb in Indian mustard by soil-applied chelating agents. *Environmental Science & Technology, 31*, 860–865.

Boufadel, M. C., Suidan, M. T., & Venosa, A. D. (2006). Tracer studies in laboratory beach simulating tidal influences. *Journal of Environmental Engineering, 132*, 616–623.

Congress US. (1991). *Office of technology assessment* (Bioremediation for marine oil spills-background paper). Government Printing Office OTA-BP-O-70: Washington, DC.

Cybulski, Z., Dzuirla, E., Kaczorek, E., & Olszanowski, A. (2003). The influence of emulsifiers on hydrocarbon biodegradation by Pseudomonadacea and Bacillacea strains. *Spill Science and Technology Bulletin, 8*, 503–507.

Das, N., & Chandran, P. (2011). Microbial degradation of petroleum hydrocarbon contaminants: An overview. *Biotechnology Research International, 2011*, 941810.

Dave, D., & Ghaly, A. E. (2011). Remediation technologies for marine oil spills: A critical review and comparative analysis. *American Journal of Environmental Sciences, 7*, 423–440.

Davis, L. C., Erickson, L. E., Narayanan, N., & Zhang, Q. (2003). Modeling and design of phyto remediation. In S. C. McCutcheon & J. L. Schnoor (Eds.), *Phytoremediation: Transformation and control of contaminants* (pp. 663–694). New York: Wiley.

Dean-Ross, D., Moody, J., & Cerniglia, C. E. (2002). Utilization of mixtures of polycyclic aromatic hydrocarbons by bacteria isolated from contaminated sediment. *FEMS Microbiology Ecology, 41*, 17.

Dias, R. L., Ruberto, L., Calabro´, A., Balbo, A. L., Del-Panno, M. T., & Mac-Cormack, W. P. (2015). Hydrocarbon removal and bacterial community structure in on-site biostimulated biopile systems designed for bioremediation of diesel-contaminated Antarctic soil. *Polar Biology, 38*, 677–687.

Dubey, R. C. (2004). *A text book of biotechnology* (3rd ed., p. 365375). New Delhi: S. Chand and Company Ltd.

Folch, A., Vilaplana, M., Amado, L., Vicent, R., & Caminal, G. (2013). Fungal permeable reactive barrier to remediate groundwater in an artificial aquifer. *Journal of Hazardous Materials, 262*, 554–560.

Frascari, D., Zanaroli, G., & Danko, A. S. (2015). In situ aerobic cometabolism of chlorinated solvents: A review. *Journal of Hazardous Materials, 283*, 382–399.

Freeze, R. A., & Cherry, J. A. (1979). *Groundwater*. Englewood Cliffs: Prentice-Hall.

Gadd, G. M. (2001). *Fungi in bioremediation*. Cambridge: Cambridge University Press.

Gidarakos, E., & Aivalioti, M. (2007). Large scale and long term application of bioslurping: The case of a Greek petroleum refinery site. *Journal of Hazardous Materials, 149*, 574–581.
Glazer, A. N., & Nikaido, H. (2007). *Microbial biotechnology: Fundamentals of applied microbiology* (2nd ed., p. 510528). Cambridge/New York: Cambridge University Press.
Gomez, F., & Sartaj, M. (2014). Optimization of field scale biopiles for bioremediation of petroleum hydrocarbon contaminated soil at low temperature conditions by response surface methodology (RSM). *International Biodeterioration & Biodegradation, 89*, 103–109.
Hamer, G. (1993). Bioremediation: A response to gross environmental abuse. *Trends in Biotechnology, 11*(8), 317–319.
Hess, A., Zarda, B., Hahn, D., Häner, A., Stax, D., Höhener, P., & Zeyer, J. (1997, June). In situ analysis of denitrifying toluene-and m-xylene-degrading bacteria in a diesel fuel-contaminated laboratory aquifer column. *Applied and Environmental Microbiology, 63*(6), 2136–2141.
Hillel, D. (1998). *Environmental soil physics*. Waltham: Academic.
Hughes, J. B., Shanks, J., Vanderford, M., Lauritzen, J., & Bhadra, R. (1997). Transformation of TNT by aquatic plants and plant tissue cultures. *Environmental Science & Technology, 31*, 266–271.
Husain, Q., Husain, M., & Kulshrestha, Y. (2009). Remediation and treatment of organopollutants mediated by peroxidases: A review. *Critical Reviews in Biotechnology, 29*(2), 94–119.
Ijah, U. J. J. (2002). Accelerated crude oil biodegradation in soil by inoculation with bacterial slurry. *Journal of Environmental Sciences, 6*(1), 3847.
Jogdand, S. N. (1995). *Environmental biotechnology* (1st ed., p. 1041). Bombay: Himalaya Publishing House.
Kamaludeen, S. P. B. K., Arunkumar, K. R., Avudainayagam, S., & Ramasamy, K. (2003). Bioremediation of chromium contaminated environments. *The Indian Journal of Experimental Biology, 41*, 972–985.
Kim, S., Krajmalnik-Brown, R., Kim, J. O., & Chung, J. (2014). Remediation of petroleum hydrocarbon-contaminated sites by DNA diagnosis-based bioslurping technology. *Science of the Total Environment, 497*, 250–259.
Kvesitadze, G., Khatisashvili, G., Sadunishvili, T., & Ramsden, J. J. (2006). Uptake, translocation and effects of contaminants in plants. In *Biochemical mechanisms of detoxification in higher plants: Basis of phytoremediation* (pp. 55–102). Berlin/New York: Springer.
Lal, B., & Khanna, S. (1996). Degradation of crude oil by Acinetobacter calcoaceticus and Alcaligenes odorans. *Journal of Applied Bacteriology, 81*(4), 355–362.
Lee, S. H., Lee, S., Kim, D. Y., & Kim, J. G. (2007). Degradation characteristics of waste lubricants under different nutrient conditions. *Journal of Hazardous Materials, 143*(1–2), 65–72.
Macek, T., Mackova, M., & Káš, J. (2000). Exploitation of plants for the removal of organics in environmental remediation. *Biotechnology Advances, 18*(1), 23–34.
Mbhele, P. P. (2007). *Remediation of soil and water contaminated by heavy metals and hydrocarbons using silica encapsulation*. Doctoral dissertation, University of Witwatersrand, Johannesburg.
Mesa, J., Rodrı´guez-Llorente, J. D., Pajuelo, E., Piedras, J. M. B., Caviedes, M. A., Redondo-Go´mez, S., & Mateos-Naranjo, E. (2015). Moving closer towards restoration of contaminated estuaries: Bioaugmentation with autochthonous rhizobacteria improves metal rhizoaccumulation in native Spartina maritima. *Journal of Hazardous Materials, 300*, 263–271.
Mohan, S. V., Sirisha, K., Rao, N. C., Sarma, P. N., & Reddy, S. J. (2004). Degradation of chlorpyrifos contaminated soil by bioslurry reactor operated in sequencing batch mode: Bioprocess monitoring. *Journal of Hazardous Materials, 116*(1–2), 39–48.
Morel, J. L., Echevarria, G., & Goncharova, N. (Eds.). (2002). *Phytoremediation of metal-contaminated soils* (Vol. 68). Dordrecht: Springer Science and Business Media.
National Research Council, Division on Engineering and Physical Sciences, Commission on Engineering and Technical Systems, Committee on In Situ Bioremediation. National Academies Press, February 01, 1993.

Newman, L. A., & Reynolds, C. M. (2004). Phytodegradation of organic compounds. *Current Option in Biotechnology, 15*, 225–230.
Nikolopoulou, M., Pasadakis, N., Norf, H., & Kalogerakis, N. (2013). Enhanced *ex-situ* bioremediation of crude oil contaminated beach sand by supplementation with nutrients and rhamnolipids. *Marine Pollution Bulletin, 77*, 37–44.
Obiri-Nyarko, F., Grajales-Mesa, S. J., & Malina, G. (2014). An overview of permeable reactive barriers for in situ sustainable groundwater remediation. *Chemosphere, 111*, 243–259.
Ojuederie, O., & Babalola, O. (2017). Microbial and plant-assisted bioremediation of heavy metal polluted environments: A review. *International Journal of Environmental Research and Public Health, 14*(12), 1504.
Paniagua-Michel, J., & Rosales, A. (2015). Marine bioremediation-A sustainable biotechnology of petroleum hydrocarbons biodegradation in coastal and marine environments. *Journal of Bioremediation and Biodegredation, 6*(2), 1.
Park, A. J., Cha, D. K., & Holsen, T. M. (1998). Enhancing solubilization of sparingly soluble organic compounds by biosurfactants produced by Nocardia erythropolis. *Water Environment Research, 70*(3), 351–355.
Paudyn, K., Rutter, A., Rowe, R. K., & Poland, J. S. (2008). Remediation of hydrocarbon contaminated soils in the Canadian Arctic by landfarming. *Cold Regions Science and Technology, 53*(1), 102–114.
Philp, J. C., & Atlas, R. M. (2005). Bioremediation of contaminated soils and aquifers. In R. M. Atlas & J. C. Philp (Eds.), *Bioremediation: Applied microbial solutions for real-world environmental cleanup* (pp. 139–236). Washington: American Society for Microbiology (ASM) Press.
Pichtel, J. (2007). *Fundamentals of site remediation: For metal and hydrocarbon-contaminated soils*. New York: Government Institutes.
Prasad, M. N. (2004a). Phytoremediation of metals in the environment for sustainable development. *Proceedings-Indian National Science Academy Part B, 70*(1), 71–98.
Prasad, M. N. V. (2004b). *Heavy metal stress in plants: From biomolecules to ecosystems* (2nd ed., 462 pp). Heidelberg: Springer.
Prescott, L. M., Harley, J. P., & Klein, D. A. (2002). *Microbiology* (5th ed., p. 1014). New York: McGrawHill.
Raskin, I., Kumar, P. B. A. N., Dushenkov, S., & Salt, D. E. (1994). Bioconcentration of heavy metals by plants. *Current Opinion in Biotechnology, 5*, 285–290.
Reichenauer, T. G., & Germida, J. J. (2008). Phytoremediation of organic contaminants in soil and groundwater. *ChemSusChem, 1*, 708–717.
Rike, A. G., Schiewer, S., & Filler, D. M. (2008). Temperature effects on biodegradation of petroleum contaminants in cold soils. In *Bioremediation of petroleum hydrocarbons in cold regions* (pp. 84–108). Cambridge/New York: Cambridge University Press.
Roy, M., Giri, A. K., Dutta, S., & Mukherjee, P. (2015). Integrated phytobial remediation for sustainable management of arsenic in soil and water. *Environment International, 75*, 180–198.
Salt, D. E., Smith, R. D., & Raskin, I. (1998). Phytoremediation. *Annual Review of Plant Biology, 49*(1), 643–668.
Sara, M. N. (2003). *Site assessment and remediation handbook*. Boca Raton: Lewis Publishers/ CRC Press.
Shanahan, P. (2004, Spring). *Bioremediation; waste containment and remediation technology*. Massachusetts Institute of Technology.
Sharma, S. (2012). Bioremediation: Features, strategies and applications. *Asian Journal of Pharmacy and Life Science, 2*(2), 4423.
Silva-Castro, G. A., Uad, I., Rodríguez-Calvo, A., González-López, J., & Calvo, C. (2015). Response of autochthonous microbiota of diesel polluted soils to land-farming treatments. *Environmental Research, 137*, 49–58.
Strong, P. J., & Burgess, J. E. (2008). Treatment methods for wine related distillery wastewaters: A review. *Bioremediation Journal, 12*, 7087.

Suthersan, S. S., Horst, J., Schnobrich, M., Welty, N., & McDonough, J. (2016). *Remediation engineering: Design concepts*. Boca Raton: CRC Press.

Tang, C. Y., Fu, Q. S., Criddle, C. S., & Leckie, J. O. (2007). Effect of flux (transmembrane pressure) and membrane properties on fouling and rejection of reverse osmosis and nanofiltration membranes treating perfluorooctane sulfonate containing wastewater. *Environmental Science and Technology, 41*(6), 2008–2014.

Thiruvenkatachari, R., Vigneswaran, S., & Naidu, R. (2008). Permeable reactive barrier for groundwater remediation. *Journal of Industrial and Engineering Chemistry, 14*(2), 145–156.

Tyagi, M., da Fonseca, M. M., & de Carvalho, C. C. (2011). Bioaugmentation and biostimulation strategies to improve the effectiveness of bioremediation processes. *Biodegradation, 22*(2), 231–241.

Van Aken, B. (2009). Transgenic plants for enhanced phytoremediation of toxic explosives. *Current Opinion in Biotechnology, 20*(2), 231–236.

Vázquez, S., Agha, R., Granado, A., Sarro, M. J., Esteban, E., Penalosa, J. M., & Carpena, R. O. (2006). Use of white lupin plant for phytostabilization of Cd and As polluted acid soil. *Water, Air, and Soil Pollution, 177*(1–4), 349–365.

Verma, J. P., & Jaiswal, D. K. (2006). Book review: Advances in biodegradation and bioremediation of industrial waste. *Frontiers in Microbiology, 6*, 1555.

Verma, P., George, K. V., Singh, H. V., Singh, S. K., Juwarkar, A., & Singh, R. N. (2006). Modeling rhizofiltration: Heavy-metal uptake by plant roots. *Environmental Modeling and Assessment, 11*(4), 387–394.

Vidali, M. (2001). Bioremediation an overview. *Pure and Applied Chemistry, 73*(7), 1163–1172.

Volpe, A., D'Arpa, S., Del Moro, G., Rossetti, S., Tandoi, V., & Uricchio, V. F. (2012). Fingerprinting hydrocarbons in a contaminated soil from an Italian natural reserve and assessment of the performance of a low-impact bioremediation approach. *Water, Air, & Soil Pollution, 223*(4), 1773–1782.

Watanabe, K., Watanabe, K., Kodama, Y., Syutsubo, K., & Harayama, S. (2000). Molecular characterization of bacterial populations in petroleum-contaminated groundwater discharged from underground crude oil storage cavities. *Applied and Environmental Microbiology, 66*(11), 4803–4809.

Whelan, M. J., Coulon, F., Hince, G., Rayner, J., McWatters, R., Spedding, T., & Snape, I. (2015). Fate and transport of petroleum hydrocarbons in engineered biopiles in polar regions. *Chemosphere, 131*, 232–240.

Zahed, M. A., Aziz, H. A., Isa, M. H., & Mohajeri, L. (2010). Effect of initial oil concentration and dispersant on crude oil biodegradation in contaminated seawater. *Bulletin of Environmental Contamination and Toxicology, 84*(4), 438–442.

Biosorption as Environmentally Friendly Technique for Heavy Metal Removal from Wastewater

Saraswati Saini, Jaskaran Kaur Gill, Jagdeep Kaur, Hridoy Ranjan Saikia, Navdeep Singh, Inderpreet Kaur, and Jatinder Kaur Katnoria

Abstract Water is an essential element of all the life forms and is a universal solvent that may contain miscellany of toxic as well as non toxic substances. Due to increasing population and urbanization, there has been substantially a great burden on the water ecosystem. Apart from these, water ecosystems are also exposed to significant quantity of contaminants released from agricultural and industrial practices which consequently cause serious health problems. Presence of contaminants in ground water along with surface water is of serious concern. Heavy metals, considered as non-degradable pollutants, are responsible to induce various types of diseases in human beings on consumption of contaminated water. Many techniques such as membrane filtration, reverse osmosis, chemical precipitation, physical methods (boiling and sand bed filtration), carbon/activated carbon adsorption, phytoremediation and biosorption have been extensively used for treatment of wastewater. Biosoption, among various types of treatments, is recognized as an environmental friendly tool to remediate the wastewater. The present review focuses on removal of heavy metals using different biosorbents.

Keywords Heavy metals · Remediation · Biosorbents · Adsorption process · Environment and health

---

S. Saini · J. K. Gill · J. Kaur · H. R. Saikia · N. Singh · J. K. Katnoria (✉)
Department of Botanical and Environmental Sciences, Guru Nanak Dev University, Amritsar, India

I. Kaur
Department of Chemistry, Centre for advance Studies, Guru Nanak Dev University, Amritsar, India

H. Qadri et al. (eds.), *Fresh Water Pollution Dynamics and Remediation*,
https://doi.org/10.1007/978-981-13-8277-2_10

1 Introduction

1.1 Sources of Water Pollution

Water is the most precious resource of nature as there can be no life without water on the earth. It is well recognized that the life on earth has originated in water itself and the survival of life is impossible, if the water is polluted beyond certain limits. The polluted water not only is fatal for aquatic animals but also causes severe toxicities including carcinogenicity and death of the terrestrial animals upon consumption of polluted water (Mouchet et al. 2006). Over the decades, it is a common observation that almost all surface water systems like canals, lakes, rivers, ground reservoirs or oceans are so much loaded with continuous discharges from households, municipals, agricultural runoffs, industries that their water is no more fit for even bathing or cloth washing purposes (Chen et al. 2004; Chandra et al. 2005; Mathur and Bhatnagar 2005; Samuel et al. 2010). Some water bodies have been polluted to the extent that even with costliest and tedious methods of treatment cannot make the water fit for domestic purposes (Garg et al. 2006). Pollution of water bodies on account of heavy metals is from both natural and anthropogenic sources. However, anthropogenic sources have now surpassed the later and many man-made activities like mining, welding, textiles, plumbing, electroplating, enameling, dyeing, manufacturing of batteries, painting and varnishing, plasticizing, canning and usage of fertilizers and pesticides are prime reasons for release of various heavy metals into the aquatic systems. Different heavy metals released from various sources with their toxic effects have been shown in Fig. 1.

In order to monitor the water quality of surface water sources with respect to different possible uses and for their proper classification, even the BIS (Bureau of India Standards) vide IS 2296-1982 has classified the water bodies into five categories as water that is (a) used as drinking water sources without any conventional treatment yet, after disinfection (b) used for outdoor bathing (c) used as drinking water sources with conventional treatment followed by disinfection (d) used for fish culture and wild life propagation (e) used for irrigation, industrial cooling and controlled waste disposal (Garg 2010). Despite the categorization of water systems, single surface water resource in most of the villages and cities of India are used for multiple purposes like drinking, washing, irrigation and last but not least inlets of wastes.

1.2 Consequences of Water Pollution

Water pollution due to heavy metals has become a serious problem throughout the world for the past decades. Different heavy metal ions including lead, cadmium, mercury, copper, chromium and zinc have been released into aquatic ecosystem through mining activities and effluent discharges from various industries such as

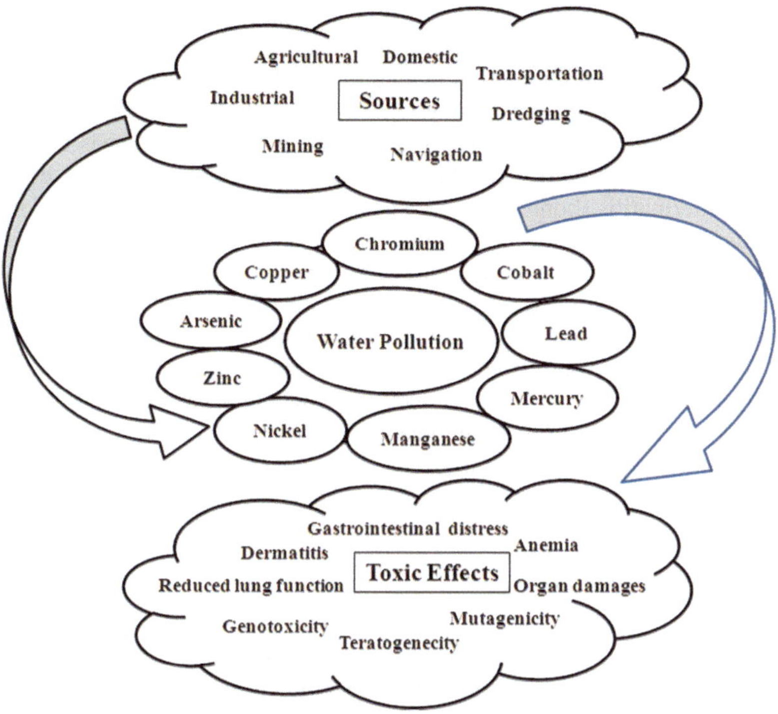

Fig. 1 Sources and toxic effects of heavy metals contamination

electroplating, photographic, steel/iron production and tanneries (Tsezos 1999; Ibrahim et al. 2006; Asberry et al. 2014). Toxic heavy metal ions, not only contaminate surface water such as lake, sea and ponds, but also contaminate ground water and ultimately posing threat to all forms of life including human beings. These non-biodegradable heavy metal ions have the potential to accumulate in different soft tissues upon entering the human body through the drinking water, food chain or dermal contact.

Among different heavy metals, copper and cadmium metal ions are highly toxic in nature and can cause gastrointestinal problems, insomnia, Wilson's disease, hypertension, prostate cancer, improper testicular function and various other reproductive as well as teratogenic effects. (Zhu et al. 2008; Farooq et al. 2010; Baraket 2011; Sirilamduan et al. 2011). Exposure to cadmium metal ions can lead to "Itai-Itai" disease, (Klaassen 2001; El-Sayed 2012). Maximum concentration limit of copper and cadmium in industrial effluents was reported to be 3.0 mg/l and 2.0 mg/l, respectively according to US Environment Pollution Agency Standards (USEPA 1997). Due to its toxicity, cadmium has been included in black list of European Economic Community (Council Directive EEC 1976) and red list by Department of Environment, U.K (U.K. Red list substances 1991).

Lead has been reported to induce anemia, anorexia, loss of appetite, damages to organs like kidney, liver and bladder, gastrointestinal damages, effects on central nervous system, mental retardation in children and induction of tumors while mercury induced damages to nervous system, liver and kidney damages, protoplasm poisoning, dermatitis and corrosive to skin eyes and mucosa (Abbas et al. 2014). Arsenic has been shown to have the potential to induce mutagenicity, hemolysis, bronchitis, dermatitis, bone marrow depression and hepatomegaly (Gadd 2010). Chromium is well recognized as a carcinogen due to its chronic toxicities like mutagenicity, genotoxicity, carcinogenicity and teratogenicity (Gadd 2010). Acute symptoms of chromium exposure through its dust inhalation, dermal contact or ingestion include gastrointestinal pains, nausea, vomiting, diarrhea, tremors and muscle contraction. Apart from effects on human beings, heavy metals induce aquatic toxicity where various organisms get affected. The entry of heavy metals into food chains results in biomagnifications, ultimately causing severe toxicity to the ecosystem as a whole.

1.3 Remedial Measures

Due to high toxicities of metals, it is essential to remove metal ions from water bodies. The best method for the same is to remove metal ions from the sewage/wastewater before its disposal to the surface water bodies. Nowadays, many technologies have been developed to remove the pollutants/heavy metals from wastewater. Various technologies such as membrane filtration, precipitation, coagulation, solvent extraction, ion exchange, neutralization, electro-dialysis, ultra filtration and reverse osmosis have been used for removal of heavy metals from aqueous solutions. Although these technologies are competent but also experience certain limitations such as high operation cost, heavy instrumentation and can operate well only for small scale water treatments (Metcalf 2003; Cardoso et al. 2004; Kandah 2004; Gupta and Ali 2008). On the other hand, sorption process has been found to be a promising technique because of its feasibility, simplicity, cost effectiveness and eco-friendly nature for removal of heavy metals from different aqueous solutions (Melckova and Ruzovic 2010; Sirilamduan et al. 2011; Gupta and Rastogi 2009; Gupta et al. 2013).

2 Biosorption Techniques

Biosorption is a physio-chemical process naturally occurring in biomass of some inactive or dead plants/plant parts as well as microbes that result in binding and concentrating heavy metals from aqueous solutions. During this binding property, biomass acts just as a chemical substance (an ion exchanger of biological origin) that passively bind and concentrate various contaminants including heavy metals onto its cellular structure. Biosorption has been considered as a promising alternative technology for removal of heavy metal ions from wastewater due to easy availability, low cost and high uptake capacity of biosorbents (Kotrba 2011; Soares 2010).

2.1 Microorganisms as Biosrbents

Heavy metal uptake using microorganisms is considered to be a complex process and is dependent on the cell physiology of microorganism, chemistry of metal ions, surface properties of biosorbent and influence of the physical as well as chemical parameters such as pH and metal ion concentration of solutions, temperature and water/moisture content (Volesky et al. 1993; Goyal et al. 2003). Various microorganisms such as yeast, algae, bacteria and fungi have been used to remove the heavy metal ions from aqueous solutions following biosorption technique (Gadd 1990; Holan and Volesky 1995; Volesky and Holan 1995; Vieira and Volesky 2000; Wang and Chen 2006; Svecova et al. 2006; Fiol et al. 2006).

Among different micro-organisms, *Saccharomyces cerevisiae* has been widely used due to its inexpensiveness, easy availability and great ability to remove heavy metals such as cobalt, cadmium, copper, zinc and lead from aqueous solutions (Huang et al. 1990). Biosorption of various heavy metals *viz.*, chromium (Hlihor et al. 2013), copper (Jianlong 2002; Zan et al. 2012), cadmium (Soares et al. 2002; Dai et al. 2008; Zan et al. 2012), lead (Suh et al. 1999a, b; Mapolelo and Torto 2004; Chen and Wang 2008), silver (Chen and Wang 2008), and zinc (Can and Jianlong 2008) onto the *Saccharomyces cerevisiae* has been well documented. Investigations conducted by several researchers demonstrated that percentage removal of heavy metals was higher for flocculent strain of *Saccharomyces cerevisiae* than non-flocculent strain (Soares et al. 2002). Moreover, flocculent strain had ability to accumulate the heavy metal ions on the surface of cells (Avery and Tobin 1992; Ferraz and Teixeira 1999; Marques et al. 1999, 2000; Ferraz et al. 2004).

Özer and Özer (2003) reported the biosorption of lead (II), nickel (II) and chromium (III) ions using an inactive form of *Saccharomyces cerevisiae.* Maximum uptake of lead, nickel and chromium was observed at temperature (25 °C), yeast solution (100 ml/l) and contact time (24 h). The authors reported that biosorption of Pb (II), Ni (II) and Cr (VI) ions onto *Saccharomyces cerevisiae* was a physical adsorption and exothermic in nature. Comparative studies on *Streptococcus equisimilis*, *Saccharomyces cerevisiae* and *Aspergillus niger* for removal of chromium (VI) and ferric (III) was reported by Goyal et al. (2003). Maximum uptake of chromium and ferric ions was observed at pH (2), temperature (30 °C) and biomass (0.75 g/l). Maximum biosorption capacity was observed to be 80.13, 34.5 and 100.3 mg/g for chromium and 19.73, 16.90 and 22.27 mg/g for ferric onto *Streptococcus equisimilis*, *Saccharomyces cerevisiae* and *Aspergillus niger*, respectively. *Aspergillus niger* was found to be the most effective biosorbent for removal of chromium. Contrary to this, Ferraz et al. (2004) reported that chromium uptake reached 80% after 24 h at temperature 30 °C using live form of *Saccharomyces cerevisiae.*

2.2 Plant/Plant Materials as Biosorbents

Biosorption removal of toxic metals from aqueous solutions using plant/plant parts has been observed as an alternate technique to many commercially available processes of adsorption processes. A number of studies have been conducted by researchers to explore different parts of the plant such as *Calotropis procera* roots (Ramalingam et al. 2013), *Terminillia catappa* leaf powder (Rao 2013), *Azadirachta indica* leaf powder (Bhattacharyya et al. 2009), *Tectona grandis* leaf powder (Kumar et al. 2006), *Acacia leucophala* bark and pods (Subbaiah et al. 2009; Dar et al. 2013), *Psidium guajana* bark (Lohani et al. 2008), Salvadora persica branches (Ileri et al. 2014) and *Larrea tridentate* roots, stem and leaves (Gardea-Torresdey et al. 1998) as biosorbents. Banana and orange peels was used as a low cost adsorbent material for removal of heavy metals (Cu, Co, Ni, Zn and Pb) from aqueous solution. The adsorption capacity was found to be 7.97 (Pb), 6.88 (Ni), 5.80 (Zn), 4.75 (Cu) and 2.55 mg/g (Co) onto banana peel and was 7.75 (Pb), 6.01 (Ni), 5.25 (Zn), 3.65 (Cu) and 1.82 mg/g (Co) onto orange peel.

The studies on use of different low cost biosorbents such as rice husk (Sharma and Singh 2008), corn stalk (Zhu et al. 2008), bamboo leaf powder (Mondal et al. 2013), bamboo charcoal (Wang et al. 2012a and b; Zheng et al. 2010), bamboo activated carbon (Khan et al. 2015), cassava root husks (Jorgetto et al. 2014), *Ficus carica* (Gupta et al. 2013) for removal of heavy metals have been well documented. Copper was biosorbed onto *Carica* papaya leaf powder (Varma and Mishra 2016); Cobalt and nickel were removed from aqueous solution using *Tectona grandis* (Vilvanathan and Shanthakumar 2016); biosorption of mercury was done by leaves of *Ricinus communis* L. (Al Rmalli et al. 2008); *Cimmamomum camphora* was used to biosorb copper metal ions (Chen et al. 2010); Zinc was removed from aqueous solution using *Moringa oleifera* Lam. Biomass (Bhatti et al. 2007) where as cadmium using *Moringa oleifera* Lam. Leaves (Ali et al. 2015); Cadmium removal was done using *Syzygium cumini* leaf powder (Rao et al. 2010).

Various natural adsorbents like sunflower stalks (Sun and Shi 1998), maize bran and cob (Singh et al. 2006; Muthusamy et al. 2012), Salvinia biomass (Dhir and Kumar 2010; Dhir et al. 2009), peanuts (Li et al. 2007), wheat straw and bran (Dhir and Kumar 2010; Bulut and Baysal 2006), rice straw and husk (Dhir and Kumar 2010; Mohan et al. 2008), Akhtar et al. 2010), and eucalyptus bark (Sarin and Pant 2006; Ghodbane and Hamdaoui 2008) have been used for removal of various heavy meals from aqueous solution.

2.3 Miscellaneous

Biosorption of some heavy metals like Fe, Ag, Cr and Cd from textile wastewater was carried out using green seaweed biomass by Latinwo et al. (2015). Biosorption of textile waste was carried out using activated and non activated marine algae

(*Gracilaria corticita*) by Sharmila et al. (2016). Cadmium and lead were biosorbed from aqueous solution using mushrooms (Vimala and Das 2009). Agricultural waste has been reported to be used for treatment of waste water containing various heavy metals *viz.*, copper, lead, nickel, chromium and zinc (Mohammed et al. 2014). Efficiency of flyash and commercial activated charcoal were used to remove chromium by Vasanthy et al. 2004). Seven isolates tolerant fungi *viz.*, *Aspergilus versicolor, A. fumigates, Paecilomyces* sp. 9, *Paecilomyces* sp. G, Terichaderma sp., *Microsporum* sp., *Cladosporium* sp. were used to remove cadmium and *Aspergilus versicolor* was observed to be the most efficient isolate for Cd removal (Fazli et al. 2015).

Biosorption/adsorbent capacity of adsorbents has been documented to be enhanced after the treatment with different compounds such as acids, alkaline solutions and chelating agents (Kaewsarn and Yu 2001; Salatnia et al. 2004; Yang and Chen 2008; Zhu et al. 2008). Documentation exists on the usage of various chelating agents *viz.,* thio-urea, polythioether, mercapto, EDTA, citric acid for the modification of biomass to enhance the biosorption capacity of different biosorbents for removal of heavy metal ions from aqueous solutions (Ni et al. 2001). However, affinity of chelating agents towards biosorbates is reported to be dependent upon its physico-chemical characteristics *viz.*, ionic radius, ionic charge and ligand bonding with functional groups on the surface of adsorbents (Szlag and Wolf 1999).

Several researchers observed that modified biosorbents had better sorption capacity than unmodified biosorbents. Different agents to modify the biosorbents were explored in order to enhance their adsorption capacities. It is well documented that heavy metals have capacity to form a complex with different functional groups such as humic or fulvic acids, ligno sulfonates, organic acids and protein (amino acid) onto most of the biosorbents. Various modified/chelating agents such as organic compounds (formaldehyde, Ethylene diamine, methanol, epichlorohydrin), dyes (reactive orange 13), oxidizing agents (hydrogen peroxide), acid solutions (hydrochloric acid, sulfuric acid, nitric acid, tartaric acid, citric acid) and base solution (sodium hydroxide, sodium carbonates, calcium hydroxide) were used for modification of adsorbents to enhance their adsorption capacities (Acar and Eren 2006; Özer 2006; Özer and Pirinc 2006; Reddy et al. 1997; Wong et al. 2003a, b).

3 Mechanism of Biosorption

Various factors like temperature, nature of biomass, pH, dose of biosorbent, initial metal ion concentration, physical-chemical interactions/affinities of biosorent to biosorbate (metal ions) and surface area play the key role to define the quality and efficiency of the biosorbents. Normally, the first step is the physical process i.e. simple surface adsorption which then is followed by the chemical interactions of metal ions with biosorbents. In most of the studies, the biosorption capacity was found to be enhanced with the increased temperature, however, the temperature beyond 40–45 °C resulted in killing of microorganisms, thereby reducing its adsorption rate (Ahalya et al. 2003; White et al. 1997). Nature or characteristics of the biosorbent can be modified by different physical (boiling, drying, autoclaving) and chemical (alkali or acidic) treatments which can enhance the surface porosity to provide the binding sites for different metals.

pH is important parameter for biosorption as it directly influences the nature of biosorbent binding sites as well as metal solubility and thus availability for biosorption. Many microorganisms including algae and bacteria are pH dependent and behave differently with changes in pH of aqueous solutions. The decline in biosorption is observed with decrease in pH from 6 to 2.5. Negligible biosorption of metals was observed at pH less than 2 (Abbas et al. 2014). Initial metal ion concentration acts as a driving force for mass transfer resistance of metal ions between solid and liquid phases (Abbas et al. 2014). Although increase in initial concentration increases the biosorption rate but too high initial concentration results in blocking of the surface binding sites ultimately decreasing the biosorption. The biosorption process involves both physical and chemical interactions of metal ions with biomass. This process can be further enhanced following different treatments like activation following various chemical treatments. The surface area of the adsorbent also plays a very important role in defining efficiency of biosorbent. With increased surface area, the biosorption can be increased (Fig. 2).

Slaiman et al. (2010) carried out a study on adsorption behavior of bamboo for removal of copper and zinc metal ions from aqueous solution. Adsorption behavior of bamboo was conducted using batch adsorption experiments at different conditions *viz.*, pH (3, 4, 4.5, 5 and 6), effect of dosage (0.5, 1 and 1.5 g), mixing speed (90, 111, 131, 156 and 170 rpm), temperature (20, 25, 30 and 35 °C) and metal ions concentration (10, 50, 70, 90 and 100 mg/L). Maximum adsorption capacity of bamboo for copper and zinc was observed to be 8.7 and 8.5 mg/g, respectively. Adsorption effectiveness of bamboo charcoal and iron modified bamboo charcoal (BC-Fe) was investigated for the removal of As (III & V) form aqueous solution by

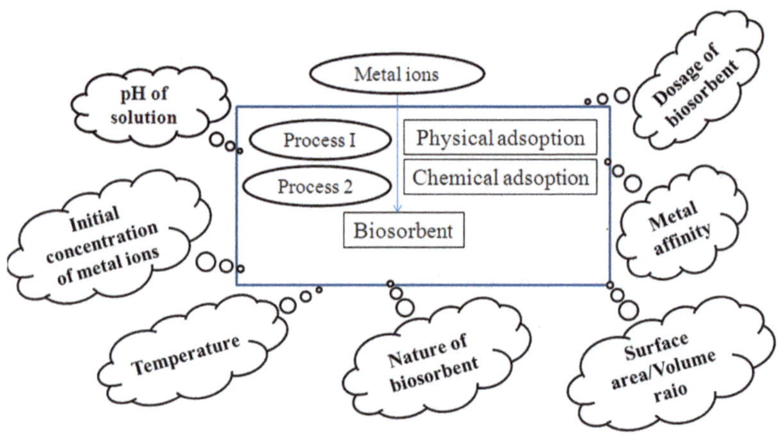

Fig. 2 Process of biosorption and factors affecting the biosorption

Liu et al. (2012). Ferric salt solution was used for the modification of bamboo charcoal. From surface analysis, it was observed that surface of bamboo charcoal and modified bamboo charcoal was highly porous which provided a large surface area for accumulation of arsenic metal ions. As (III) and As (V) removal was maximum at ranging from pH 4–5 and 3–4, respectively.

Wang et al. (2012b) reported the removal of lead from aqueous solution using $KMnO_4$ modified bamboo charcoal. $KMnO_4$ modified bamboo charcoal was prepared with microwave irradiation. The surface studies of modified bamboo charcoal were characterized by N_2 adsorption, XRD, FTIR, SEM, EDS and pH_{zpc}. Surface area of modified bamboo charcoal was observed to be higher (172.3 m^2/g) than the unmodified bamboo charcoal (15.5 m^2/g). The adsorption behavior of lead using modified bamboo charcoal was found to be maximum at pH (5), contact time (600 min), and temperature (298 K) of the solution. Maximum adsorption capacity of lead was found to be 25.03 mg/g for bamboo charcoal and 55.56 mg/g for modified bamboo charcoal.

4 Application

Wide applicability of biosorption technique has been witnessed in the field of remediation for water pollution. The technique is not only applicable for removal of heavy metals but also for pesticides, dyes and other inorganic pollutants. Hameed et al. (2007) studied adsorption of methylene blue using bamboo-based activated carbon (850 °C for 2 h). Hameed and El-Khaiary (2008) studied the adsorption of malachite green using activated carbon of bamboo. Activated bamboo was prepared following the physical and chemical activation processes using CO_2 and K_2CO_3. Wang (2012) studied the adsorption of azo disperse dye using activated charcoal derived from "waste" bamboo culms. Maximum removal of dye was observed at pH (1).

5 Conclusion

The present review deals with potential of different adsorbents of biological origin towards the removal of various heavy metals. Although there are many conventional and commercial methods available for removal of heavy metals from the aqueous solutions yet, biosorption has been designated as one of the most economic, viable, potent last but not least, eco-friendly technique for remediation of polluted water.

References

Abbas, S. H., Ismail, I. M., Mostafa, T. M., & Sulaymon, A. H. (2014). Biosorption of heavy metals: A review. *Journal of Chemical Science and Technology, 3*, 74–102.

Acar, F. N., & Eren, Z. (2006). Removal of Cu (II) ions by activated poplar sawdust (*Samsun clone*) from aqueous solutions. *Journal of Hazardous Materials B, 137*, 909–914.

Ahalya, N., Ramachandra, T. V., & Kanamadi, R. D. (2003). Biosorption of heavy metals. *Research Journal of Chemistry and Environment, 7*, 71–79.

Akhtar, M., Iqbal, S., Kausar, A., Bhanger, M. I., & Shaheen, M. A. (2010). An economically viable method for the removal of selected divalent metal ions from aqueous solutions using activated rice husk. *Colloids and Surfaces B: Biointerfaces, 75*, 149–155.

Al Rmalli, S. W., Gahmani, A. A., Abuein, M. M., & Gleza, A. A. (2008). Biosorption of mercury from aqueous solutions by powdered leaves of castor tree (*Ricinus communis* L.). *Journal of Hazardous Materials, 152*, 955–959.

Ali, E. N., Alfarra, S. R., Yusoff, M. M., & Rahman, M. L. (2015). Environmentally friendly biosorbent from *Moringa oleifera* leaves for water treatment. *International Journal of Environmental Svcience and Development, 6*, 165–169.

Asberry, H. B., Kuo, C., Gung, C., Conte, E. D., & Suen, S. (2014). Characterization of water bamboo husk biosorbents and their application in heavy metals ion trapping. *Microchemical Journal, 113*, 59–63.

Avery, S. V., & Tobin, J. M. (1992). Mechanism of strontium uptake by laboratory and brewing strains of *Saccharomyces cerevisiae*. *Applied Environmental Microbiology, 58*, 3883–3889.

Baraket, M. A. (2011). New trends in removing heavy metals from industrial wastewater. *Arabian Journal of Chemistry, 4*, 361–377.

Bhattacharyya, K. G., Sarma, J., & Sarma, A. (2009). *Azadirachta indica* leaf powder as a biosorbent for Ni (II) in aqueous medium. *Journal of Hazardous Materials, 165*, 271–278.

Bhatti, H. N., Mumtaz, B., Hanif, M. A., & Nadeem, R. (2007). Removal of Zn (II) ions from aqueous solution using *Moringa oleifera* Lam. (horseradish tree) biomass. *Process Biochemistry, 42*, 547–553.

Bulut, Y., & Baysal, Z. (2006). Removal of Pb (II) from wastewater using bran. *Journal of Environmental Management, 78*, 107–113.

Can, C., & Jianlong, W. (2008). Investigating the interaction mechanism between zinc and *Saccharomyces cerevisiae* using combined SEM – EDX and XAFS. *Applied Microbiology and Biotechnology, 79*, 293–299.

Cardoso, V. A., de Souza, A. G., Sartoratto, P. P. C., & Nunes, L. M. (2004). The ionic exchange process of cobalt, nickel and copper (II) in alkaline and acid – layered titanates. *Colloids and Surface A: Physicochemical and Engineering Aspects, 248*, 145–149.

Chandra, S., Chauhan, L. K. S., Murthy, R. C., Saxena, P. N., Pande, P. N., & Gupta, S. K. (2005). Comparative biomonitoring of leachates from hazardous solid waste of two industries using *Allium test. Science of the Total Environment, 347*, 46–52.

Chen, C., & Wang, J. (2008). Removal of Pb^{+2}, Ag^{+}, Cs^{+} and Sr^{+2} from aqueous solution by brewery's waste biomass. *Journal of Hazardous Materials, 151*, 65–70.

Chen, Y., Wang, Z., & Huang, S. (2004). Assessment of the contamination and Genotoxicity of soil irrigated with wastewater. *Plant and Soil, 56*, 189–196.

Chen, H., Dai, G., Zhao, J., Zhong, A., Wu, J., & Yan, H. (2010). Removal of copper (II) ions by a biosorbent – *Cinnamomum camphora* leaves powder. *Journal of Hazardous Materials, 177*, 228–236.

Council Directive: Pollution caused by certain dangerous substances discharged into the aquatic environment of the Community, 1976. (76/464/EEC).

Dai, S., Wei, D., Zhou, D., Jia, C., Wang, Y., & Liu, W. (2008). Removing cadmium from electroplating wastewater by waste *Saccharomyces cerevisiae*. *Transactions of Nonferrous Metals Society of China, 18*, 1008–1013.

Dar, B. A., Taher, A., Wani, A., & Farooqui, M. (2013). Isotherms and thermodynamic studies on adsorption of copper on powder of shed pods of *Acacia nilotica*. *Journal of Environmental Chemistry and Ecotoxicology, 5*, 17–20.

Dhir, B., & Kumar, R. (2010). Adsorption of heavy metals by *Salvinia* biomass and agricultural residues. *International Journal of Environmental Research, 4*, 427–432.

Dhir, B., Sharmila, P., Pardha Saradhi, P., & Nasim, S. A. (2009). Physiological and antioxidant responses of *Salvinia natans* exposed to chromium rich wastewater. *Ecotoxicology and Environmental Safety, 72*, 1790–1797.

El-Sayed, M. T. (2012). The use of *Saccharomyces cerevisiae* for removing cadmium (II) from aqueous waste solutions. *African Journal of Microbiology Research, 6*, 6900–6910.

Farooq, U., Kosinzki, J. A., Khan, M. A., & Athar, M. (2010). Biosorption of heavy metal ions using wheat based biosorbents – A review of the recent literature. *Bioresource Technology, 101*, 5043–5053.

Fazli, M. M., Soleimani, N., Mehrasbi, M., Darabian, S., Mohammadi, J., & Ramazani, A. (2015). Highly cadmium tolerant fungi: Their tolerance and removal potential. *Journal of Environmental Health, 13*, 19–27.

Ferraz, A. I., & Teixeira, J. A. (1999). The use of flocculating brewer's yeast for Cr (III) and Pb (II) removal from residual wastewaters. *Bioprocess Engineering, 21*, 431–437.

Ferraz, A. I., Tavares, T., & Teixeira, J. A. (2004). Cr (III) removal and recovery from *Saccharomyces cerevisiae*. *Chemical Engineering Journal, 105*, 11–20.

Fiol, N., Villaescusa, I., Martinez, M., Miralles, N., Poch, J., & Serarols, J. (2006). Sorption of Pb (II), Ni (II), Cu (II) and Cd (II) from aqueous solution by olive stone waste. *Separation and Purification Technology, 50*, 132–140.

Gadd, G. M. (1990). Heavy metal accumulation by bacteria and other microorganisms. *Experientia, 46*, 834–840.

Gadd, G. M. (2010). Metals, minerals and microbes: Geomicrobiology and bioremediation. *Microbiology, 156*, 609–643.

Gardea-Torresdey, J. I., Hernandez, A., Tiemann, K. J., Bibb, J., & Rodriguez, O. (1998). Adsorption of toxic metal ions from solution by inactivated cells of *Larrea tridentate* (Creosote Bush). *Journal of Hazardous Substance Research, 1*, 13–16.

Garg, S. K. (2010). *Ecology and environmental studies*. New Delhi: Khanna Publishers.

Garg, S. K., Garg, R., & Garg, R. (2006). *Ecological and environmental studies*. New Delhi: Khanna Publishers.

Ghodbane, I., & Hamdaoui, O. (2008). Removal of mercury (II) from aqueous media using *eucalyptus* bark: Kinetic and equilibrium studies. *Journal of Hazardous Materials, 160*, 301–309.

Goyal, N., Jain, S. C., & Banerjee, U. C. (2003). Comparative studies on the microbial adsorption of heavy metals. *Advances in Environmental Research, 7*, 311–319.

Gupta, V. K., & Ali, I. (2008). Removal of endosulfan and methoxychlor from water on carbon slurry. *Environmental Science and Technology, 42*, 766–770.

Gupta, V. K., & Rastogi. (2009). Biosorption of hexavalent chromium by raw and acid-treated green alga Oedogonium hatei from aqueous solutions. *Journal of Hazardous Materials, 163*, 396–402.

Gupta, V. K., Pathania, D., Agarwal, S., & Sharma, S. (2013). Removal of Cr (IV) onto *Ficus carica* biosorbent from water. *Environment Science and Pollution Research, 20*, 2632–2644.

Hameed, B. H., & El-Khaiary, M. I. (2008). Equilibrium and kinetics mechanism of malachite green adsorption on activated carbon prepared from bamboo by K_2CO_3 activation and subsequent gasification with CO_2. *Journal of Hazardous Materials, 157*, 344–351.

Hameed, B. H., Din, A. T. M., & Ahmad, A. L. (2007). Adsorption of methylene blue onto bamboo – Based activated carbon: Kinetics and equilibrium studies. *Journal of Hazardous Materials, 141*, 819–825.

Hlihor, R. M., Diaconu, M., Fertu, D., Chelaru, C., Sandu, I., Tavares, T., & Gavrilescu, M. (2013). Bioremediation of Cr (VI) polluted wastewaters by sorption on heat inactivated *Saccharomyces cerevisiae* biomass. *International Journal of Environmental Research, 7*, 581–594.

Holan, Z., & Volesky, B. (1995). Accumulation of cadmium, lead and nickel by fungal and wood biosorbents. *Applied Biochemistry and Biotechnology, 53*, 133–146.

Huang, C., Huang, C. P., & Morehart, A. L. (1990). The removal of Cu (II) from dilute aqueous solutions by *Saccharomyces cerevisiae*. *Water Research, 24*, 433–439.

Ibrahim, S. C., Hanafiah, M. A. K. M., & Yahya, M. Z. A. (2006). Removal of cadmium from aqueous solutions by adsorption onto sugarcane bagasse. *American – Eurasian Journal of Agriculture and Environmental Science, 1*, 179–184.

Ileri, O., Cay, S., Uyanik, A., & Erduran, N. (2014). Removal of common heavy metals from aqueous solutions by waste *Salvadora persica* L. branches (Miswak). *International Journal of Environmental Research, 8*, 987–996.

Jianlong, W. (2002). Biosorption of copper (II) by chemically modified biomass of *Saccharomyces cerevisiae*. *Process Biochemistry, 37*, 847–850.

Jorgetto, A. O., Silva, R. I. V., Saeki, M. J., Barbosa, R. C., Martines, M. A. U., Jorge, S. M. A., Silva, A. C. P., Schneider, J. F., & Castro, G. R. (2014). Cassava root husks powder as green adsorbent for the removal of Cu (II) from natural river water. *Vérsila Biblioteca Digital, 288*, 356–362.

Kaewsarn, P., & Yu, Q. (2001). Cadmium (II) removal from aqueous solutions by pre-treated biomass of marine alga *Padina* sp. *Environmental Pollution, 112*, 209–213.

Kandah, M. I. (2004). Zinc and cadmium adsorption on low grade phosphate. *Separation and Purification Technology, 35*, 61–70.

Khan, M. A., Alemayehu, A., Duraisamy, R., & Berekete, A. K. (2015). Removal of lead ions from aqueous solution by bamboo activated carbon. *International Journal of Water Research, 5*, 33–46.

Klaassen, E. D. (2001). (2001). Heavy metal and heavy metal antagonists. In J. G. Hardman, L. E. Limbird, & A. G. Gilman (Eds.), *Goodmna and Gilman's: The pharmacological basis of therapeutices* (9th ed., pp. 1851–1875). New York: McGraw Hill.

Kotrba, P. (2011). Microbial biosorption of metals-general introduction. In P. Kotrba, M. Mackova, & T. Macek (Eds.), *Microbial biosorption of metals* (pp. 1–6). Dordrecht: Springer.

Kumar, Y. P., King, P., & Prasad, V. S. R. K. (2006). Equilibrium and kinetic studies for the biosorption system of copper (II) ion from aqueous solution using *Tectona grandis* L.f. leaves powder. *Journal of Hazardous Materials B, 137*, 1211–1217.

Latinwo, G. K., Jimoda, L. A., Aggary, S. E., & Adeniran, J. A. (2015). Biosorption of some heavy metals from textile wastewater by green seaweed biomass. *Universal Journal of Environmental Research and Technology, 5*, 210–219.

Li, Q., Zhai, J., Zhang, W., Wang, M., & Zhou, J. (2007). Kinetic studies of adsorption of Pb (II), Cr (III) and Cu (II) from aqueous solution by sawdust and modified peanut husk. *Journal of Hazardous Materials, 141*, 163–167.

Liu, X., Ao, H., Xiong, X., Xiao, J., & Liu, J. (2012). Arsenic removal from water by iron-modified bamboo charcoal. *Water, Air & Soil, 223*, 1033–1044.

Lohani, M. B., Singh, A., Rupainwar, D. C., & Dhar, D. N. (2008). Studies on efficiency of guava (*Psidium guajava*) bark as biosorbent for removal of Hg (II) from aqueous solutions. *Journal of Hazardous Materials, 159*, 626–629.

Mapolelo, M., & Torto, N. (2004). Trace enrichment of metal ions in aquatic environments by *Saccharomyces cerevisiae*. *Talanta, 64*, 39–47.

Marques, P. A., Pinheiro, H. M., Teixeira, J. A., & Rosa, M. F. (1999). Removal efficiency of Cu (II), Cd (II) and Pb (II). By waste brewery biomass: pH and cation association effects. *Desalination, 124*, 137–144.

Marques, P. A. S. S., Rosa, M. F., & Pinheiro, H. M. (2000). pH effects on the remove of Cu^{2+}, Cd^{2+} and Pb^{2+} from aqueous solution by waste brewery biomass. *Bioprocess Engineering, 23*, 135–141.

Mathur, N., & Bhatnagar, P. (2005). Mutagenicity assessment of textile dyes from Sanganer (Rajasthan). *Journal of Environmental Biology, 28*, 123–126.

Melckova, I., & Ruzovic, T. (2010). Biosorption of zinc from aqueous solution using algae and plant biomass. *Nova Biotechnoligica, 10*, 33–43.
Metcalf, E. (2003). *Wastewater engineering: Treatment, disposal and reuse*. New York: McGraw-Hill.
Mohammed, M. A., Shitu, A., Tadda, M. A., & Ngaura, M. (2014). Utilization of various agricultural waste materials in the treatment of industrial wastewater containing heavy metals: A review. *International Research Journal of Environmental Sciences, 3*, 62–71.
Mohan, S., Gandhimathi, R., & Sreelakshmi, G. (2008). Isotherm studies for heavy metal adsorption on rice husk. *Asian Journal of Water Environment and Pollution, 5*, 71–78.
Mondal, D. K., Nandi, B. K., & Purkait, M. K. (2013). Removal of mercury (II) from aqueous solution using bamboo leaf powder: Equilibrium, thermodynamic and kinetic studies. *Journal of Environmental Chemical Engineering, 1*, 891–898.
Mouchet, F., Gauthier, L., Mailhes, C., Jourdian, M. J., Ferrier, V., & Triffault, G. (2006). Biomonitoring of the Genotoxic potential of aqueous extracts of solids and bottom ash resulting from municipal solid waste incineration using the comet and micronucleus tests on amphibian (Xenopus laevis) larvae and bacterial assays (Mutatox R and Ames tests). *Science of the Total Environment, 355*, 232–246.
Muthusamy, P., Murugan, S., & Smitha, M. (2012). Removal of nickel ions from industrial waste water using maize cob. *Journal of Biological Sciences, 1*, 7–11.
Ni, C., Yi, C., & Feng, Z. (2001). Studies of synthesis and adsorptive properties of chelating resin from thiourea and formaldehyde. *Journal of Applied Polymer Science, 82*, 3127–3132.
Ozer, A. (2006). Removal of Pb (II) ions from aqueous solutions by sulfuric acid treated wheat bran. *Journal of Hazardous Materials B, 141*, 753–761.
Ozer, A., & Ozer, D. (2003). Comparative study of biosorption of lead (II), nickel (II) and chromium (VI) ions onto *S. cerevisiae*: Determination of biosorption heats. *Journal of Hazardous Materials B, 100*, 219–229.
Ozer, A., & Pirinc, c. i. H. B. (2006). The adsorption of cadmium (II) ions on sulfuric acid treated wheat bran. *Journal of Hazardous Materials B, 137*, 849–855.
Ramalingam, S. J., Khan, T. H., Pugazhlenthi, M., & Thirumurugan, V. (2013). Removal of Pb (II) and Cd (II) ions from industrial wastewater using *Calotropis procera* roots. *International Journal of Engineering Science Invention, 2*, 1–6.
Rao, L. N. (2013). Adsorption of lead and zinc from aqueous solution using *Terminalia catappa* L. as adsorbent. *International Journal of Engineering Research and Science and Technology, 2*, 65–76.
Rao, K. S., Anand, S., & Venkateshwarlu, P. (2010). Cadmium removal from aqueous solutions using biosorbent *Syzygium cumini* leaf powder: Kinetic and equilibrium studies. *Korean Journal of Chemical Engineering, 27*, 1547–1554.
Reddy, B. R., Mirghaffari, N., & Gaballah, I. (1997). Removal and recycling of copper from aqueous solutions using treated Indian barks. *Resources Conservation Recycling, 21*, 227–245.
Salatnia, A., Madani, A., Bakhti, M. Z., Kertous, L., Mansouri, Y., & Yous, R. (2004). Biosorption of Ni^{2+} from aqueous solution by a NaOH-treated bacterial dead *Streptomyces rinosus* biomass. *Minerals Engineering, 17*, 903–911.
Samuel, O. B., Osuala, F. I., & Odeigah, G. C. (2010). Cytogenotoxicity evaluation of two industrial effluents using *Allium cepa* assay. *African Journal of Environmental Science and Technology, 4*, 21–27.
Sarin, V., & Pant, K. K. (2006). Removal of chromium form industrial waste by using eucalyptus bark. *Bioresources Technology, 97*, 15–20.
Sharma, N., & Singh, J. (2008). Removal of Zn^{+2} ions from aqueous solution using rice (*Oryza sativa*) husk in sequential bed adsorption column. In *Proceeding of Taal 2007: The 12th World Lake Conference* (pp. 944–951).
Sharmila, S., Amaraselvum, K., Rebecca, L. J., & Kowsalya, E. (2016). Biosorption of textile effluent using marine algae. *International Journal of Pharmaceutical Sciences Review and Research, 39*, 108–111.

Sirilamduan, C., Umpuch, C., & Kaewsarn, P. (2011). Removal of copper from aqueous solutions by adsorption using modify *Zalacca edulis* peel modify. *Songklanakarin Journal of Science and Technology, 33*, 725–732.
Singh, K. K., Talat, M., & Hasan, S. H. (2006). Removal of lead from agqueous solutions by agricultural waste maize bran. *Bioresource Technology., 97*, 2124–2130.
Slaiman, Q. J. M., Haweel, C. K., & Abdulmajeed, Y. R. (2010). Removal of heavy metal ions from aqueous solutions using biosorption onto bamboo. *Iraqi Journal of Chemical and Petroleum Engineering, 11*, 23–32.
Soares, E. V. (2010). Flocculation in *Saccharomyces cerevisiae*: A review. *Journal of Applied Microbiology., 110*, 1–18.
Soares, E. V., Coninck, G. D., Duarte, F., & Soares, H. M. V. M. (2002). Use of *Saccharomyces cerevisiae* for Cu^{+2} removal from solution: The advantages of using a flocculent strain. *Biotechnology Letters, 24*, 663–666.
Subbaiah, M. V., Vijaya, Y., Kumar, N. S., Reddy, A. S., & Krishnaiah, A. (2009). Biosorption of nickel from aqueous solutions by *Acacia leucocephala* bark: Kinetics and equilibrium studies. *Colloids and Surfaces B: Biointerfaces, 74*, 260–265.
Suh, J. H., Yun, J. W., & Kim, D. S. (1999a). Cation (K^+, Mg^{2+}, Ca^{2+}) exchange in Pb^{+2} accumulation by *Saccharomyces cerevisiae. Process Biochemistry, 21*, 383–387.
Suh, J. H., Yun, J. W., & Kim, D. S. (1999b). Effects of pH on Pb^{+2} accumulation in *Saccharomyces cerevisiae* and *Aureobasidium pullulans. Bioprocess Engineering, 20*, 471–474.
Sun, G., & Shi, W. (1998). Sunflower stalks as adsorbents for removal of metal ions from wastewater. *Industrial Engineering and Chemical Research, 37*, 1324–1328.
Svecova, L., Spanelova, M., Kubal, M., & Guibal, E. (2006). Cadmium, lead and mercury biosorption on waste fungal biomass issued from fermentation industry: Equilibrium studies. *Separation and Purification Technology, 52*, 142–153.
Szlag, D. C., & Wolf, N. J. (1999). Recent advances in ion exchange materials and processes for pollution prevention. *Clean Products and Processes, 1*, 117–131.
Tsezos. (1999). Biosorption of metals. The experience accumulated and the outlook for technology development. *Process Metallurgy, 9*, 171–173.
UK Red List Substances: Environmental Protection (Prescribed Processes and Substances) Regulations, 1991 (SI 1991/472).
USEPA (US Environmental Protection Agency). (1997). *Exposure factors handbook–general factors. EPA/600/P-95/002Fa, vol. I. Office of Research and Development. National Center for Environmental Assessment.* Washington, DC: US Environmental Protection Agency.
Varma, G. V., & Misra, A. K. (2016). Equilibrium and kinetic studies on adsorption of copper on to carca papaya leaf powder. *J Colloid Interface Sci, 7*, 403–416.
Vasanthy, M., Sangeetha, M., & Kalaiselvi, R. (2004). A comparative study on the chromium removal efficiency of flyash and commercial activated carbon. *Journal of Industrial Pollution Control, 20*, 37–44.
Vieira, R. H. S. F., & Volesky, B. (2000). Biosorption: A solution to pollution. *International Microbiology, 3*, 17–24.
Vilvanathan, S., & Shanthakumar, S. (2016). Removal of Ni(II) and Co(II) ions from aqueous solution using teak (Tectona grandis) leaves powder: Adsorption kinetics, equilibrium and thermodynamics study. *Desalination and Water Treatment, 57*, 3995–4007.
Vimala, R., & Das, N. (2009). Biosorption of cadmium (II) and lead (II) from aqueous solutions using mushrooms: A comparative study. *Journal of Hazardous Materials, 168*, 376–382.
Volesky, B., & Holan, Z. R. (1995). Biosorption of heavy metals. *Biotechnology Progress, 11*, 235–250.
Volesky, B., May, H., & Holan, Z. R. (1993). Cadmium bosorption by *Saccharomyces cerevisiae. Biotechnology and Bioengineering, 41*, 826–829.
Wang, L. (2012). Application of activated carbon derived from 'waste' bamboo culms for the adsorption of azo disperse dye: Kinetic, equilibrium and thermodynamic studies. *Journal of Environmental Management., 102*, 79–87.

Wang, J. L., & Chen, C. (2006). Biosorption of heavy metals by *Saccharomyces cerevisiae*: A review. *Biotechnology Advance, 24*, 427–451.

Wang, Y., Wang, X., Wang, X., Liu, M., Yang, L., Wu, Z., Xia, S., & Zhao, J. (2012a). Adsorption of Pb (II) in aqueous solutions by bamboo charcoal modified with $KMnO_4$ via microwave irradiation. *Colloids and Surfaces A: Physicochemical and Engineering Aspects, 414*, 1–8.

Wang, Y., Wang, X. J., Liu, M., Wang, X., Wu, Z., Yang, L. Z., Xia, S. Q., & Zhao, J. F. (2012b). Cr (VI) removal from water using cobalt – Coated bamboo charcoal prepared microwave heating. *Industrial Crops and Products, 39*, 81–88.

White, C., Sayer, J. E., & Gadd, G. M. (1997). Microbial solubilization and immobilization of toxic metals: Key biogeochemical processes for treatment of contamination. *FEMS Microbiology Reviews, 20*, 503–516.

Wong, K. K., Lee, C. K., Low, K. S., & Haron, M. J. (2003a). Removal of Cu and Pb by tartaric acid modified rice husk from aqueous solutions. *Chemosphere, 50*, 23–28.

Wong, K. K., Lee, C. K., Low, K. S., & Haron, M. J. (2003b). Removal of Cu and Pb from electroplating wastewater using tartaric acid modified rice husk. *Process Biochemistry, 39*, 437–445.

Yang, L., & Chen, J. P. (2008). Biosorption of hexavalent chromium onto raw and chemical modification *Sargassun* sp. *Bioresource Technology, 99*, 297–307.

Zan, F., Huo, S., Xi, B., & Zhao, X. (2012). Biosorption of Cd^{+2} and Cu^{+2} on immobilized *Saccharomyces cerevisiae*. *Frontier Environmental Science and Engineering, 6*, 51–58.

Zheng, L., Dang, Z., Yi, X., & Zhang, H. (2010). Equilibrium and kinetic studies of adsorption of Cd (II) from aqueous solution using modified corn stalk. *Journal of Hazardous Materials, 176*, 650–656.

Zhu, B., Fan, T., & Zhang, D. (2008). Adsorption of copper ions from aqueous solution by citric acid modified soybean straw. *Journal of Hazardous Materials, 153*, 300–308.

Biotechnological Intervention as an Aquatic Clean Up Tool

Moonisa Aslam Dervash, Rouf Ahmad Bhat, Sadiqa Shafiq, Dig Vijay Singh, and Nighat Mushtaq

Abstract Although three quarters of Earth is occupied by water but quantity of available fresh water is limited. In a vast arena of environmental issues during the present era, aquatic pollution is one of the major problems. In order to curb the growing concern of aquatic pollution, biotechnological interventions provide distinguished avenues in the form of novel techniques of remediation (biodegradation, biostimulation, blastofiltration, cyanoremediation, biosparging and mycoremediation). And in order to hold back effluence of pollutants into aquatic environs, biotechnological gadgets (biological fuel cells and biosensors) are quite helpful to achieve sustainable development.

Keywords Biosensor · Heavy metals · Remediation · Biosorption · Blast filtration · Biofuels

1 Introduction

Biotechnological interventions are one of the vital options that offer the possibility to destroy or render various contaminants into harmless entities using natural biological activity (Gupta and Mahapatra 2003). It includes all the innovative progressions utilized in enrichment and mediation of activities of biological entities.

M. A. Dervash (✉) · R. A. Bhat
Division of Environmental Science, Sher-e-Kashmir University of Agricultural Sciences and Technology, Shalimar, India

S. Shafiq
Institute of Home Science, University of Kashmir, Srinagar, Jammu and Kashmir, India

D. V. Singh
School of Environmental Science, DES, Babasaheb Bhimrao Ambedkar University, Lucknow, India

N. Mushtaq
Division of Vegetable Science, Sher-e-Kashmir University of Agricultural Sciences and Technology, Shalimar, India

H. Qadri et al. (eds.), *Fresh Water Pollution Dynamics and Remediation*,
https://doi.org/10.1007/978-981-13-8277-2_11

Aquatic environs are facing tremendous pressures due to continuous inputs of pollutants; oil spillage and solid waste pollution (Bhat et al. 2012, 2014), heavy metal pollution (Mehmood et al. 2019), plastic pollution (Bhat et al. 2018) which alter the physicochemical (Bhat et al. 2017a, b) and biological characteristics as well as aesthetic values across the globe. In order to curb the growing aquatic pollution, biotechnological clean up techniques present various strategies broadly summed up into single consortium called 'Bioremediation'. Bioremediation can operate in situ or ex situ by either removing the contaminants from the substratum (decontamination or cleanup techniques) or reducing exposure (stabilization techniques) which can otherwise pose risk through contamination (Vangronsveld et al. 2009). There are various bio-techniques which include biosorption (Mustapha and Halimoon 2015), biodegradation (Adams and Reddy 2003), biostimulation (Tang et al. 2007), blastofiltration (Conesa et al. 2012), cyanoremediation (Fiset et al. 2008), biosparging (Adams and Reddy 2003) and mycoremediation (Rhodes, 2014). And in order to achieve sustainable development, various innovative biotechnological interventions illuminate the present day world with the advent of bioelectricity through biological fuel cells (Wang and Ren 2013; Logan and Rabaey 2012) and production of biodiesel by exploitation of various microbes (Xiong et al. 2008). And in order to detect pollution levels in aquatic ecosystems, novel gadgets 'biosensors' (Nigam and Shukla 2015) are emerging as innovative biotechnological intervention.

2 Microbial Biosorption

Biosorption can be defined as the removal of metal or metalloid species, compounds and particulates from contaminated aquatic stream by low cost biological materials (Wang and Chen 2009; Mustapha and Halimoon 2015). All biological materials can be useful biosorbents for metals sequestration with the exception of mobile alkali metal cations like sodium and potassium ions, and this can be a significant passive process in living and dead organisms (Gadd 2010; Mustapha and Halimoon 2015). Several cheap biosorbents for the removal of metals mainly arrive under the following categories: bacteria, fungi, algae, plants, industrial wastes, agricultural wastes and other polysaccharide materials (Schiewer and Patil 2008; Kumar et al. 2014; Mustapha and Halimoon 2015). The major advantages of biosorption over conventional treatment methods include small expenditure, high competence, and regeneration of biosorbents and recovery of metals (Azouaou et al. 2008; Sud et al. 2008).

Bacterial biosorption is chiefly used for the elimination of pollutants from effluents contaminated with recalcitrant pollutants (metals ions and dyes). However, their isolation, screening and harvesting on a larger scale may be complicated but still remain one of the efficient ways of remediating pollutants. Different bacterial strains possess different sorption capabilities (Mustapha and Halimoon 2015). Algae are proficient and cheap biosorbents as the nutrient requirement by algae is minute. Based on statistical analysis on algae potentiality in biosorption, it has been reported that algae absorb about 15.3–84.6% which is higher as compared to other microbial biosorbents (Mustapha and Halimoon 2015). Biosorption of metal ions occurs on the cell surface by means of ion exchange method. Brown marine algae has the capacity to absorb metals like Cd, Ni and Pb through chemical groups on their surface such as carboxyl, Sulfonate, amino, as well as sulfhydryl (Mustapha and Halimoon 2015). Likewise, capability of the many type of fungi to produce extracellular enzymes for the assimilation of complex carbohydrates for former hydrolysis makes capable the degradation of various degrees of pollutants with the benefit of being relatively uncomplicated to grow in fermenters, therefore being appropriate for large scale production. In comparison to yeasts, filamentous fungi are less sensitive to variations in nutrients, aeration, pH, temperature and have a lower nucleic content in the biomass (Leitao 2009; Li et al. 2015; Mustapha and Halimoon 2015) (Table 1).

Table 1 Metal biosorption capacity of microorganisms

Metal	Microbial species	Biosorption capacity (mg/g)	References
Zn	*Pseudomonas putida*	17.7	Mustapha and Halimoon (2015), Freitas et al. (2008), Fan et al. (2008), and Ghosh et al. (2016)
	Sargassum muticum	34.10	
	Penicillium simpliccium	65.60	
Cu	*Enterobacter* sp. J1	32.5	Lu et al. (2006), Celekli et al. (2010), Dursun (2006), and Infante et al. (2014)
	Spirulina platensis	67.93	
	Aspergillus niger	28.7	
	Penicillium chrysogenum	92.0	
Cd	*Pseudomonas fluorescence*	40.8	Uzel and Ozdemir (2009), Li et al. (2010), Infante et al. (2014), Katsumata et al. (2003), and Tian et al. (2014)
	Chlorella miniata	34.60	
	Penicillium purpurogenum	36.5	
Ni	*E. coli*	6.9	Quintelas et al. (2009), Uzel and Ozdemir (2009), and Tan et al. (2004)
	Pseudomonas fluorescence	40.8	
	Penicillium chrysogenum	260	
Cd	*Enterobacter* sp. J1	46.2	Quintelas et al. (2009), Bulgariu et al. (2013), and Fan et al. (2008)
	Ulva lactuca sp.	43.02	
	Penicillium simpliccium	52.50	
Pb	*Aspergillus Niger*	34.4	Zeng et al. (2015) and Mustapha and Halimoon (2015)
	Penicillium chrysogenum	204	

3 Biodegradation

Biodegradation of wastes in aerobic environment (Adams and Reddy 2003; Bhat et al. 2018) involves activity of microbes after minutely examining and understanding biogeochemical processes and genomics relevant to microbial consortium (Chauhan and Jain 2010; Jeffries et al. 2012; Rayu et al. 2012; Tyagi et al. 2011).

4 Biostimulation

Microorganisms (particularly bacteria and fungi) are natural bio-transformers with unique capability to transform variety of xenebiotics into sources of energy (Tang et al. 2007). Biostimulation (in situ bioremediation) involves injection of nutrients and allied components in order to induce propagation at a hastened rate of accidental spills and unceasingly polluted ecosystems across the world (Cheng et al. 2009; Tyagi et al. 2011). Bioremediation potential of biologically activated microbial culture is possible to be isolated from heavy metals waste disposal contaminated sites (Fulekar et al. 2012). Supplementing nitrogen, phosphorus (Bhat et al. 2017a, b) oxygen as electron acceptor, and various substrates, in addition to introducing microorganisms with desired catalytic capabilities basically designs framework for bio-stimulation (Ma et al. 2007; Baldwin et al. 2008). In order to bioremediate chromium, heterogeneous group of bacteria isolated from various polluted environs possess unique plasmid-mediated chromate resistance which eventually result in enzymatically mediated reduction Kanmani et al. (2012). With the advent of molecular engineering, strains with improved traits are derived which can withstand stressful conditions. In this context, bacterial merC gene (a potential molecular tool) was isolated for improving the efficiency of cadmium phytoremediation Kiyono et al. (2012). Similarly, sulfate-reducing bacteria (SRB) were incorporated into a column reactor containing sulfate and heavy metals (As, Cd, Cr, Cu and Zn) which resulted in 50% abatement of sulfate, whereas, heavy metals were totally removed from the bioreactor (Viggi et al. 2010).

5 Blast Filtration

Blastofiltration is a biological clean up technique in which aerated water precipitate and heavy metal contaminated effluent is allowed to pass through selected seedlings (Conesa et al. 2012). Through blastofiltration, it is possible to decrease the concentration of 100 mg/L Pb to 5 mg/L Pb in 72 h using various crop seedlings (Lin et al. 2002). Extensive research on accumulative power of various crops is demanded to remediate heavy metal contaminated aquatic ecosystems.

6 Cyanoremediation

Incorporation of cyanobacteria to remediate heavy metals from polluted aquatic ecosystems and wetlands as well (Fiset et al. 2008), is known as cyanoremediation (Kumar and Singh 2017; Khanday et al. 2016). Cyanoremediation involves wild stains, mutant or genetically designed stains of cyanobacteria (Yin et al. 2012). In this context, *Synechocysis* sp. PCC6803 (blue alga) possess ability to accumulate arsenic (Yin et al. 2012). Likewise, *Spirogyra* sp. bioaccumulate Cr (98.23%), Cu (89.6%), Fe (99.73%), Mn (99.6%), Se (98.16%) and Zn (81.53%) (Mane and Bhosle 2012), whereas, *Spirulina* sp. possess potential to bioremove Cr (98.3%), Cu (81.2%), Fe (98.93%), Mn (99.73%), Se (98.83%) and Zn (79%), at 5 mg/L initial metal concentration (Mane and Bhosle 2012). *Hydrodictylon*, *Oedogonium* and *Rhizoclonium* (algal species) efficiently uptake heavy metals from waste water which eventually contributes to their dry weight (Saunders et al. 2012).

7 Biosparging

In addition to surface water resources, ground water pollution is of much concern. And to deal with hectic ground water contamination, biosparging paves a handsome way to deal with the problem. Biosparging is simply injection of air below the pressure of water table to increase oxygen enrichment in ground water which can result in enhancing the biodegradation of contaminations (Adams and Reddy 2003). The momentous intensification in biodegradation pace is function of air injection speed which is the main factor behind biodegradation efficiency in heavily contaminated areas (Machackova et al. 2012).

8 Mycoremediation

Employing fungi to sequester various pollutants from contaminated environs is referred as Mycoremediation (Rhodes 2014). Fungal species possess remarked ability to clean up contaminated aquatic ecosystems and, in this context, *Aspergillus niger*, *Aureobasidium pullulans*, *Cladosporium resinae*, *Funalia trogii*, *Ganoderma lucidum*, *Penicillium* spp. (Loukidou et al. 2003; Say et al. 2003), *Rhizopus arrhizus* and *Trametes versicolor* are known to recover heavy metals from the polluted environment. Bio-accumulative removal yield of *Aspergillus versicolor* for various heavy metals Cr, Ni and Cu ions in wastewater effluents was 99.89, 30.05 and 29.06%, respectively (Tastan et al. 2010). Similarly, *Aspergillus fumigates* can efficiently remove Pb ions from the aqueous solution of electronic waste (containing Pb 100 mg/L) with 85.41% adsorption rate (Ramasamy et al. 2011). Besides, Arbuscular mycorhiza has great potential to degrade heavy metal pollution (Khanday et al. 2016; Bhat et al. 2017a, b).

9 Bioelectricity Through Biofuel Cells

Microbial fuel cells (MFCs) are the biotechnological interventions that characteristically contain two domains, i.e. the negative terminal (anode) and the positive terminal (cathode) which together make up the circuit, and in addition microbes (bacteria) act as biological catalysts which stems up the reaction by oxidation of organic substrate to carbon dioxide which consequently leads to generation of electric current which is referred as Bioelectricity (Wang and Ren 2013, Logan and Rabaey 2012) and the bacteria has a well established capability to produce protein based filaments known as 'microbial nanowires' which possess unique attribute to conduct and transfer electrons (Malvankar and Lovley 2012). Schematic representation of a microbial fuel cell is shown in Fig. 1, Table 2.

10 Production of Biodiesel from Wastes

Many microbes (especially yeast, fungi and microalgae) possess potential biosynthesizing machinery (large amounts of fatty acids) which eventually results in production of biodiesel (Xiong et al. 2008). By the cultivation of *Trichosporon fermentans*, microbial oil can be produced from sulphuric acid rice straw hydrolysate (SARSH) using waste rice straw as substrate (Huang et al. 2009). Using the

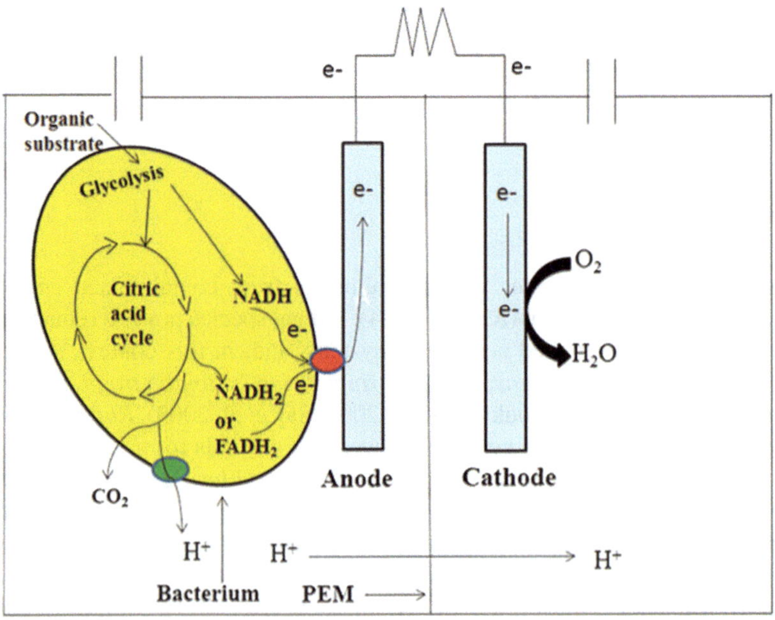

Fig. 1 General principle of a microbial fuel cell

Table 2 Various micro organisms with potential to be exploited for generation of Bioelectricity (through MFCs)

Micro organisms	References
Bacteria	
Aeromonas spp. and *Actinomycetes*	Malvankar and Lovley (2012), and Bhatti et al. (2017)
Geobacter spp.	Bond and Lovley (2003)
Shewanella spp.	Gorby et al. (2006)
Rhodoferax ferrireducens	Chaudhuri and Lovley (2003)
Aeromonas hydrophila	
Pseudomonas aeruginosa	Rabaey et al. (2004) and Jayapriya and Ramamurthy (2012)
Thermincola ferriacetica	Wrighton et al. (2011)
Clostridium butyricum	Park et al. (2001)
Enterococcus gallinarum	Chisti (2007)
Escherichia coli	Qiao et al. (2008)
Saccharomyces cerevisiae	Raghavulu et al. (2011)
Lysinibacillus sphaericus	Nandy et al. (2013)
Citrobacter sp.	Xu and Liu (2011)
Dechlorospirillum anomalous, *Acinetobacter calcoaceticus*, *Staphylococcus carnosus*, *Streptococcus mutans*, *Enterococcus faecalis*, *Shigella flexneri* and *Lactobacillus farciminis*	Thrash et al. (2007), Aulenta et al. (2010), and Cournet et al. (2010)
Klebsiella pneumonia	Lifang et al. (2010)
Desulfovibrio desulfuricans	Kang et al. (2014)
Ochrobactrum sp.	Xin et al. (2014)
Yeast	
Saccharomyces cerevisiae	Prasad et al. (2007)
Candida melibiosica	Hubenova and Mitov (2010)
Hansenula anomala	Prasad et al. (2007)
Algae	
Scenedesmus	Cui et al. (2014)
Arthrospira maxima	Inglesby et al. (2012)
Chlorella vulgaris	González et al. (2013)
Coriolus versicolor	Wu et al. (2012)

same strain at optimized conditions (pH 6.5 and temperature 25 °C), microbial culture cultivation is allowed to flourish for 7 days which eventually result in exploitation of waste molasses (from sugar industry) through production of lipids in microbial biomass and can be utilized as biodiesel (Xiong et al. 2008; Zhu et al. 2008). Ethanol production through biomass utilization technology can be termed as 'Bio-ethanol' (Nigam and Singh 2011). The various processes involved in Bio-ethanol production from biomass are: (a) pretreatment of substrates (separation of cellulose, hemicellulose and lignin in order to facilitate their enzyme-catalysed hydrolysis into their constituent simple sugars), (b) saccharification process to release the fermentable sugars from polysaccharides, (c) fermentation of released sugars to ethanol (Barron et al. 1996) and (d) distillation step to separate ethanol. Fermentation of cellulose directly to ethanol is specialized attribute of few microbe species like *Neurospora*, *Monilia*, *Paecilomyces* and *Fusarium* (Lynd et al. 2005) and the process is said to be simultaneous saccharification and fermentation (SSF). Consolidated bioprocessing (CBP) facilitating cellulase production, cellulose hydrolysis and fermentation in a single master stroke is actually an alternative approach with exceptional potential (Saxena et al. 1992).

11 Biosensors for Environmental Monitoring

Biosensors are highly sensitive tools of biological origin which comprise of biological detection element, a transducer and a signal processing system (Nigam and Shukla 2015) and recognition of a specific analyte is made by a bioreceptor which is basically a biomolecule, and transducer converts the detected response into a quantifiable signal.

12 Conclusion

Biotechnological interventions are promising avenues to deal with amalgam of environmental pollutions and can curb contamination of aquatic environs. Microbes are cheapest natural industries which can be exploited through means of biosorption, biodegradation, biostimulation, blastofiltration, cyanoremediation, biosparging and mycoremediation. In addition to aforementioned remarkable functions, microbes can generate electricity through biological fuel cells and biodiesel. And in order to deal with the growing aquatic pollution, biosensors can be employed to detect presence and concentration of various pollutants in the system.

References

Adams, J. A., & Reddy, K. R. (2003). Extent of benzene biodegradation in saturated soil column during air sparging. *Ground Water Monitoring and Remediation, 23*(3), 85–94.

Aulenta, F., Reale, P., Canosa, A., Rossetti, S., Panero, S., & Majone, M. (2010). Characterization of an electro-active biocathode capable of dechlorinating trichloroethene and cis-dichloroethene to ethane. *Biosensors & Bioelectronics, 25*, 1796–1802. https://doi.org/10.1016/j.bios.2009.12.033.

Azouaou, N., Sadaoui, Z., & Mokaddem, H. (2008). Removal of Cadmium from aqueous solution by adsorption on vegetable wastes. *Journal of Applied Sciences, 8*(24), 4638–4643.

Baldwin, B. R., Peacock, A. D., Park, M., Ogles, D. M., et al. (2008). Multilevel samplers as microcosms to assess microbial response to biostimulation. *Ground Water, 46*, 295–304.

Barron, N., Brady, D., Love, G., Marchant, R., Nigam, P., McHale, L., & McHale, A. P. (1996). Alginate immobilized thermotolerant yeast for conversion of cellulose to ethanol. In R. H. Wijffels, R. M. Buitelaar, C. Bucke, & J. Tramper (Eds.), *Progress in biotechnology e immobilized cells: Basics and applications* (pp. 379–383). Amsterdam: Elsevier Science BV.

Bhat, R. A., Dar, G. H., Jehangir, A., Bhat, B. A., & Yousuf, A. R. (2012). Municipal solid waste generation and present scenario of waste management during Yatra season in Pahalgam: A tourist health resort of Kashmir valley. *International Journal of Current Research, 4*, 004–009.

Bhat, R. A., Nazir, R., Ashraf, S., Ali, M., Bandh, S. A., et al. (2014). Municipal solid waste generation rates and its management at Yusmarg forest ecosystem, a tourist resort in Kashmir. *Waste Management and Research, 32*, 165–169.

Bhat, R. A., Dervash, M. A., Mehmood, M. A., Skinder, B. M., Rashid, A., Bhat, J. I. A., Singh, D. V., & Lone, R. (2017a). Mycorrhizae: A sustainable industry for plant and soil environment. In A. Varma et al. (Eds.), *Mycorrhiza – Nutrient uptake, biocontrol, ecorestoration* (pp. 473–502). https://doi.org/10.1007/978-3-319-68867-1_25.

Bhat, R. A., Shafiq-ur-Rehman, Dervash, M. A., Mushtaq, N., Bhat, J. I. A., & Dar, G. H. (2017b). Current status of nutrient load in Dal Lake of Kashmir Himalaya. *Journal of Pharmacognosy and Phytochemistry, 6*(6), 165–169.

Bhat, R. A., Dervash, M. A., Mehmood, M. A., & Hakeem, K. R. (2018). Municipal solid waste generation and its management, a growing threat to fragile ecosystem in Kashmir 756 Himalaya. *American Journal of Environmental Sciences, 2*(11), 145–167.

Bhatti, A. A., Haq, S., & Bhat, R. A. (2017). Actinomycetes benefaction role in soil and plant health. *Microbial Pathogenesis, 111*, 458–467.

Bond, D. R., & Lovley, D. R. (2003). Electricity production by *Geobacter sulfurreducens* attached to electrodes. *Applied and Environmental Microbiology, 69*, 1548–1555. https://doi.org/10.1128/AEM.69.3.1548-1555.2003.

Bulgariu, L., Lupea, M., Bulgariu, D., Rusu, C., & Macoveanu, M. (2013). Equilibrium study of Pb (II) and Cd (II) biosorption from aqueous solution on marine green algae biomass. *Environmental Engineering and Management Journal, 12*, 183–190.

Celekli, A., Yavuzatmaca, M., & Bozkurt, H. (2010). An eco-friendly process: Predictive modelling of copper adsorption from aqueous solution on Spirulina platensis. *Journal of Hazardous Materials, 173*, 123–129.

Chaudhuri, S. K., & Lovley, D. R. (2003). Electricity generation by direct oxidation of glucose in mediatorless microbial fuel cells. *Nature Biotechnology, 21*, 1229–1232. https://doi.org/10.1038/nbt867.

Chauhan, A., & Jain, R. K. (2010). Biodegradation: Gaining insight through proteomics. *Biodegradation, 21*, 861–879.

Cheng, S. S., Hsieh, T. L., Pan, P. T., Gaop, C. H., Chang, L. H., Whang, L. M., & Chang, T. C. (2009). *Study on biomonitoring of aged TPHcontaminated soil with bioaugmentation and biostimulation (Conference paper)*. 10th International in situ and on-site bioremediation symposium, Baltimore MD, May 5–8, 2009.

Chisti, Y. (2007). Biodiesel from microalgae. *Biotechnology Advances, 25*, 294–306. https://doi.org/10.1016/j.biotechadv.2007.02.001.

Conesa, H. M., Michael, W., Evangelou, H., Robinson, B. H., & Schulin, R. A. (2012). Critical view of current state of phytotechnologies to remediate soils: Still a promising tool? *ScientificWorldJournal, 2012*, 173829.

Cournet, A., Délia, M. L., Bergel, A., Roques, C., & Bergé, M. (2010). Electrochemical reduction of oxygen catalyzed by a wide range of bacteria including Gram-positive. *Electrochemistry Communications, 12*, 505–508. https://doi.org/10.1016/j.elecom.2010.01.026.

Cui, Y., Rashid, N., Hu, N., Rehman, M. S. U., & Han, J. I. (2014). Electricity generation and microalgae cultivation in microbial fuel cell using microalgae-enriched anode and biocathode. *Energy Conversion and Management, 79*, 674–680. https://doi.org/10.1016/j.enconman.2013.12.032.

Dursun, A. Y. (2006). A comparative study on determination of the equilibrium, kinetic and thermodynamic parameters of biosorption of copper (II) and lead (II) ions onto pretreated *Aspergillus niger*. *Biochemical Engineering Journal, 28*, 187–195.

Fan, T., Liu, Y., Feng, B., Zeng, G., Yang, C., et al. (2008). Biosorption of cadmium(II), zinc(II) and lead(II) by Penicillium simplicissimum: Isotherms, kinetics and thermodynamics. *Journal of Hazardous Materials, 160*, 655–661.

Fiset, J. F., Blais, J. F., & Riverso, P. A. (2008). Review on the removal of metal ions from effluents using seaweeds, alginate derivatives and other sorbents. *Revue des Sciences de l'Eau, 21*(3), 283–308.

Freitas, O. M., Martins, R. J., Delerue-Matos, C. M., & Boaventura, R. A. (2008). Removal of Cd (II), Zn (II) and Pb (II) from aqueous solutions by brown marine macro algae: Kinetic modelling. *Journal of Hazardous Materials, 153*, 493–501.

Fulekar, M. H., Sharma, J., & Tendulkar, A. (2012). Bioremediation of heavy metals using biostimulation in laboratory bioreactor. *Environmental Monitoring and Assessment, 184*(12), 7299–7307.

Gadd, G. M. (2010). Metals, minerals and microbes: Geomicrobiology and bioremediation. *Microbiology, 156*, 609–643.

Ghosh, A., Ghosh Dastidar, M., & Sreekrishnan, T. (2016). Recent advances in bioremediation of heavy metals and metal complex dyes: Review. *Journal of Environmental Engineering, 142*(9), C4015003. https://doi.org/10.1061/(ASCE)EE.1943-7870.0000965.

González del Campo, A., Cañizares, P., Rodrigo, M., Fernández, F. J., & Lobato, J. (2013). Microbial fuel cell with an algae-assisted cathode: A preliminary assessment. *Journal of Power Sources, 242*, 638–645. https://doi.org/10.1016/j.jpowsour.2013.05.110.

Gorby, Y. A., Yanina, S., McLean, J. S., Rosso, K. M., Moyles, D., Dohnalkova, A., Beveridge, T. J., Chang, I. S., Kim, B. H., Kim, K. S., Culley, D. E., Reed, S. B., Romine, M. F., Saffarini, D. A., Hill, E. A., Shi, L., Elias, D. A., Kennedy, D. W., Pinchuk, G., Watanabe, K., Ishii, S., Logan, B., Nealson, K. H., & Fredrickson, J. K. (2006). Electrically conductive bacterial nanowires produced by *Shewanella oneidensis* strain MR-1 and other microorganisms. *Proceedings of the National Academy of Sciences of the United States of America, 103*, 11358–11363. https://doi.org/10.1073/pnas.0604517103.

Gupta, R., & Mahapatra, H. (2003). Microbial biomass: An economical alternative for removal of heavy metals from wastewater. *Indian Journal of Experimental Biology, 41*, 945–966.

Huang, C., Zong, M. H., Hong, W., & Liu, Q. P. (2009). Microbial oil production from rice straw hydrolysate by *Trichosporon fermentans*. *Bioresource Technology, 100*, 4535–4538.

Hubenova, Y., & Mitov, M. (2010). Potential application of *Candida melibiosica* in biofuel cells. *Bioelectrochemistry, 78*, 57–61. https://doi.org/10.1016/j.bioelechem.2009.07.005.

Infante, J., De Arco, R., & Angulo, M. (2014). Removal of lead, mercury and nickel using the yeast Saccharomyces cerevisiae. *Revista MVZ Córdoba, 19*, 4141–4149.

Inglesby, A. E., Beatty, D. A., & Fisher, A. C. (2012). Rhodopseudomonas palustris purple bacteria fed *Arthrospira maxima* cyanobacteria: Demonstration of application in microbial fuel cells. *RSC Advances, 2*, 4829–4838. https://doi.org/10.1039/C2RA20264F.

Jayapriya, J., & Ramamurthy, V. (2012). Use of non-native phenazines to improve the performance of *Pseudomonas aeruginosa* MTCC 2474 catalysed fuel cells. *Bioresource Technology, 124*, 23–28. https://doi.org/10.1016/j.biortech.2012.08.034.

Jeffries, T. C., Seymour, J. R., Newton, K., Smith, R. J., Seuront, L., & Mitchell, J. G. (2012). Increases in the abundance of microbial genes encoding halotolerance and photosynthesis along a sediment salinity gradient. *Biogeosciences, 9*(2), 815–825.

Kang, C. S., Eaktasang, N., Kwon, D. Y., & Kim, H. S. (2014). Enhanced current production by Desulfovibrio desulfuricans biofilm in a mediator-less microbial fuel cell. *Bioresource Technology, 165*, 27–30. https://doi.org/10.1016/j.biortech.2014.03.148.

Kanmani, P., Aravind, J., & Preston, D. (2012). Remediation of chromium contaminants using bacteria. *International journal of Environmental Science and Technology, 9*, 183–193.

Katsumata, H., Kaneco, S., Inomata, K., Itoh, K., Funasaka, K., et al. (2003). Removal of heavy metals in rinsing wastewater from plating factory by adsorption with economical viable materials. *Journal of Environmental Management, 69*, 187–191.

Khanday, M., Bhat, R. A., Haq, S., Dervash, M. A., Batti, A. A., Nissa, M., & Mir, M. R. (2016). Arbuscular mycorrhizal fungi boon for plant nutrition and soil health. In K. R. Hakeem et al. (Eds.), *Soil science: Agricultural and environmental prospectives* (pp. 317–332). https://doi.org/10.1007/978-3-319-34451-5_14.

Kiyono, M., Oka, Y., Sone, Y., Tanaka, M., Nakamura, R., Sato, M. H., Pan-Hou, H., Sakabe, K., & Inoue, K. (2012). Expression of bacterial heavy metal transporter MerC fused with a plant SNARE, SYP121 in *Arabidopsis thaliana* increases cadmium accumulation and tolerance. *Planta, 235*, 841–850.

Kumar, A., & Singh, J. S. (2017). Cyanoremediation: A green-clean tool for decontamination of synthetic pesticides from agro- and aquatic ecosystems. In J. Singh & G. Seneviratne (Eds.), *Agro-environmental sustainability*. Cham: Springer. https://doi.org/10.1007/978-3-319-49727-3_4.

Kumar, R., Sharma, A. K., Singh, P., Dhir, B., & Mehta, D. (2014). Potential of some fungal and bacterial species in bioremediation of heavy metals. *Journal of Nuclear Physics, Material Sciences, Radiation and Applications, 1*, 213–223.

Leitão, A. L. (2009). Potential of *Penicillium* species in the bioremediation field. *International Journal of Environmental Research and Public Health, 6*, 1393–1417.

Li, J., Lin, Q., & Zhang, X. (2010). Mechanism of electron transfer in the bioadsorption of hexavalent chromium within *Leersia hexandra* Swartz granules by X-ray photoelectron spectroscopy. *Journal of Hazardous Materials, 182*, 598–602.

Li, X., Li, A., Long, M., & Tian, X. (2015). Equilibrium and kinetic studies of copper biosorption by dead *Ceriporia lacerata* biomass isolated from the litter of an invasive plant in China. *Journal of Environmental Health Science and Engineering, 13*, 37.

Lifang, D., Frang, L., Shungui, Z., Yin, H. D., & Jinren, N. I. (2010). A study of electron-shuttle mechanism in *Klebsiella pneumoniae* based microbial fuel cells. *Environmental Science & Technology, 55*, 99–104. https://doi.org/10.1007/s11434-009-0563-y.

Lin, Q. U. R., Sen, L. D., Qian, D. U. R., & Yao, J. M. (2002). Phytoremediation for heavy metal pollution in water II. The blastofiltration of Pb from water. *Journal of Agro-Environment Science*. http://en.cnki.com.cn/Article_en/CJFDTOTAL-NHBH200206005.htm

Logan, B. E., & Rabaey, K. (2012). Conversion of wastes into bioelectricity and chemicals by using microbial electrochemical technologies. *Science, 337*, 686–690. https://doi.org/10.1126/science.1217412.

Loukidou, M. X., Matis, K. A., Zouboulis, A. I., & Liakopoulou-Kyriakidou, M. (2003). Removal of As(V) from wastewaters by chemically modified fungal biomass. *Water Research, 37*(18), 4544–4552.

Lu, W. B., Shi, J. J., Wang, C. H., & Chang, J. S. (2006). Biosorption of lead, copper and cadmium by an indigenous isolate Enterobacter sp. J1 possessing high heavy-metal resistance. *Journal of Hazardous Materials, 134*, 80–86.

Lynd, L. R., VanZyl, W. H., McBride, J. E., & Laser, M. (2005). Consolidated bioprocessing of cellulosic biomass: An update. *Current Opinion in Biotechnology, 16*, 577–583.

Ma, X., Nonvak, P. J., Ferguson, J., Sadowsky, M., et al. (2007). The impact of H2 addition or dechlorinating microbial communities. *Bioremediation Journal, 11*, 45–55.

Machackova, J., Wittlingerova, Z., Vlk, K., & Zima, J. (2012). Major factors affecting in situ biodegradation rates of jet-fuel during largescale biosparging project in sedimentary bedrock. *Journal of Environmental Science and Health, Part A, 47*(8), 1152–1165.

Malvankar, N. S., & Lovley, D. R. (2012). Microbial nanowires: A new paradigm for biological electron transfer and bioelectronics. *Chem Sus Chem, 5*, 1039–1046. https://doi.org/10.1002/cssc.201100733.

Mane, P. C., & Bhosle, A. B. (2012). Bioremoval of some metals by living Algae Spirogyra sp. and Spirullina sp. from aqueous solution. *International Journal of Environmental Research, 6*(2), 571–576.

Mehmood, M. A., Qadri, H., Bhat, R. A., Rashid, A., Ganie, S. A., Dar, G. H., & Shafiq-ur-Rehman. (2019). Heavy metal contamination in two commercial fish species of a trans-Himalayan freshwater ecosystem. *Environmental Monitoring and Assessment Environ, 191*, 104. https://doi.org/10.1007/s10661-019-7245-2.

Mustapha, M. U., & Halimoon, N. (2015). Microorganisms and biosorption of heavy metals in the environment: A review paper. *Journal of Microbial and Biochemical Technology, 7*, 253–256. https://doi.org/10.4172/1948-5948.1000219.

Nandy, A., Kumar, V., & Kundu, P. P. (2013). Utilization of proteinaceous materials for power generation in a mediatorless microbial fuel cell by a new electrogenic bacteria *Lysinibacillus sphaericus* VA5. *Enzyme and Microbial Technology, 53*, 339–344. https://doi.org/10.1016/j.enzmictec.2013.07.006.

Nigam, V. K., & Shukla, P. (2015). Enzyme based biosensors for detection of environmental pollutants – A review. *Journal of Microbiology and Biotechnology, 11*, 1773–1781.

Nigam, P. S., & Singh, A. (2011). Production of liquid biofuels from renewable resources. *Progress in Energy and Combustion Science, 37*, 52–68.

Park, H. S., Kim, B. H., Kim, H. S., Kim, H. J., Kim, G. T., Kim, M., Chang, I. S., Park, Y. K., & Chang, H. I. (2001). A novel electrochemically active and Fe(III)-reducing bacterium phylogenetically related to *Clostridium butyricum* isolated from a microbial fuel cell. *Anaerobe, 7*, 297–306. https://doi.org/10.1006/anae.2001.0399.

Prasad, D., Arun, S., Murugesan, M., Padmanaban, S., Satyanarayanan, R. S., Berchmans, S., & Yegnaraman, V. (2007). Direct electron transfer with yeast cells and construction of a mediatorless microbial fuel cell. *Biosensors & Bioelectronics, 22*, 2604–2610. https://doi.org/10.1016/j.bios.2006.10.028.

Qiao, Y., Li, C. M., Bao, S. J., Lu, Z., & Hong, Y. (2008). Direct electrochemistry and electrocatalytic mechanism of evolved Escherichia coli cells in microbial fuel cells. *Chemical Communications, 11*, 1290–1292. https://doi.org/10.1039/B719955D.

Quintelas, C., Rocha, Z., Silva, B., Fonseca, B., Figueiredo, H., et al. (2009). Removal of Cd (II), Cr (VI), Fe (III) and Ni (II) from aqueous solutions by an *E. coli* biofilm supported on kaolin. *Chemical Engineering Journal, 149*, 319–324.

Rabaey, K., Boon, N., Siciliano, S. D., Verhaege, M., & Verstraete, W. (2004). Biofuel cells select for microbial consortia that self-mediate electron transfer. *Applied and Environmental Microbiology, 70*, 5373–5382. https://doi.org/10.1128/AEM.70.9.5373-5382.2004.

Raghavulu, S. V., Goud, R. K., Sarma, P. N., & Mohan, S. V. (2011). *Saccharomyces cerevisiae* as anodic biocatalyst for power generation in biofuel cell: Influence of redox condition and substrate load. *Bioresource Technology, 102*, 2751–2757. https://doi.org/10.1016/j.biortech.2010.11.048.

Ramasamy, R. K., Congeevaram, S., & Thamaraiselvi, K. (2011). Evaluation of isolated fungal strain from e-waste recycling facility for effective sorption of toxic heavy metals Pb(II) ions and fungal protein molecular characterization-a Mycoremediation approach. *Asian Journal of Experimental Biological Sciences, 2*(2), 342–347.

Rayu, S., Karpouzas, D. G., & Singh, B. K. (2012). Emerging technologies in bioremediation: Constraints and opportunities. *Biodegradation, 23*, 917–926.

Rhodes, C. (2014). Mycoremediation (bioremediation with fungi) – Growing mushrooms to clean the earth. *Chemical Speciation and Bioavailability, 26*, 196–198. https://doi.org/10.3184/095422914X14047407349335.

Saunders, R. J., Paul, N. A., Hu, Y., & de Nys, R. (2012). Sustainable sources of biomass for bioremediation of heavy metals in wastewater derived from coal-fired power generation. *PLoS One, 7*(5), e36470. https://doi.org/10.1371/journal.pone.0036470.

Saxena, A., Garg, S. K., & Verma, J. (1992). Simultaneous saccharification and fermentation of waste newspaper to ethanol. *Bioresource Technology, 39*, 13–15.

Say, R., Yimaz, N., & Denizli, A. (2003). Removal of heavy metal ions using the fungus *Penicillium canescens*. *Adsorption Science and Technology, 21*, 643–650.

Schiewer, S., & Patil, S. B. (2008). Pectin-rich fruit wastes as biosorbents for heavy metal removal: Equilibrium and kinetics. *Bioresource Technology, 99*(6), 1896–1903.

Sud, D., Mahajan, G., & Kaur, M. P. (2008). Agricultural waste material as potential adsorbent for sequestering heavy metals ions from aqueous solutions: A review. *Bioresource Technology, 99*(14), 6017–6027.

Tan, T. W., Hu, B., & Su, H. (2004). Adsorption of Ni^{2+} on amine-modified mycelium of *Penicillium chrysogenum*. *Enzyme and Microbial Technology, 35*, 508–513.

Tang, C. Y., Criddle, Q. S., Fu, C. S., & Leckie, J. O. (2007). Effect of flux (Transmembrane pressure) and membranes properties on fouling and rejection of reverse osmosis and nanofiltration membranes treating perfluorooctane sulfonate containing wastewater. *Environmental Science & Technology, 41*, 2008–2014.

Tastan, B. E., Ertuğrul, S., & Dönmez, G. (2010). Effective bioremoval of reactive dye and heavy metals by *Aspergillus versicolor*. *Bioresource Technology, 101*(3), 870–876.

Thrash, J. C., Van Trump, I. V., Weber, K. A., Miller, E., Achenbach, L. A., & Coates, J. D. (2007). Electrochemical stimulation of microbial perchlorate reduction. *Environmental Science & Technology, 41*, 1740–1746. https://doi.org/10.1021/es062772m.

Tian, J., Peng, X. W., Li, X., Sun, Y. J., Feng, H. M., et al. (2014). Isolation and characterization of two bacteria with heavy metal resistance and phosphate solubilizing capability. *Huan Jing Ke Xue, 35*, 2334–2340.

Tyagi, M., Fonseca, M. M. R. D., & Carvalho, C. C. C. R. D. (2011). Bioaugmentation and biostimulation strategies to improve the effectiveness of bioremediation processes. *Biodegradation, 22*, 231–241.

Uzel, A., & Ozdemir, G. (2009). Metal biosorption capacity of the organic solvent tolerant Pseudomonas fluorescens TEM08. *Bioresource Technology, 100*, 542–548.

Vangronsveld, J., Herzig, R., Weyens, N., Boulet, J., et al. (2009). Phytoremediation of contaminated soils and groundwater: Lessons from the field. *Environmental Science and Pollution Research, 16*, 765–794.

Viggi, C. C., Pagnanelli, F., Cibati, A., Uccelletti, D., Palleschi, C., & Toro, L. (2010). Biotreatment and bioassessment of heavy metal removal by sulphate reducing bacteria in fixed bed reactors. *Water Research, 44*(1), 151–158.

Wang, J., & Chen, C. (2009). Biosorbents for heavy metals removal and their future. *Biotechnology Advances, 27*, 195–226.

Wang, H., & Ren, Z. J. (2013). A comprehensive review of microbial electrochemical systems as a platform technology. *Biotechnology Advances, 31*, 1796–1807. https://doi.org/10.1016/j.biotechadv.2013.10.001.

Wrighton, K. C., Thrash, J. C., Melnyk, R. A., Bigi, J. P., Byrne-Bailey, K. G., Remis, J. P., Schichnes, D., Auer, M., Chang, C. J., & Coates, J. D. (2011). Evidence for direct electron transfer by a Gram-positive bacterium isolated from a microbial fuel cell. *Applied and Environmental Microbiology, 77*, 7633–7639. https://doi.org/10.1128/AEM.05365-11.

Wu, L., Li, Z., Akahane, I., Liu, L., et al. (2012). Effects of organic amendments on Cd, Zn and Cu bioavailability in soil with repeated phytoremediation by Sedum plumbizincicola. *International Journal of Phytoremediation, 14*(10), 1024–1038.

Xin, L., Guo-Zhen, Z., Yan, Q., Jing, H., Weihua, H., Xing-Guo, W., & Changming, L. (2014). A high performance xylose microbial fuel cell enabled by Ochrobactrum sp.575 cells. *RSC Advances, 4*, 39839–39843. https://doi.org/10.1039/C4RA05077K.

Xiong, W., Li, X., Xiang, J., & Wu, O. (2008). High-density fermentation of microalga Chlorella protothecoides in bioreactor for microbiodiesel production. *Applied Microbiology and Biotechnology, 78*, 29–36.

Xu, S., & Liu, H. (2011). New exoelectrogen Citrobacter sp. SX-1 isolated from a microbial fuel cell. *Journal of Applied Microbiology, 111*, 1108–1115. https://doi.org/10.1111/j.1365-2672.2011.05129.x.

Yin, X. X., Wang, L. H., Bai, R., Huang, H., & Sun, G. X. (2012). Accumulation and transformation of arsenic in the blue-green alga *Synechocysis* sp. PCC6803. *Water, Air, and Soil Pollution, 223*(3), 1183–1190.

Zeng, X., Wei, S., Sun, L., Jacques, D. A., Tang, J., et al. (2015). Bioleaching of heavy metals from contaminated sediments by the *Aspergillus niger* strain SY1. *Journal of Soils and Sediments, 15*, 1029–1038.

Zhu, L. Y., Zong, M. H., & Wu, H. (2008). Efficient lipid production with T. fermentas and its use for biodiesel preparation. *Bioresource Technology, 99*, 7881–7885.

Analysis of Hydrology, Sediment Retention, Biogenic- Calcification and -Scavenging as Self-Remediative Lacustrine Functions.

Umar Nazir Bhat and Anisa Basheer Khan

Abstract Urban water bodies are indicators of anthropogenic intrusion surfacing mutability in intrinsic homeostasis. Ecological assessment of various bio-physicochemical variables at periodic intervals is vital for eventual implementation of management and conservation practices in lakes. An inter-annual monitoring of surface-waters, surface-sediments and dominant macrophytes for standard variables at 50 sampling sites in 5 zones (10 each) of Anchar and Dal lakes is carried out to assess their spatio-temporal heterogeneity under human pressures. Temperature, pH, conductivity and ionic composition of the epilimnion show $p < 0.01$ and $R^2 > 0.5$. The trophic range for total-P exceeds critical eutrophic index (≤ 0.05 mgL^{-1}) but nitrate-N persists beneath it (≤ 0.5 mgL^{-1}) normally. Conductivity maintains superior solute richness though autotrophic assimilation and biocalcification episodes subsidize it towards summer. The anionic predominance of HCO_3^-(BIC) and Cl^- exist alongside cationic progression of Ca > Mg > Na > K. Lime-catchment adds to Ca ascendancy and hard-waters. Agricultural runoff links with K while Cl to faunal organic pollution. Superior nitrate concentration is accumulative of human actions (agriculture, farming, sewage, factories, etc.), spring fed lake-basin, preferential NH^{4+} autotrophic assimilation, geogenic N-pockets and forest surface runoff. Significant Coefficient of Determination (R^2) for pH versus temperature, conductivity versus pH and temperature substantiate biological uptake and calcite co-precipitation. An equation with average worldwide stream abundance (mgL^{-1}) of recorded Ca (> 15), Mg (> 4), K (> 2.3) and Na (> 6.3) besides observed average epilimnion trace element concentration (μgL^{-1}) for As (> 2), Cd (> 1), Cr (> 1), Co (> 0.2), Cu (> 10), Fe (> 700), Pb (> 3), Mn (> 7), Ni (> 1), Se (> 0.2), Sn (> 0.1) and Zn (> 20) acclaim their anthropogenic origins. However, all priority pollutants (As, Cd, Cr, Cu, Pb, Ni and Se) continued below USEPA chronic levels. Fe and Zn exceed maximum permissible limits for irrigation. The flushing-out of harmful nutrient- and contaminant-levels due to semi-drainage hydrology recuperated the aqueous volume. Sediment assessment identifies Ca-Si domination with temporal gradients in pH, bicarbonate, conductivity, Organic Carbon (OC), Organic Matter

U. N. Bhat (✉) · A. B. Khan
Department of Ecology and Environmental Sciences, Pondicherry Central University, Puducherry, India

H. Qadri et al. (eds.), *Fresh Water Pollution Dynamics and Remediation*,
https://doi.org/10.1007/978-981-13-8277-2_12

(OM), Total Nitrogen (TN) and C/N. Almost no outliers in box-plots across the select sites suggest their tranquil nature. Element composition revealed the order of Si > Ca > Mg > K > Na > P > S > Cl. Micro and trace element quantification denote the descending series of Fe > Al > Zn > Mn > Cu > Cr > Ni > Co > As > Sn > Pb > Cd while Hg and Se remained Below Detection Level (BDL). Sediment pH stayed on the basic side but slender acidic nature is noticed during late summer. Significant correlation for conductivity with OC and OM ($p < 0.01$) establish the latter a source for nutrient ions. Total-N is complementary to OC and OM of sediments too. Active/Passive-bioaccumulation or anoxic release from sediments tends to slight gradual decline in nutrient concentration till culmination of macrophytic growth phases. Enrichment Factor (EF), Geo-accumulation Index (I_{geo}) and Contamination Factor (CF) expound the contaminants to be largely anthropogenic. Integrated Pollution Index (IPI) and Pollution Load Index (PLI) catalog the lakes to have moderate metal contamination. Sediment Quality Guidelines (SQG's) point to pollution status and associated ecological risks involved. Cr, Ni and Zn exceed SQG's but Cd and Pb don't transcend them. As is below Effects Range Low (ERL) and Cu lags in Probable Effect Concentration (PEC). The typical $C/N < 10$ infers autochthonous sediment OM with low decomposition rates. Upgraded [N]:[P] ratios parallel chronic nitrogen influx. Higher temperature and lower [N]:[P] ratio during summer develop internal loading of P. But higher Al, Ca and Fe proportions in sediments inactivate P mobilization. Curbing of external N and P loads is effective in remediation but the internal supplement compensates the loss. OM or Fe/Mn- oxide decomposition and reductive dissolution respectively separate bound trace-metals near hypolimnion-sediment overlap. Lower [Ca]:[Al] sponsor exsitu human Potentially Toxic Element (PTE) transport. Nonetheless, OM enriched sediments and calcite co-precipitation together curtails PTE mobility. Macrophytes optimize ambient water quality and sediment medium. The peak biomass (gm^{-2}) values on dry weight basis are 880.2, 678.4, 182.4 and 45 for *Myriophyllum aquaticum*, *Nelumbo nucifera*, *Ceratophyllum demersum* and *Salvinia natans* respectively. Dry Weight, Productivity, Net Primary Productivity (NPP) and Specific Growth Rate institute affiliated variations but species Turn-Over is highest in case of *S. natans* and lowest for *C. demersum*. The species differ in tissue nutrient and trace element concentrations but correlate with ambient water-sediment medium. The peak nutrient uptake and bioconcentration coincide with peak biomass in summer and autumn. Bioconcentration Factor (BCF) indicates hyperaccumulation for most of the metals in case of *C. demersum* and *S. natans*. Removal Potential for different elements is divergent but the pattern is related which suggests unselective absorption. Turn-over Rates for elements closer to the reference value of 1 is significant. Bioavailability of nutrients and toxins becomes fractional conjointly by flushing hydrology, biological scavenging and biocalcification. An insitu self-reclaimed nutrient balance and eco-restoration is conceivable in the region of anthro-urban intensification by limiting human perturbations, practicing periodic dredging, sediment trapping, scaled-cum-selective deweeding and construction of vegetation buffer strips.

Keywords Assimilation · Bicarbonates · Conductivity · Contamination · Nutrients Resilience · Spectrometer

1 Introduction

Natural ecosystems maintain complex heterogeneous form and function on exposure to disturbances by shifting their equilibrium. The dynamic homeostasis simultaneously operative between physical, chemical and biological mechanisms tend towards stabilization to counter internal and external strains. Resilience enables the system to recover or adjust to the change inorder to retain essential structure, function, identity and feedbacks. However, frequent, intense and even irreversible impairment is evident due to accelerated anthropogenic stresses. The differentiated activities have led to the global spread of spectra of contaminants intoxicating environment (Karanlik et al. 2011) to pose potential risks for biota and humans (Zhang et al. 2012). The ever-evolving human actions are contributing to observable changes in aquatic ecosystems in terms of their distribution, size and composition. Freshwater sources are overplayed and at high peril to curtail the elixir of life. Lakes also possess a dynamic equilibrium among and between physical, chemical and biological attributes responding to natural and anthropogenic fluctuations. The changeover from oligotrophy to eutrophy steered by nutrient enrichment, sedimentation and higher productivity eventually transforms them into dystrophic marshes (Søndergaard et al. 2010). Accordingly, their natural setting and homeostasis is continuously and significantly altered by progressive socioeconomic, urban, industrial and agricultural developments (EEA 2012; UNEP 2007; Vörösmarty et al. 2010). These influences on the surface-water quality devaluate it for any designated purpose besides manipulating the biotic and abiotic subsystems (MEA 2005).

Sediments are a major nutrient pool archiving historical hydroclimatic records (Dean et al. 2015). They act as pollutant monitoring indicators (Nilin et al. 2013) mineralized either in organic or inorganic form. Sediments retain nutrients (Rothwell et al. 2010) as well as release contaminants (Manap and Voulvoulis 2015). The surface sediments conduct both ascendant and descendent biogeochemical exchange with the overlying water column (Lenzi et al. 2012). The bottom sediments act as a sink (reservoir) of nutrients instead (Akin et al. 2010).

Hydrophytes include phytoplankton, periphyton and multicellular macrophytes occupying various ecological niches in the water environment. Macrophytes consist principally of aquatic vascular flowering plants but also include the aquatic mosses, liverworts, ferns and the larger macro-algae (Chambers et al. 2008). Macrophytes are predominant community of primary producers performing roles in energy input, nutrient budgeting and recycling (Algesten et al. 2003), biofiltration (Dhote 2007), sedimentation, furnishing habitat to micro and macro water-organisms (Gurnell et al. 2012). Aquatic weeds not only act as "biological sinks" for minerals and organics (Gottschall et al. 2007) but their quality and quantity facilitate insight of environmental state (Lenzi et al. 2012). Extreme eutrophication modifiesmacrophyte-dominated clear water state to turbid phytoplankton-dominated state of "algal blooms"(Scheffer 1989) and loss of submerged macrophytes (Xiangcan 2003). The frequent abundance of macrophytes in aquatic ecosystems prescribes them as preferred and potent bioagents for biomonitoring and pollution abatement. 'Green Clean' or phytoremediation detoxify, remove, sequester and stabilize persistent pollutants in aesthetic manner (Ibanez et al. 2016) both by physical obstruction (biomass adsorption) and uptake from soil solution (Lukacs et al. 2009). It is natural, spontaneous, efficient both in-situ and ex-situ, ecofriendly, aesthetically pleasing, cost effective, solar driven, sustainable, least site invasive, applicable on large scale, demands minimal maintenance and amenable to amelioration of a broad range of contaminants (Tian et al. 2007).

The endeavor in the current study is to evaluate the comparative spontaneous efficacy of principal lake compartments for their in-situ utilityin self-remediation. Continuous monitoring and characterisation to understand contribution of each contaminant responsible for deterioration is necessary to evolve sustainable eco-conservation strategies, practices and technologies (Allan et al. 2006; Bierman et al. 2009; Moreno et al. 2011).

2 Materials and Methods

2.1 Study Area

Jammu and Kashmir (≥ 1500 m a.s.l) embeds in the north-west Himalayan biogeographic zone. The temperate climatic Kashmir valley with mountainous physiography (about 64%) has the Great Himalayas in its North and Pir Panjal range in its South. Kashmir owns a collection of natural lacustrine water bodies upholding utmost ecological, cultural, historical and socio-economic substance. The estimated geographic area occupied by lakes in J&K is 13,762 ha that approximates to 3.52% of all its aquatic ecosystems (National Wetland Atlas 2010). Srinagar (33°59′14″ - 34°12′37" N latitudes and 74°41′06″ - 74°57′27″ E longitudes)- the summer capital city of J&K sited on the banks of the Jhelum (*Vyeth*) spreads across the plains of Kashmir vale. It has a moderate physiography ≥1580 m a.s.l. representing hill-topography. The general climatic conditions resemble the sub-Mediterranean characterized by seasonal and diurnal temperature extremes alongside frequent precipitation episodes except for a few dry-periods during summer and autumn. The study periods' monthly maximum, mean and minimum temperatures (°C) and average rainfall (mm) is illustrated in Fig. 1.

Srinagar has a lake area cover of 2194 ha equivalent to its 21.76% of wetland area. The very existence of the urban lacustrine water bodies is under continuously intensifying multiple stressors of nutrient loading, siltation, waste disposal, sewerage and agro-chemical residue receivers, expanding floating-garden area, encroachment, blockade and narrowing of drainage channels, hydrological alterations, catchment perturbations and so on (Abubakr and Kundanger 2009; Badar et al. 2012). The focus of present limnological research remain two fluvial urban valley lakes namely Anchar and Dal situated in the summer capital (Srinagar) of J&K. Anchar is a single basined lake located in the north-west of Srinagar city within the geographical coordinates of 34° 20′ – 34° 26' N latitudes and 74° 82′ –

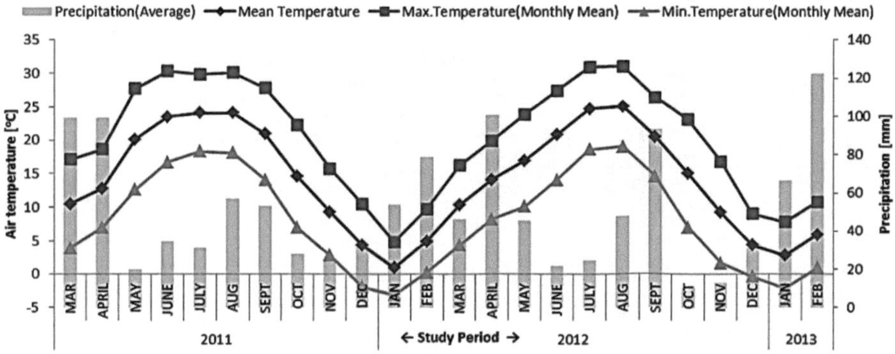

Fig. 1 Climate profile of Srinagar 2011–2013. (Data source: India Meteorological Department)

74° 85′ E longitudes. It has fluvial origin and occupies about 6 km^2 water surface area at an altitude of 1584 m a.s.l. (Zargar et al. 2012). Its mean depth is about 1.6 m and has a catchment cover of ~ 66 km^2. It has an open drainage feature reflected by a feeding network of streams from river Sindh; Gilsar-Khushalsar lake connections besides natural basin springs while its exit contributes to the Jhelum waters. While Dal lake occupying the coordinates of 34° 04′ N to 34° 09′ N and 74° 49′ E to 74° 52′ E is positioned in north-east of the city at 1587 m a.s.l. altitude (National Wetland Atlas 2010). It is a tetra-basined ox-bow type lake formed by meandering course of river Jhelum. Its main feeding source is a perennial Dachigam creek (splitted into 3 streams of Telbal, Pishpu and Meerakshah) and several lake-bed springs, however, the key exit discharges via Dalgate into the Jhelum. Alternate Hydro Energy Center (AHEC), Roorkee (now IIT, Roorkee) designates an area of 337 km^2 as its catchment while holding a water volume of 15 × 10^6 m^3. The satellite imagery database of Lakes and Waterways Development Authority (LWDA) reveals its total area as 24.6 Km^2 with an open water expanse of 15.41 Km^2 and the remaining occupied by floating gardens, emergent vegetation zone, house boats and human settlement area. A recent assessment intimates the open water spread restricted to just 10.5 Km^2 (Rashid et al. 2016). The mean depth averages to <3 m while the maximum depth equals 6 m. A preliminary comparison indicates Anchar to be unmanaged while Dal as fairly managed with substantial differences in their physical appearance and open-water surface area. However, both the lakes are eutrophied and majorly engulfed in urban and agricultural transformations as evident from satellite imagery (Fig. 2).

In defining the prevailing environmental conditions of two temperate urban lacustrine ecosystems of Srinagar, Kashmir, selection of appropriate bulk descriptors - water, sediment and macrophytes is applied. Due consideration is given to collect and preserve the sub-samples of water, sediment and hydrophytes from the same spot. The assessment plan implicates spatio-temporal distribution and dynamics quantification of major-, minor- and trace- elements in the epilimnion in combination with the characterisation of surface-sediments for minerals and metals in

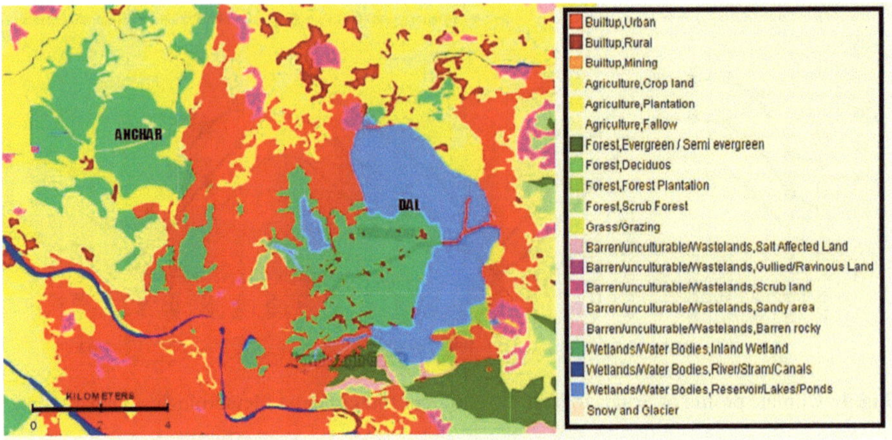

Fig. 2 Visualisation of land- cover/use for site catchment. (Source: NRSC/ISRO- Bhuvan)

comparison with geochemical indices and sediment quality criteria followed by in-situ bioconcentration, nutrient storage, removal potential and turnover of selective macrophyte species from sediment-water continuum as self-remediative lacustrine functions. The inter-annual survey, sample collection and hydrobiochemical study of the lakes is based on their 5 select stations. The site selection criterion implemented here represents the distinct spatial features within the lakes but concurrent availability of water, sediment and macrophytes. The 5 locations designated as A1, A2, D1, D2 and D3 were each further fragmented into ten sampling stations (Fig. 3) in order to obtain composite representative samples in triplicate for enhanced precision. A1 adjoins the urban locale highly infested with vegetation and diffuse sewage inputs; A2 is situated closer to the lakes' exit and recipient of run-off from paddy fields; D1 vicinity is contributed with key inlet and STP discharges; D2 typifies the floating garden area; and D3 nearby the outlet has hotel and house-boat zone.

2.2 *Physico-Chemical Analysis of Water*

The collection, preservation, preparation, storage and estimation of water samples follow standard methods of (APHA 2005; Morford et al. 2005; Mazej and Germ 2009; Parker and Bloom 2005; USEPA 1994; Matusiewicz 2003). The water samples were analysed for temperature, pH, conductivity, bicarbonates (BIC), chloride, calcium, magnesium, sodium, potassium, nitrate nitrogen and total phosphorus on monthly basis while trace elements of Al, As, Cd, Co, Cr, Cu, Fe, Hg, Mn, Ni, Pb, Se, Sn and Zn were measured for spring, summer and winter seasons.

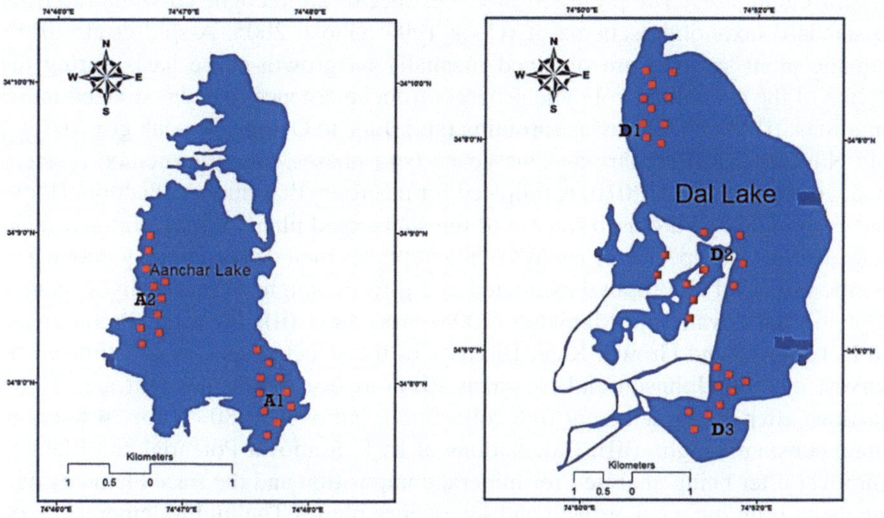

Fig. 3 Sampling stations of A1, A2, D1, D2 and D3

2.3 Chemical Analysis of Sediments

Surface sediment samples are collected with the help of shovel and Ekman dredge (~ 0 to 25 cm depth) at the selected sites from both the lakes on seasonal basis for 2 year study period. In order to reduce heterogeneous nature of sediments due to changes in hydrological regime and catchment area geomorphology, composite samples are constituted by mixing at least 5 sub-samples taken from each sampling station. Except pH and electrical conductivity, sediment parameters are carried out on oven dried samples following standard recommended methods from Ryan et al. (2007), Gupta (2004), Radojevic and Bashkin (2006) and Estefan et al. (2013). Their analysis includes H-ion concentration (pH), electrical conductivity, bicarbonates, organic-C, organic matter, total-N, C/N ratio, elemental composition of Ca, Cl, K, Mg, Na, P, S, Si and trace elements for each season.

2.4 Bio-Chemical Analysis of Macrophytes

Macrophytes employed in the study includes select species from emergent, free-floating, rooted-floating and submerged categories namely *Myriophyllum aquaticum* (Parrot-feather Watermilfoil), *Salvinia natans* (Floating-Moss or -Fern), *Nelumbo nucifera* (Asian Lotus) and *Ceratophyllum demersum* (Hornwort or Coontail) respectively. Their selection represents omnipresent abundance at each sampling location besides being high biomass yielding, eutrophication tolerant and hyperaccumulators. The preferred producer species are recognized while referring to standard taxonomic scheme of (Cook 1996; Ghosh 2005; Arshid et al. 2011). Aquatic plant samples are obtained manually on growth-phase basis during the stretch of the annual macrophytic progress from emergence to decay divided into 3 durations: (i) March to May as sprouting (sp.), June to October as peak growth (pg.) and November to February as senescence (sn.) phases. Quadrat method (Havens et al. 2004; Gunn et al. 2010) is followed for macrophytic sample collection. Hooks and Ekman dredge are used in case of the submerged plants falling within a float-able quadrat. The specific bioassays of macrophytes include **(i):** Photosynthetic pigments, viz., total chlorophyll estimated in 1 g fresh weight of macrophytic sample applying Sadasivam and Manikam (2005) procedure. **(ii):** Biomass, Productivity, NPP, Turnover and Growth Rate: Biomass of the selected species is estimated by harvest method (Johnson and Newman 2011) as dry weight per unit area (1 m^2 quadrat) after drying adherent free collection in an oven at 80 °C for 24 hours to attain constant weight. **(iii):** Calculations of BCF, Removal Potential and Element Turnover after being analysed for mineral composition and the trace elements during their sprouting, peak growth and senescence phases.The multi-elemental quantification of Ca, Cl, K, Mg, Na, P, S and Si in sediments and plant biomass is performed via Wavelength Dispersive X-ray Fluorescence Spectrometer (WD-XRF) wherein 1 g of oven dried, finely ground and sieved (0.425 mm/420 micron mesh) of such sample is used. A homogeneous sample is prepared with 1:4 ratio borate flux (Alloway 1995) for the quantitative analysis on WD- XRF spectrometer of Bruker S4 Pioneer make and model. However, the trace elements were determined on ICP-OES (Perkin Elmer Optima 5300 DV) instrument entirelyusing pre-digested extracts concentrated with HNO_3 and HCl refluxing. The simultaneous quantification follow manufacturer's standard operating procedures and conditions. But the analyte concentrations in sediment and biomass extract solutions are reported as $mgKg^{-1}$ on dry weight basis using the following W/V correction equation:

$$\text{Sample concentration}\left(\frac{\text{mg}}{\text{Kg}}\right) = \text{C} \times \frac{\text{W}}{\text{V}}$$

where C = concentration in the extract (mgL^{-1}); V = volume of extract (100 mL or 0.1 L) and W = weight of sample (1 g or 0.001 Kg).

2.5 *Statistical Methods*

The descriptive and illustrative statistical analysis including Means and Standard Deviation, Box plots, Correlation, Regression, Analysis of Variance (ANOVA), Cluster Analysis, Principal Components Analysis (PCA) and Biplotsis performed using MS-Excel 2010, PAST 3 and SPSS 19.The extrapolation, graphic visualization and interpretation of statistical procedures and scores are based on Reimann et al. (2008) and Greenacre and Primicerio (2013).

3 Results

3.1 Surface Waters

The inter-annual quantitative analysis of water quality parameters presents a vivid variation at the given sites (Fig. 4).

The mean epilimnion temperatures recorded for the sites reflect a distinct contrast with higher values in summers and a declining trend towards the winters as a

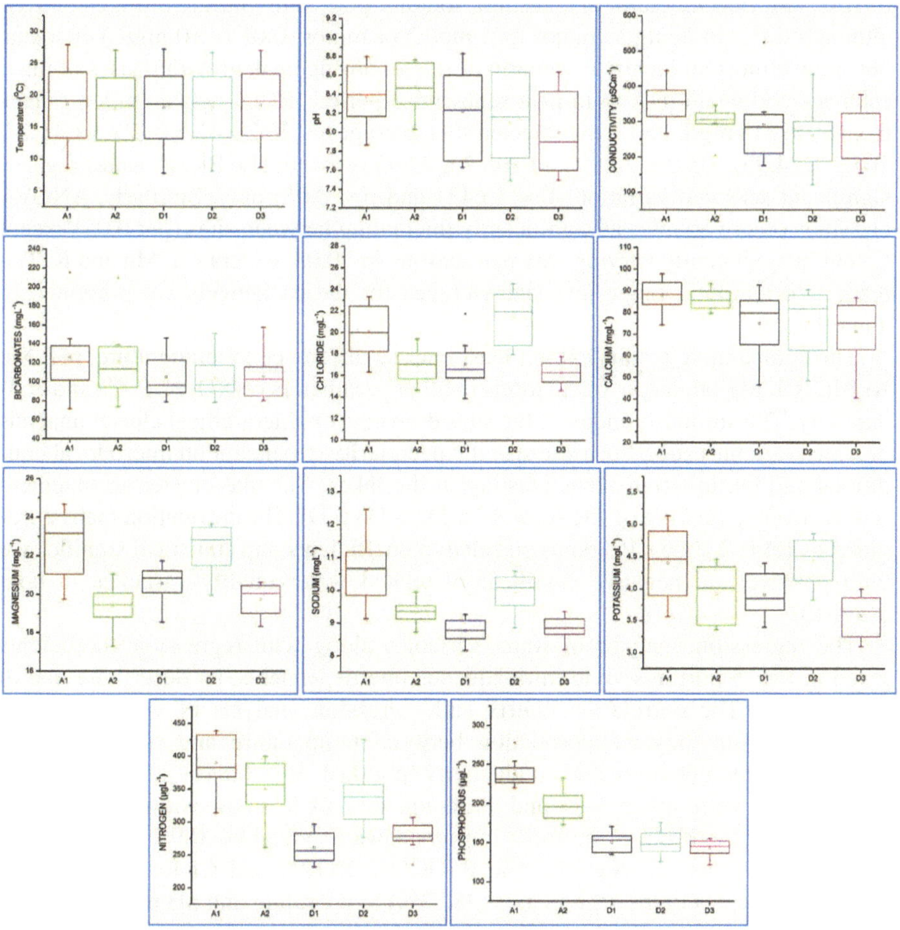

Fig. 4 Spatial variation in water quality variables

season specific feature. It ranged between 5 and 30 °C with averages of 18.6 ± 7.3, 17.9 ± 7.2, 18.1 ± 7.4, 18.1 ± 7.0, 18.5 ± 7.0 recorded at the sites. The temperature variance between sites ($F = .013$ and $p > 0.05$) is insignificant, however, seasonal variation is significant ($F = 1.39$ and $p < 0.05$). pH fluctuated between 7.2 and 8.8 symbolizing alkaline waters with comparatively higher summer peaks established by macrophytic productivity. The site-wise averages are 8.4 ± 0.32, 8.5 ± 0.28, 8.2 ± 0.43, 8.1 ± 0.43 and 8.0 ± 0.41. ANOVA for pH has significance between lakes ($F = 7.807$ and $p < 0.05$) and between seasons ($F = 9.782$ and $p < 0.05$) as well. Conductivity exhibited a fall towards the spring and summer months presumably due to dilution and ion uptake by flora present in the water-column. Conductivity has minimum value of $100 \mu SCm^{-1}$ and maximum $565 \mu SCm^{-1}$ besides having average site values of 353.9 ± 57.46, 306.4 ± 48.88, 294.7 ± 113.83, 266.8 ± 100.87 and 271.3 ± 87.60. Seasonal ANOVA of conductivity is more significant ($F = 7.829$ and $p < 0.05$) than lake comparisons ($F = 4.526$ and $p < 0.05$). The water characteristics of BIC, chloride, calcium, magnesium, sodium, potassium, nitrate nitrogen and TP although differ in being as major (> 5 mg/L) or minor (0.01 to 10 mg/L) elements but show almost an identical scenario of retreat during peak growth phases of macrophytes and again pickup in post senescence period. HCO_3 concentration ranges from 60 to 218 mg/L and at the selected sites average to 126.2 ± 42.57, 121.8 ± 41.77, 105.2 ± 24.93, 105.0 ± 25.89 and 105.7 ± 27.95 respectively. Bicarbonates register significant seasonal variation ($F = 35.431$ and $p < 0.05$) only. Similarly, ANOVA classifies other variables as significantly fluctuating between sites ($p < 0.05$) except Ca but insignificantly varying between seasons ($p > 0.05$) except Ca, Mg and K. The respective temporal range and site averages for these elements are presented in Table 1.

The comparative results exhibit least inter se difference in temperature, pH, Na, K, BIC, Cl, Mg but larger fluctuations in other parameters of NO_3^-N, P, Ca and conductivity. The similarity index of the sites derived from hierarchical cluster analysis of water variables suggests a site specific peculiarity dependent on micro local conditions and perturbation types existing in the lakes. The site cluster arrangement reflects trophic ranking of site A1 > A2 > D2 > D3 > D1.The correlation matrix provided in Table 2 shows Pearsons correlation coefficients and statistical significance with positive and negative significantly related water quality variables in bold asterisks.

The regression analysis of water variables along with regression coefficients given in Fig. 5 portrays vivid intra-aqueous interdependence by describing line of best-fit (R^2). The correlation matrix and regression analysis of water variables deduce a strong positive correlation between temperature and pH ($p < 0.01$, $R^2 = 0.8617$); temperature and conductivity ($p < 0.01$, $R^2 = 0.8419$); temperature or conductivity with other ions and intra-aqueous ionic composition ($p < 0.01$, $R^2 > 0.5$). PCA disclose the most important water variables (89% contribution) being conductivity, K, Na, Mg, Ca, P, HCO_3^-, NO_3-N and Cl for the first axis (60.27%) whereas in the second axis (28.73%) temperature and pH predominated.

The graphical presentation in Figs. 6 and 7 illustrate average trace element concentration in lake waters and their seasonal changes respectively. The trace elements

Table 1 Temporal range and site averages for elements in surface waters

Element	Range	A1	A2	D1	D2	D3
Cl (mg/L)	11.6–28.9	20.1 ± 3.40	17.0 ± 1.92	17.1 ± 3.06	21.4 ± 3.07	16.3 ± 1.44
Ca (mg/L)	38–102	88.1 ± 8.39	86.2 ± 7.56	75.1 ± 17.34	75.6 ± 17.09	71.4 ± 16.2
Mg (mg/L)	16.8–26	22.6 ± 2.26	19.4 ± 1.28	20.6 ± 1.36	22.4 ± 1.59	19.8 ± 1.05
Na (mg/L)	7.5–13.4	10.6 ± 1.23	9.4 ± 0.66	8.8 ± 0.47	10.0 ± 0.72	8.9 ± 0.52
K (mg/L)	2.6–5.3	4.4 ± 0.70	3.9 ± 0.52	3.9 ± 0.45	4.4 ± 0.49	3.6 ± 0.42
NO_3-N (μg/L)	220–557	390.3 ± 42.6	351.3 ± 45.7	261.4 ± 25.1	338.4 ± 57.4	279.3 ± 23.5
P (μg/L)	110–290	238.17 ± 20.36	197.96 ± 20.68	150.67 ± 13.70	150.67 ± 18.46	143.42 ± 12.66

do not follow any definite seasonal variability. But ANOVA between sites confirm significance for Al ($F = 49.14$ and $p < 0.05$), Cr ($F = 3.854$ and $p < 0.05$), Cu ($F = 28.64$ and $p < 0.05$) and Ni ($F = 7.05$ and $p < 0.05$). Besides as per correlation analysis among trace elements in surface- waters, Al shows significant correlations with Co, Cr, Cu and Ni while as association of As is significant with Cd and Ni. Similar is the case of Cd versus Pb;Coversus Cu and Ni; Cr versus Cu; Cu versus Ni atsignificance level of "$p < 0.01$ or 0.05".

3.2 *Surface Sediments*

Sediment characterisation featuring seasonal quantification marks a distinctive outcome. The boxplots of analysed parameters pronounce site wise trend in Fig. 8. The observed range of H-ion concentration in surface sediments is 6.16 to 8.35 and the respective mean site values include 7.22 ± 0.60, 7.30 ± 0.64, 7.58 ± 0.42, 7.20 ± 0.58, 7.49 ± 0.56. Sediment conductivity ranged between 234 and 498 μScm^{-1} at 25 °C and the average inter-site comparison is as 395.4 ± 55.7, 312.6 ± 47.2, 293.3 ± 36.4, 363.3 ± 50.9 and 323.3 ± 38.4 μScm^{-1} respectively. The sediment bicarbonate content has minima of 116 and maxima of 264 $mgKg^{-1}$ and varied between the designated sites as 187.4 ± 25.3, 193.5 ± 37.2, 169.8 ± 41.5, 172.6 ± 41.9 and 189.6 ± 42.3 $mgKg^{-1}$ correspondingly. The ranges and average site wise percentage dry-weight variations in OC, OM and C/N ratio are visualised in Fig. 9. The assessed average percentage elemental composition on dry-weight (DW) basis (Fig. 10) reveal a slight inter-seasonal variance and follow the dominance order of Si > Ca > Mg > K > Na > P > S > Cl. Besides, the average micro- and trace- element quantification (Fig. 11) remains seasonally almost uniform and represent the descending series of Fe > Al > Zn > Mn > Cu > Cr > Ni > Co > As > Sn > Pb > Cd while Hg and Se as BDL. The above acquired data for sediment metal profile paved way for Pollution Indices and Sediment Quality Criteria adoption as follows.

Table 2 Correlations among water variables

		T	pH	Conductivity	Bicarbonates	Cl	Ca	Mg	Na	K	Nitrate-N	P
T	Pearson correlation	1										
	Sig. (2-tailed)											
pH	Pearson correlation	.918**	1									
	Sig. (2-tailed)	.000										
Conductivity	Pearson correlation	−.917**	−.932**	1								
	Sig. (2-tailed)	.000	.000									
Bicarbonates	Pearson correlation	−.736**	−.507*	.585**	1							
	Sig. (2-tailed)	.000	.012	.003								
Cl	Pearson correlation	.266	.159	−.308	−.228	1						
	Sig. (2-tailed)	.209	.459	.143	.285							
Ca	Pearson correlation	−.531**	−.568**	.617**	.222	−.114	1					
	Sig. (2-tailed)	.008	.004	.001	.296	.595						
Mg	Pearson correlation	−.803**	−.715**	.800**	.613**	−.307	.329	1				
	Sig. (2-tailed)	.000	.000	.000	.001	.144	.116					
Na	Pearson correlation	−.641**	−.672**	.641**	.299	.173	.532**	.590**	1			
	Sig. (2-tailed)	.001	.000	.001	.156	.420	.007	.002				
K	Pearson correlation	−.661**	−.726**	.709**	.249	.258	.677**	.497*	.813**	1		
	Sig. (2-tailed)	.000	.000	.000	.240	.223	.000	.014	.000			
Nitrate- N	Pearson correlation	.356	.382	−.312	−.408*	.295	.210	−.124	.083	.073	1	
	Sig. (2-tailed)	.088	.065	.138	.048	.162	.324	.563	.700	.736		
P	Pearson correlation	−.485*	−.530**	.531**	.187	.343	.614**	.414*	.745**	.784**	.378	1
	Sig. (2-tailed)	.016	.008	.008	.381	.101	.001	.044	.000	.000	.069	

**Correlation is significant at 0.01 level (2-tailed), *Correlation is significant at 0.05 level (2-tailed)

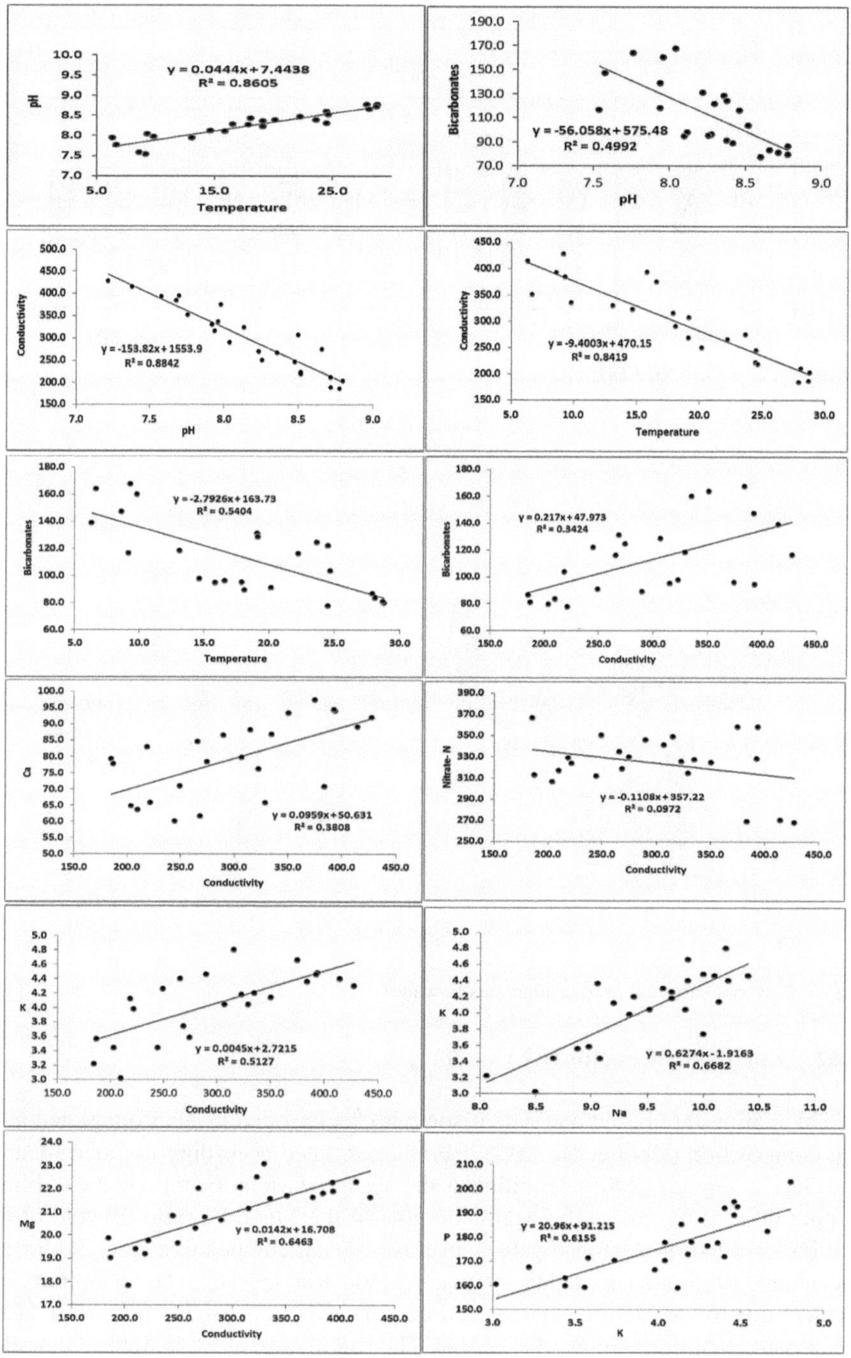

Fig. 5 Regression analysis of water variables

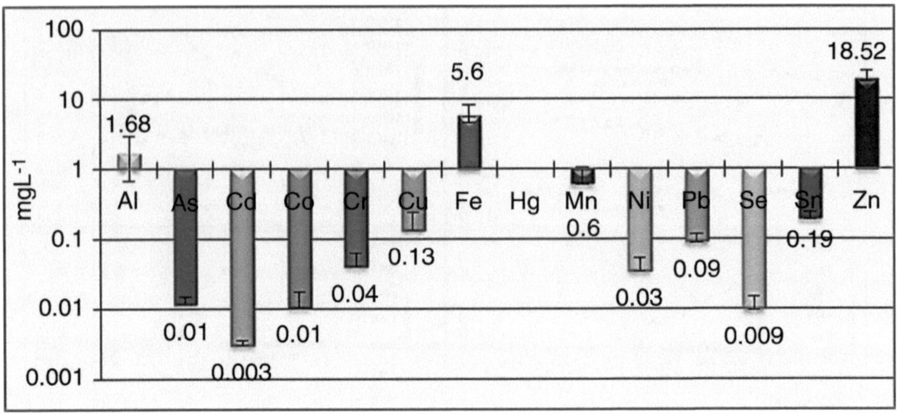

Fig. 6 Average trace metal concentration in waters

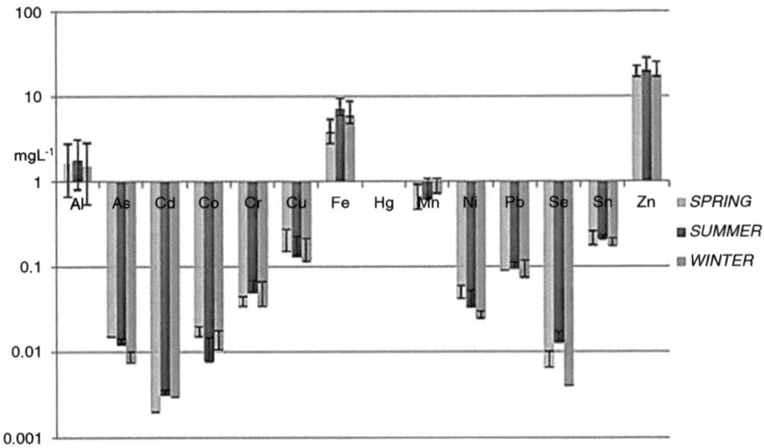

Fig. 7 Seasonal variation in trace elements in waters

3.2.1 Enrichment Factor (EF)

EF calculations are based on comparison with background levels from global average composition (Choi et al. 2012). EF is determined according to Guerra-Garcia and Garcia- Gomez (2005) definition as the observed Metal/Fe ratio in the sediment sample divided by natural background value of the Metal/Fe ratio. Devoid of any real background or reference values, average crustal composition from Taylor and Mclennan (1995) are adopted to serve as preindustrial levels. Fe being an immobile element due to vast natural sources and dominant input is used to normalize heavy metal contamination (Goher et al. 2014). The calculation of EF in Table 3 classifies each trace element into 'No Enrichment' to 'Very High Enrichment' classes of sediment samples collected from the study stations in accordance to Han et al. (2006)

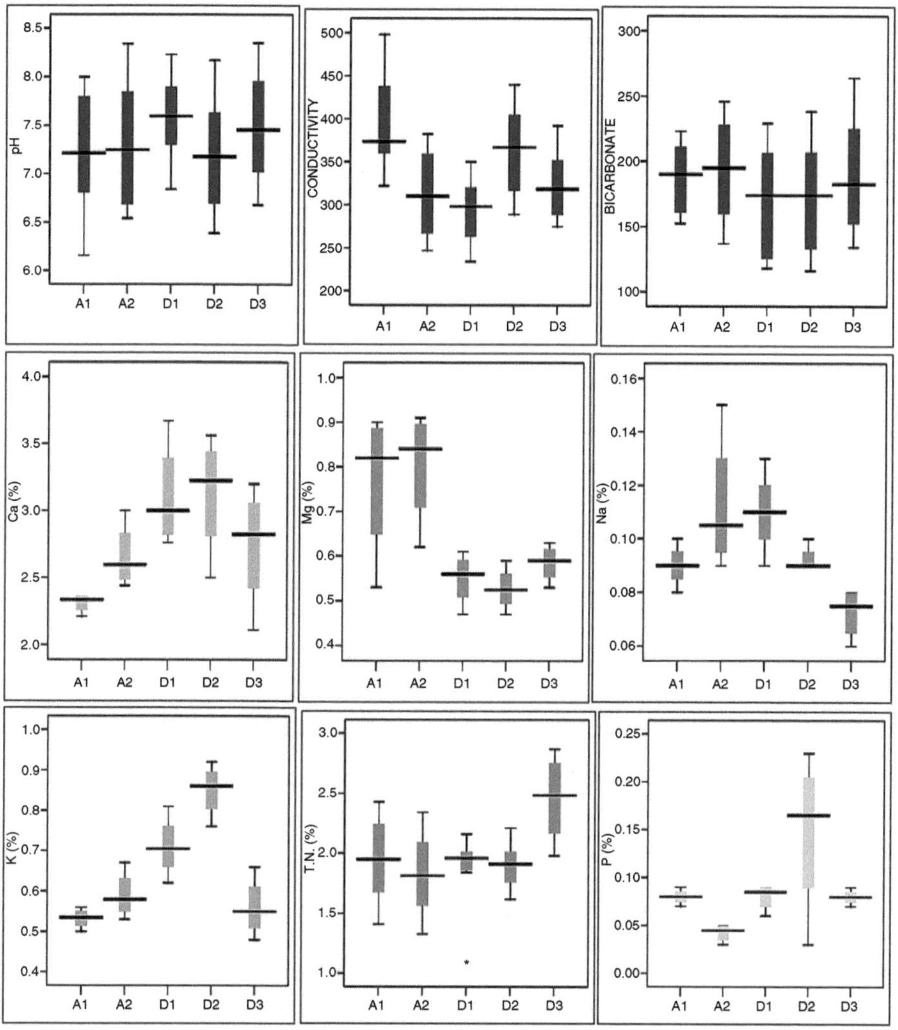

Fig. 8 Boxplots of sediment parameters

interpretation. Besides, Percentage Enrichment Factor (%EF) for each element is calculated as per (Chandrasekaran et al. 2015) by applying the equation: $\%\ EF = C - C_{min} / C_{max} - C_{min} \times 100$ where C = mean total concentration of an element; C_{min} = the minimum metal concentration and C_{max} = the maximum metal concentration.

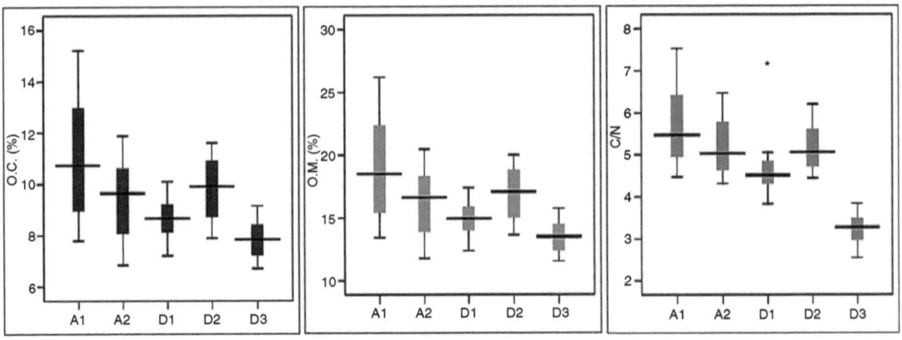

Fig. 9 Boxplots of OC, OM and C/N

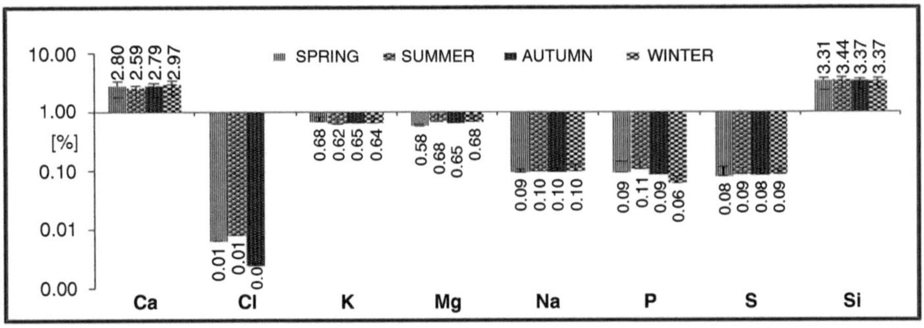

Fig. 10 Average percentage elemental composition of surface sediments

3.2.2 Geo-Accumulation Index (I_{geo})

$I_{geo}$is a quantitative criterion originally propounded by Muller (1979) to measure pollution intensity on a given qualitative scale (Macias et al. 2006; Muller 1981) as $I_{geo} = \log_2 (C_n / 1.5B_n)$ where C_n = measured concentration of an examined element (n) in the enriched sediment; B_n = geochemical background concentration of the element (n) and 1.5 is the matrix correction factor to minimize possible lithogenic differences in the control values (Mediolla et al. 2008). Here the world's crustal surface average composition by Taylor and Mclennan (1995) is used as the geochemical background. The measures of I_{geo} scale in Table 4 labels 'Unpolluted' to 'Highly Polluted' element groups in sediments (Macias et al. 2006).

3.2.3 Contamination Factor (CF)

CF connotes to a fraction of mean content of an element (C_i) as antecedent and its preindustrial level (C_n) as consequent, viz., $CF = C_i / C_n$. The construal of CF differentiates between 'Contaminated' and 'Uncontaminated' score of elements in Table 5 as per Raj and Jayaprakash (2007).

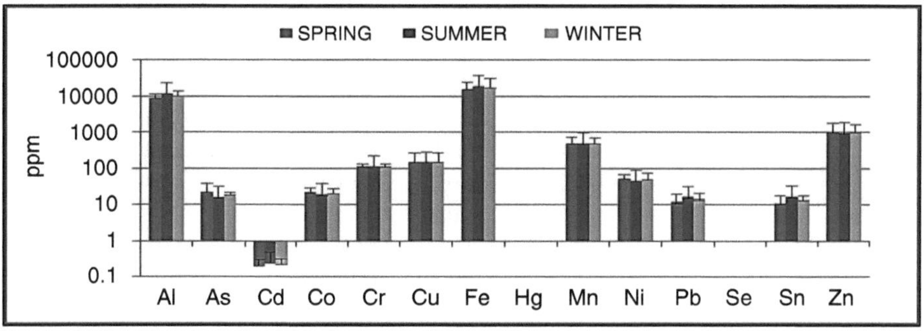

Fig. 11 Mean trace element quantification in sediments

Table 3 Enrichment factor (EF) and % EF classification of elements

Symbol	$EF_{[COMBINED]}$	% EF	EF $_{[ANCHAR]}$	%EF	EF $_{[DAL]}$	% EF	EF Interpretation
Al	0.2	57.7	0.2	59.0	0.25	57.37	No enrichment
As	22.0	20.9	24.0	23.6	20.43	30.77	Very high
Cd	4.5	42.4	3.7	33.3	5.02	48.52	Moderate
Co	2.3	41.7	2.3	47.1	2.29	38.72	Moderate
Cr	2.6	64.4	2.6	75.7	2.56	55.94	Moderate
Cu	10.9	29.5	10.0	24.8	11.53	30.73	Significant
Fe	1.0	37.6	1.0	37.1	1.00	35.24	Minor
Mn	1.4	43.5	1.5	51.0	1.34	38.63	Minor
Ni	1.9	40.3	1.9	43.0	1.93	38.07	Minor
Pb	1.6	54.5	1.3	47.4	1.79	57.79	Minor
Sn	4.5	51.7	5.9	60.0	3.41	27.30	Anchar significant/Dal moderate
Zn	26.4	34.7	27.0	32.1	26.03	35.52	Very high

3.2.4 Integrated Pollution Index (IPI) and Pollution Load Index (PLI)

IPI is the mean value of single factor pollution index (PI) for each element which in turn denotes the fraction of C_n (observed metal concentration) and B_n (baseline metal concentration) (Wei and Yang 2010). Again, Pollution Load Index (PLI) provides an integrated and site-wise comparative assessment of cumulative heavy metal pollution (Qingjie et al. 2008). It is given as the nth root of the product of n contamination factor values, viz., $PLI = (CF_1 \times CF_2 \times CF_3 \times \ldots \times CF_n)^{1/n}$ where CF is the contamination factor and n designates the number of metals examined. IPI sums upthe lakes have 'High Level of Pollution'while PLI enlists them as 'Unpolluted to Moderately Polluted'on their corresponding scales in accordance to Wei and Yang (2010) and Zhang et al. (2011) respectively.

Table 4 Geochemical Index (I_{geo}) measure of elements

Element	$I_{geo\ [COMBINED]}$	$I_{geo\ [ANCHAR]}$	$I_{geo\ [DAL]}$	Interpretation
Al	−3.59	−3.58	−3.59	Unpolluted
As	2.93	3.14	2.76	Moderately Tohighly Polluted
Cd	0.61	0.44	0.74	Unpolluted Tomoderately POLLUTED
Co	−0.34	−0.25	−0.39	Unpolluted
Cr	−0.17	−0.06	−0.23	Unpolluted
Cu	1.91	1.88	1.94	Moderately polluted
Fe	−1.51	−1.45	−1.59	Unpolluted
Mn	−1.06	−0.89	−1.16	Unpolluted
Ni	−0.58	−0.5	−0.64	Unpolluted
Pb	−0.86	−1.03	−0.75	Unpolluted
Sn	0.62	1.1	0.18	Unpolluted to moderately polluted
Zn	3.19	3.31	3.11	Highly polluted

Table 5 Contamination Factor (CF) of elements

Element	$CF_{[COMBINED]}$	$CF_{[ANCHAR]}$	$CF_{[DAL]}$	Interpretation
Al	0.12	0.13	0.12	Uncontaminated
As	11.41	13.23	10.19	Contaminated
Cd	2.35	2.04	2.51	Contaminated
Co	1.20	1.26	1.14	Contaminated
Cr	1.34	1.44	1.28	Contaminated
Cu	5.66	5.52	5.75	Contaminated
Fe	0.52	0.55	0.50	Uncontaminated
Mn	0.72	0.81	0.67	Uncontaminated
Ni	1.00	1.06	0.96	Uncontaminated
Pb	0.83	0.73	0.89	Uncontaminated
Sn	2.31	3.23	1.70	Contaminated
Zn	13.73	14.86	12.99	Contaminated

3.2.5 Sediment Quality Guidelines (SQG's)

SQG's including Threshold Effect Level (TEL), Effects Range Low (ERL), Lowest Effect Level (LEL), Probable Effect Concentrations (PEC) and Threshold Effect Concentrations (TEC) benchmark connotations include the final corollaries. SQG's include chemical specific numerical criterion or standard or guideline or value or indicator or alert/action/threshold range levels above which site contamination or ecological effect occurrence are expected. Such quantitative index or characteristic concentration when evaluated demonstrates the current state, quality or ecological trend of a system for its strategic regulatory management practices. Adverse effects are rarely observed due to elements < TEL or < ERL and no potential ecological risks are associated below PEC or TEC while acceptable effects occur at < LEL. These SQG's are compared to the observed mean values for various elements in Table 6 in order to ascertain their pollution status and associated ecological risks (MacDonald et al. 2000; Maanan et al. 2015).

Table 6 Sediment quality guidelines

Sediment quality guidelines[$mgKg^{-1}$ or ppm or μgg^{-1}]						
Element	TEL	ERL	LEL	PEC	TEC	Observed Mean
As	5.9	33	6	*	*	17.1
Cd	0.6	5	0.6	5	1	0.2
Cr	37.3	80	26	110	43	114.1
Cu	35.7	70	16	150	32	141.5
Pb	35	35	31	130	36	13.3
Hg	0.17	0.15	0.2	*	*	*
Ni	18	30	16	49	23	50.0
Zn	123	120	120	460	120	975.2

(*indicates unfamiliar value)
TEL Threshold Effect Level, *ERL* Effects Range Low, *LEL* Lowest Effect Level, *PEC* Probable Effect Concentrations, *TEC* Threshold Effect Concentrations

3.3 Macrophytes

The biochemical assays of select macrophytes for suitable parameters carried out in tandem with water and sediment associates divulge their cohesive and codependent networking. A marked difference in mean total chlorophyll content (mgg^{-1}) of fresh weight across the hydrophyte species and their growth-phases of sprouting (sp.), peak growth (pg.) and senescence (sn.) is shown in Fig. 12. The maximum total chlorophyll concentration is noted in *C. demersum* (0.92 mgg^{-1}) during its peak growth at site D3 while a minimum 0.02 mgg^{-1} is recorded both in case of *M. aquaticum* at site A2 and *S. natans* at site D3 throughout senescence. The individual ranges of chlorophyll content (mgg^{-1}) for *M. aquaticum, N. nucifera, C. demersum* and *S. natans* include 0.02 to 0.37, 0.05 to 0.72, 0.10 to 0.92 and 0.02 to 0.87 respectively.

The productive capacity of the investigated macrophytes calculated on the basis of dry weight mean biomass (gm^{-2}) for each experimental species in the three major growth phases is presented in Fig. 13. The biomass calculations reflect subsequent ranges for each species as 107 to 960 gm^{-2} for *M. aquaticum*, 66 to 750 gm^{-2} for *N. nucifera*, 39 to 231 $gm^{-2}$for *C. demersum*and 1 to 55gm^{-2} for *S. natans*. In terms of mean peak dry weight biomass values, *M. aquaticum* (880.2 gm^{-2}) dominated *N. nucifera* (678.4 gm^{-2}), *C. demersum* (182.4 gm^{-2}) and *S. natans* (45 gm^{-2}). A similar fashion low occurred in dry weight (gm^{-2}) during their senescence period ranking by way of 152.4 > 75.4 > 49 > 4.2 respectively. The percentage contribution of various life form classes among the studied macrophytes towards production (dry weight biomass, gm^{-2}) as well as specific percent relative difference during sprouting, peak growth and senescence phases is described in pie-chart diagrams (Fig. 14). *M. aquaticum* accumulated 49% of the dry weight biomass annually whereas for *N. nucifera* amounted to 38%, *C. demersum* to 10% and *S. natans* contributed only 3%.

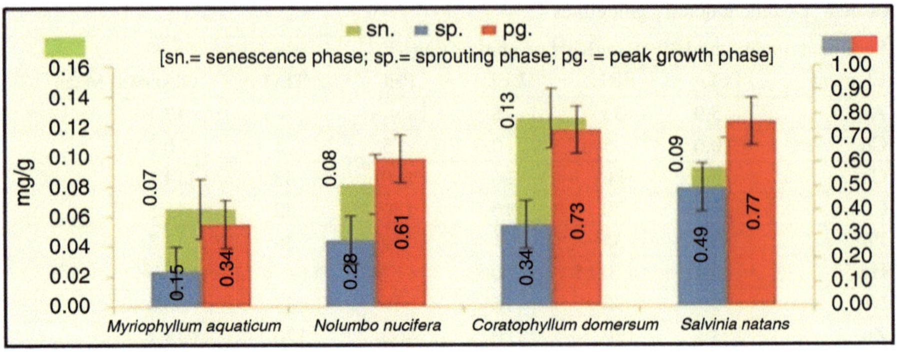

Fig. 12 Average total chlorophyll concentration of the macrophytes

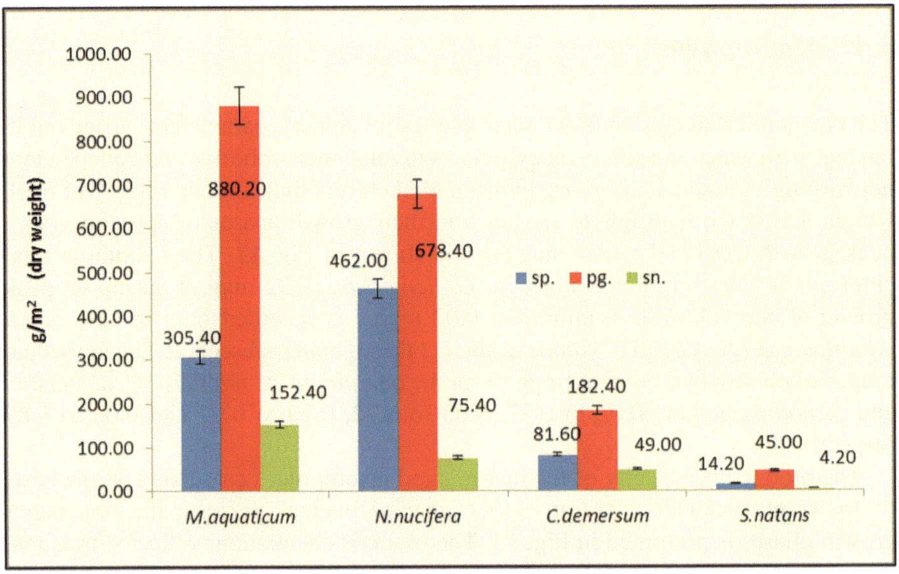

Fig. 13 Average biomass values in the growth phases (gm^{-2} dry weight)

Other biomass parameters like Productivity, NPP and Specific Growth Rate establish uniform variations in the experimental species of the order: emergent *M. aquaticum* > rooted floating *N. nucifera* > submerged *C. demersum* > free floating *S. natans* but Species Turnover is highest in case of *S. natans* and lowest for *C. demersum*.The average tissue concentration of analysed elements fluctuated not only between the select species (Fig. 15) but also depicted growth-phase variations within the same species. However, no significant site contrasts were noticed. The ANOVA test conducted between the species ('df' = 11) for the mineral composition

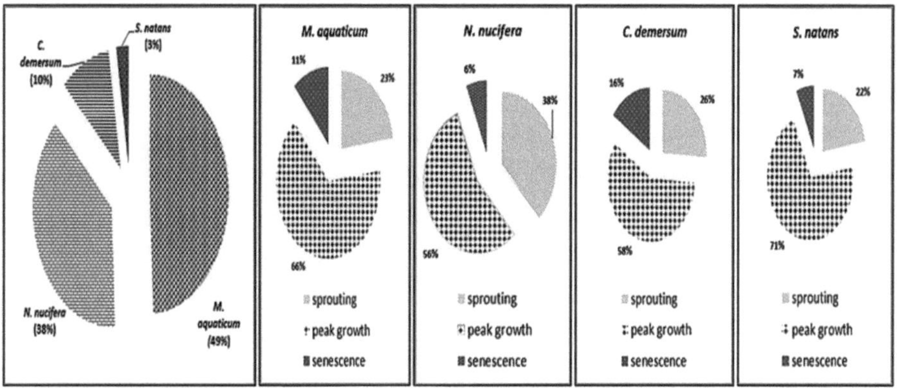

Fig. 14 Species-wise percentage biomass (gm^{-2}) during growth-phases

of macrophytes convey significance for Cu ($p < 0.05$), K ($p < 0.001$), Na ($p < 0.05$), Se ($p < 0.01$) and Si ($p < 0.01$) only. Whereas ANOVA for their growth-phases (‘df’ = 11) reflect significance in case of Al ($p < 0.05$), As ($p < 0.001$), Ca ($p < 0.05$), Co ($p < 0.05$), Cr ($p < 0.05$), Fe ($p < 0.05$), Mg ($p < 0.05$), Mn ($p < 0.001$), Ni ($p < 0.01$), Pb ($p < 0.01$) and Zn ($p < 0.001$).

The BCF criterion adopted indicates hyperaccumulation phenomenon for most of the micro- and trace- elements in *C. demersum* and *S. natans* except for Pb, Sn and Zn. And *M. aquaticum* hyperaccumulates Cd and Mn while *N. nucifera* Mn only. The Removal Potential of the macrophytes clearly stipulates species specific variation ensuing a generalized arrangement of emergent *M. aquaticum* > rooted floating *N. nucifera* > submerged *C. demersum* > free floating *S. natans*. Nevertheless, element-wise Removal Potential of each studied macrophyte for accumulation of nutrients and traces respectively follow the following pattern:

M. aquaticum: Si > N > Ca > K > S > Cl > Mg > P > Na and Fe > Al > Mn > Zn > Cu > Cr > Ni > As > Co > Sn > Pb > Se > Cd.

N. nucifera: K > Ca > N > Si > Cl > P > Mg > S > Na and Fe > Al > Mn > Zn > C u > Ni > Cr > As > Pb > Sn > Co > Cd > Se.

C. demersum: Ca > N > Si > K > Cl > P > Mg > S > Na and Fe > Al > Mn > Cu > Zn > Ni > Cr > Sn > As > Co > Pb > Cd > Se.

S. natans: Ca > N > Si > K > Mg > Cl > P > S > Na and Fe > Al > Mn > Zn > Cu > Cr > Ni > As > Sn > Co > Pb > Cd > Se.Significant Element Turnover rate is observed for almost all minerals except Mg and Na in case of *M. aquaticum*. However, *N. nucifera* has significant Turnover for Ca, Cl, Mg, As, Cd, Co, Cr, Cu, Ni, Pb, Se, Sn and Zn. Similarly, less significance is reflected for K, Mg, Na, Al and Se Turnover by *C. demersum*. And *S. natans* has lesser Turnover rate for Mg, N and Se only.

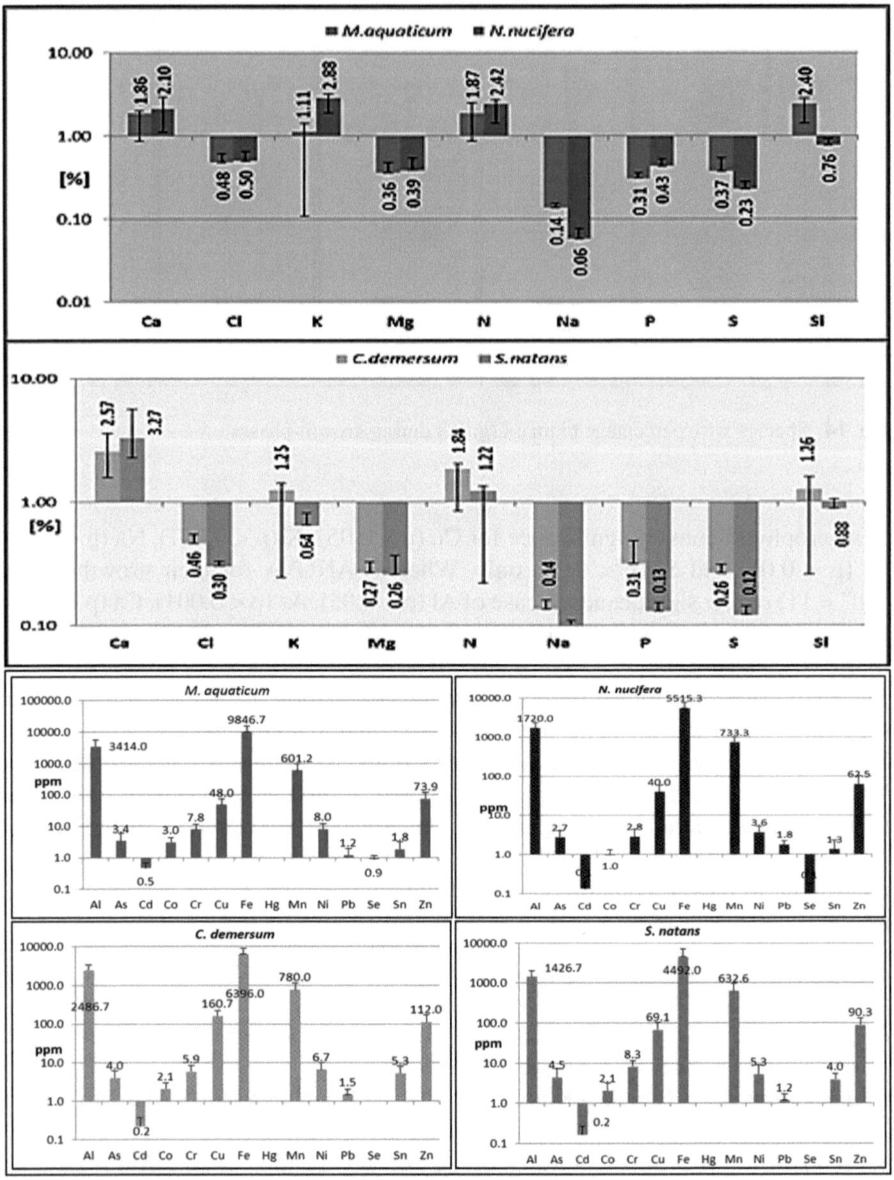

Fig. 15 Bioconcentration (%) in *M. aquaticum*, *N. nucifera*, *C. demersum* and *S. natans*

4 Discussion

4.1 Aquatic Nutrient Dynamics

The results of the study establish interdependent and interactive compartmentalization of inorganic elements in lacustrine components of water, sediment and macrophytes on time and space scales. Water quality (ANZECC/ARMCANZ 2000), sediment characteristics (Batley 2000), biological indicators (Wright et al. 2000) and even key ecological processes like primary production (Bunn and Davies 2000) are essential to assess the integrity and health of the ecosystem. The temporal variation in ionic composition of natural waters along with its spatial distribution in a region aids significantly to distinguish the source (Gupta et al. 2015). The limnochemical statistics that evolved during the study retro expound unambiguous and unavoidable evolving physico-biochemical configuration of the sites in both lakes under multiple anthropogenic stress (de Jonge et al. 2002). The chemical composition of water closely related to dual factors of concerned catchment and human intervention thereof (Bu et al. 2014). The lakes studied deem to be classified under eutrophic category with considerable differences in bio-physicochemical milieu due to varied nature of operative anthropogenic stress. But for the semi-drainage type hydrology flushing via regular outflow channels supplemented by ephemeral and sub-aqueous inputs is vital for their self-reclaimed homeostasis since the basin holding-time of water keeps low ($2^{1/2}$ months). The statistical insignificance in recorded epilimnion temperature variance as well as significant temporal discrepancy discloses similar geo-climatic setting of the lakes. Although the entire recorded pH profile (> 7) reflects buffered waters but has prominent variance both site- and time-wise suggesting position and period specific photosynthetic activity that withdraws HCO_3^- thereby elevating diurnal pH (Boyd et al. 2016). The waters are well buffered having insignificant site variation but there is gradual increase in summers due to higher proton uptake during peak photosynthetic activity thereby shifting CO_2 and HCO_3 equilibrium (Maberly 1996). Contrarily, lower pH during winter supposedly on account of excess carbonic acid generation is due to higher [Respiration: Photosynthesis] proportion. A higher pH associated with elevated temperature and alkaline conditions is reported to favour co-precipitation of carbonate and phosphate due to rapid carbon assimilation from dissolved bicarbonates (Pandit 1999). Despite a temporary inverse relationship of bicarbonate concentration and pH during peak-growth summer days, the enriched alkaline pH switch ascribes to equivalent carbonate dominance (Wetzel 2001). Conductivity describes the trophic status in terms of total nutritive ionic strength of the water column. Its decline in peak growth season implies ionic depletion by photoautotrophic consumption and biogenic calcification (precipitation of $CaCO_3$) commonly referred to as seasonal whiting or clouding of lakes (Thompson et al. 1997). The ionic strength (conductivity) measures vindicate superior eutrophy (solute richness) besides autotrophic assimilation and biocalcification contributing towards its summer falls (Wiik et al. 2015). The cationic content revealed a progression order of

Ca > Mg > Na > K (Pandit 2002) all along with anionic predominance of HCO_3^- and Cl^-(Kalff 2002; Njenga 2004). The lime grade catchment characteristic contributed to Ca dominance phenomenon and hard water type classification (Jeelani and Shah 2006; Singh and Singh 2010). K features base association with agricultural runoff and Cl to faunal organic pollution (Berzas-Nevado et al. 2009; Khan and Ansari 2005). Although a range of human actions (agriculture, farming, sewage, factories, etc.) result in N species inflow (Yu et al. 2012), the spring fed basins of both the lakes and preferential reduced NH_4^+ autotrophic assimilation instead of oxidised NO_3^- also greater concentration of nitrate– N in water (Kalff 2002). Additionally, isolated geogenic N pockets are ascribed as evaporative deposits in phyllite, schist and carbonate bed rocks in Kashmir (Singh and Singh 2010). Forest surface runoff too contain ≤20 mgL^{-1} nitrate as a natural source to lakes (Feichtinger et al. 2002). Summer decreases of the least abundant macronutrient (P) limits bio-productivity owing to its utilization and carbonate coprecipitation (Hayakawa et al. 2015). The overall total P limit surpassed critical eutrophic index (≤ 0.05 mgL^{-1}) but nitrate N persisted underneath it (≤ 0.5 mgL^{-1}) mostly (Srebotnjak et al. 2012). The amount of N and P forms is intimately related to biological productivity of aquatic ecosystems (Kalff 2002). NO_3– N shows the only negative association with HCO_3^- concluding eutrophication steers on C assimilation for it serves as the main inorganic N source (Olsen et al. 2014). The close covariance of P with Ca, Mg, Na and K depict their analogous limited absorbance constraint by the biota. The supportive understanding of water variable interaction is offered by significant coefficient of determination (R^2) alongside simple least square regression lines for pH versus temperature (0.8), conductivity versus pH (0.9), conductivity versus temperature (0.8) and BIC versus temperature (0.5) (Nnaji and Agunwamba 2014). This ascertains that higher temperatures favour bio-production and thereby elevate pH significantly. The diminishing conductivity at improved temperature conditions suggest higher bioaccumulation rate of nutrient ions like HCO_3, Ca, Mg, Na, K, P. The examination of the correlation matrix verily suggests temperature increments elevating pH but diminishing conductivity, HCO_3^-, Ca, Mg, Na, K and P significantly (Michard et al. 2001). However, Cl and NO_3-N remain unaffected due to temperature and pH ordeal. So, the factors of temperature and pH clearly state that improved growth environment for biological uptake and calcite co-precipitation of ions reduce their epilimnion concentrations in peak growth summer cycle. Such a systematics attributed to photosynthetically induced precipitation and utilization by autotrophs is concomitant to studies of Muller et al. (2016). An equation with the world average stream content (mgL^{-1}) of recorded Ca (> 15), Mg (> 4), K (> 2.3) and Na (> 6.3) acclaim pronounced human inputs from common uses like fertilizers, plaster, pigments, lime, alloys, pharmaceuticals, batteries, food additives, glass, baking powder, soft drinks, electroplating, caustic soda, water treatment chemicals, etc. (APHA 2005). ANOVA reveals significant seasonal variance in temperature, pH, conductivity and BIC compared to significant site contrasts for Cl, Mg, Na, K, NO_3– N and P (Belkhiri and Narany 2015; Xu et al. 2014). The bicarbonates, chloride, calcium, magnesium, sodium, potassium, nitrate nitrogen and total phosphorus show almost an identical scenario of retreat during peak growth phases of macrophytes and again pickup in post senescence period (Ouma and Mwamburi 2014; Wetzel 2001).

Anthropogenic effluents are established primary sources of traces in surface waters (Chon et al. 2012). While comparing the observed average epilimnion trace element concentration (μgL^{-1}) with the average global stream abundance in absence of any baseline data (APHA 2005), a vivid skewed hike for As (~ 10 > 2), Cd (~ 3 > 1), Cr (~ 40 > 1), Co (~ 10 > 0.2), Cu (~ 130 > 10), Fe (~ 5600 > 700), Pb (~ 90 > 3), Mn (~ 600 > 7), Ni (~ 30 > 1), Se (~ 9 > 0.2), Sn (~ 190 > 0.1) and Zn (~ 18,520 > 20) proclaim their anthropogenic geneses (Moiseenko et al. 2016). Pertinently, such proportions exceed maximum permissible limits for drinking water but are suitable as irrigation liquid ordinarily (FAO 2010; USEPA 2014). The exceptions of Fe (> 5 mgL^{-1}) and Zn (> 2 mgL^{-1}) enhance the maximal concentrations in irrigation recommendations (Tchobanoglous et al. 2003). Still, Zn has lesser toxicity in organic soils at pH > 6 and Fe contributes towards acidification and reduced P availability. Al, a non-priority pollutant at pH 6.5 to 9 is well below the freshwater Criterion Continuous Concentration (CCC) chronic exposure limit of ≥87 μgL^{-1}. Fe exceeds CCC at >1000 μgL^{-1} too. But all the priority pollutants of As, Cd, Cr, Cu, Pb, Ni and Se remained below chronic level of CCC except Zn (USEPA 2014). The complexing phenomena of heavy metals with carbonate and BIC limit their direct or indirect repercussions (Markich et al. 2001). Now, the possible source identification includes solid and liquid discards from an array of utilities namely alloys, storage batteries, pesticides, wood preservatives, electroplating, pigments, fertilizers, porcelain, glass, electrical wiring, roofing, utensils, piping, chemicals, paints, vapour lamps, mirror coatings, thermometers, catalysts, ceramics, fossil fuels, electronics, solder and so on (APHA 2005).

4.2 *Sediment Chemistry of Lakes*

The sediment characterisation and comparison provides an insight into its influences on the hydro-geomorphological setting of a lentic system. The abiotic quiescent zone of sediments is suitable long-term indicator of lake environmental conditions (Downing et al. 2008; Wagner et al. 2009). Sediment nutrient constitution corresponds to catchment land use (Knoll et al. 2014). They provide settling space for water carried elements, both acquired and innate, in the form of ions, organic/inorganic- complexes and dissolved/suspended- matter associations (Schaller et al. 2013). Sediments reflect weaker source but stronger sink behaviour and capacity (Ammar et al. 2015) depicted via transformations of adsorption/desorption, mineralization/demineralization, bioassimilation and burial phenomenon. Alkaline pH records of the sediment in general determine the catchment calcite predominance (Singh and Singh 2010) whereas slight summer diminution is temperature improved OM decomposition effect. Urban et al. (2009) labelled sediment pH as the principal factor enacting nutrient accessibility and movement. The seasonal OC changes in sediments aptly describe Gudasz et al. (2010) pattern of temperature induced decrement during summers fulfilling the mineralization promotion. The analysis of superficial sediment chemical characters reveals connections with trophic index (de Vicente et al. 2010). Besides, Lazzarino et al. (2009) and Heathcote and Downing (2012) evidenced eutrophication influence on C flow in lakes to the effect of 0.6 PgY^{-1} global OC burial estimate by Tranvik et al. (2009). The typical <10 C/N implies autochthonous OM (Wetzal 2001) besides low decomposition rates controlled by N budgeting and speciation via biochemical (enzyme and pH) fluctuations (Min et al. 2011) under alkaline conditions. It's possible reason could be NH_4^+ microbial preference instead of NO_3^- stimulating decomposition (Garland et al. 2010) despite eutrophied state. NO_3^- exacerbations also shift decomposer community structure from fungal to bacterial causing OM decomposition decline (Allison et al. 2008). Hence, eutrophied shallow lakes favour C sequestration. Surface sediments harbouring maximum OM profoundly determine biogeochemical cycling of major, minor and trace elements (Wetzal 2001). Significant correlation occurred for conductivity with OC and OM as the latter serves an established source for nutrient ions. Similar is the case with sediment OC, OM and TN as being complementary to one another.

Improved N:P ratios parallel chronic nitrogen influx in P-limited lakes (Elser et al. 2009), however, N-limited lakes suffer greater eutrophication catastrophes (Abell et al. 2010). Higher temperature and lower N:P ratio during summer develop the P internal loading process. This P recycling mediates via degradation (oxic), denitrification and sulphate reduction (Canavan et al. 2006). Nonetheless, higher Al, Ca and Fe proportions in sediments inactivate P mobilization (Smolders et al. 2006). Curbing the external N and P loads is effective in remediation but the sediment internal loading supplement compensates the loss (Jing et al. 2013). Meanwhile sediment dredging is beneficial for internal nutrient deloading only after the external inputs are checked (Jing et al. 2015). Trace metals bound to OM or Fe/Mn-oxides (Turner et al. 2004) separate on their decomposition and reductive dissolution respectively near hypolimnion-sediment overlap (Canavan et al. 2006). A comparative of sediment: water compartmentalization for different elements exhibit more retention potential for the sediment component conforming their sink-selves. Assuming the immensity and multifaceted networking in insitu conditions only a slight inconsistent inter-seasonal variance of Ca, Cl, K, Mg, Na, P and Si were observed. But BIC, conductivity, pH, OM, TN and C/N fluxed more on account of additional compartmental exchange quanta. A general gradual decline in nutrient concentration till culmination of active macrophytic growth phase can be associated with active/passive bioaccumulation or anoxic release from sediments (Selig and Schlungbaum 2003).

Devoid of any universal sediment pollution indicator or guideline, multiple approaches were applied on the recorded holistic seasonal and site contents of micro and trace elements (Lopes et al. 2014). The EF differentiated very high anthropogenic contamination of As, significant in case of Cu and none for Al (Wali et al. 2015). I_{geo} specifies Zn as highly polluting contaminant, Asto be moderate to highly polluting and Cu moderately polluting (Chandrasekaran et al. 2015). In the context of IPI and PLI, the sites and lakes exhibit moderate metal contamination (Bastami et al. 2015). CF calculations scale contamination with respect to As, Cd, Co, Cr, Cu, Sn and Zn (Iqbal et al. 2013). Furthermore, the observed mean concentration of Cr, Ni and Zn exceed the given sediment quality guidelines posing particular potential ecological risks (Ji et al. 2015). However, Cd and Pb don't transcend them. Again As is below ERL and Cu lags in PEC (Maanan et al. 2015; Sany et al. 2014). A generalization of the above indices elucidates the source of contaminants to be chiefly anthropogenic, degrading the sediment profile with hazardous consequences for the dependent biota in the offing (Yuan et al. 2014; Zhu et al. 2013). The lower [Ca]:[Al] values also suggest exsitu human cause for PTE transport to the designated lacustrine sites (Ammar et al. 2015). Nonetheless, the mobility of PTE's is effectively curtailed by OM enriched sediments and calcite co-precipitation (Paramasivam et al. 2015). And the SQG's suggest either rare or acceptable potential ecological risks from the observed mean quantum of PTE spectra.

4.3 Macrophytic Remediation Trends

The phytoassessment of aquatic environments involving vascular plants determine phytostimulation as well as phytotoxicity due to nutrient inputs from sewage, sediments and catchment surface runoff and toxins from commercial chemicals, industrial effluents, agrochemical and municipal wastes, hazardous wastes etc. respectively. The in-situ biomonitoring technique utilizing vascular aquatic vegetation gives information about environmental quality and plant responses to toxins and nutrients. The various parameters of study during phyto-assessment include measurement of biomass and pigment content, determination of bioaccumulation pattern, productivity, physiological effects and community composition. All of these diagnostic attributes reflect the trophic status and effect of all sources and types of contaminants present in the study area. The selected biomonitors among the macrophyte community manifest both contribution (command) and subservience to its abiotic allies. Macrophytes dominate mineral regulation between sediment and water (Kissoon et al. 2013). They are capable to remove nutrients even at low loading rates (Gottschall et al. 2007). Their adaptive advantage to lock up minerals from nutrient pool is helpful to check degradation primarily and restore mineral loading later on (Ismail et al. 2014). N and P are intimately related to biological productivity of aquatic ecosystems (Kalff 2002). The productivity also lies in consonance with developmental stage, ambient nutrient medium and physiognomy of a species (El-Otify 2015). Higher temperature favour bio-production and elevate carbonate driven pH significantly (Hasler et al. 2016). Among the evaluated species *M. aquaticum* accumulated 49% of the dry weight biomass annually, whereas *S. natans* contributed only 3% but it outdid in species turnover comparison. Biomass parameters like dry weight, productivity, NPP and specific growth rate establish similar variations in the experimental species but species turn-over is highest in case of *S. natans* and lowest for *C. demersum.* Substantial OM productions by macrophytes contribute towards nutrient immobilization and provide OC requirement for denitrification while decaying (Lin et al. 2002; McElarney et al. 2010). The upgraded biomass configuration property establishes an additional nutrient and trace element interception feature of hyperaccumulation from the growth medium (Shaltout et al. 2014).

Macrophytes are tested ecological indicators when compared against watershed (Sondergaard et al. 2010, 2013); catchment development (Sass et al. 2010) and even climate (Beck et al. 2014) inorder to validate ecosystem changes. Pertinently submergeds are least pollution and stress resistant species (Pulido et al. 2011; Sierszen et al. 2012). Environmental variations are more influential compared to interspecific competition in governing the community structure (Grabas et al. 2012). A combination of limnochemical and physicomorphic factors of water bodies describes the macrophyte community characterisation (Schneider et al. 2015) but nitrate commonly influences their growth, height and biomass profile as against ammonium supplements (Wersal and Madsen 2011). The gaining momentum for comprehending phyto-nutrient dynamics and their implications for water quality management or eco-restoration is a subject of site specific (habitat) application of a suitable phytoremediator (Udeigwe et al. 2015). A number of determinants like light, temperature, redox, pH, solubility, water flow, metal-compound chelation, metal concentration, metal species, exposure duration and cation exchange capacity altogether affect extent and rate of bioaccumulation for decontamination. The mere presence of nutrients is not the condition for rising or reducing uptake of toxic metals (Gothberg et al. 2004). The solubility of target element isn't a problem in aqueous media but its bioavailability matters which consequently control plant accumulation and translocation (Zaier et al. 2010). The peak nutrient uptake and bioconcentration coincided with peak biomass in summer and autumn (Garbey et al. 2004). The diminishing conductivity at improved temperature conditions suggest higher bioaccumulation rate of nutrients (Liu et al. 2014). The structural (physiognomy) difference between the species and the micro-habitat occupied within aquatic ecosystem are major factors related to bioconcentration of various elements (Lukacs et al. 2015). The removal potential pattern of the emergent accrued more Si, the rooted floating retained greater K, while the rootless submersed and free floating ones amassed Ca. Although the quantum of removal potential for different elements in the analysed species is divergent but the pattern is related which suggests unselective absorption (Sarwar et al. 2010). The elemental turn-over rates in select macrophytes closer to the reference value of 1 has significance. The vascular aquatic plants possess higher productivity (C:N and C:P ratios) with organic polymer body capacitating their slowed decomposition that improves nutrient and C capture potential and as a result net autotrophy (Cotner et al. 2009). The concurrence of peak bioaccumulation and biomass production empowers the concept of high yield harvesting (Quilliam et al. 2015). The harvesting of macrophyte biomass suitably during peak growth phase is perfect to avoid nutrient and metal remobilization from belowground vegetative parts or during senescence leaching on ultimate OM mineralization (Wang et al. 2015). Similarly, a mono annual harvesting practice improves height; shoot density and biomass of macrophytes, although leading to nominal increment for mass TN and TP removal rates (Zheng et al. 2015). The ANOVA comparing the species shows uptake preference mode for Cu, K, Na, Se and Si indicating specific role in the unlike macrophytes (Sistla et al. 2015). Contrarily, the growth-phase variance for Al, As, Ca, Co, Cr, Fe, Mg, Mn, Ni, Pb and Zn possibly implies their short-term raised presence or bioavailability during the annual developmental cycle (Wiik et al. 2015). BCF criterion indicates hyperaccumulation for most of the metals in case of *C. demersum* and *S. natans*. The BCF apportionment clearly summarize *M. aquaticum* and *N. nucifera* to dissipate the PTE in their rhizosphere zone but the rootless *C. demersum* and *S. natans* concentrate them in their foliage parts (Harguinteguy et al. 2014).

Anthropogenic disturbances arising from land-use change, altered global biogeochemistry and biodiversity interferences are clearly reflected by community structure and function with macrophytic communities being no exception. The ecological disturbances experienced by freshwater biota vary in magnitude, frequency, predictability, duration and on spatial-scale as well (Lake et al. 2000). Macrophytes react to the ambient physicochemical environment changes not only by altering species composition but also display plant biomass variations. Their presence, density and morphology is determined by factors like sediment type, water turbidity, water current, nutrient concentration, water depth, shoreline disturbance, herbivore grazing and human activities (Bornette and Puijalon 2011; Wood et al. 2012). They exhibit a stress response of depletion in diversity (genetic erosion) associated with predominantly excessive proliferation of a selected few (Chappius et al. 2011; Michelan et al. 2010). Although the tissue concentration of nutrients and trace elements differ species wise but well correlate with ambient water and sediment media (Amari et al. 2014). The biological sinks scavenge the deleterious proportions of nutrient supply from lakes preventing them from being overtaken by blooms. Even if slight phyto-extraction of inorganics would have occurred physically compared to large scale quanta of the insitu media, still the biogenic calcification evidenced in the summers is capable of lowering them mediated via temperature, pH, HCO_3^- and Ca changes (Boyd et al. 2016). Therefore, a selective and scaled deweeding practice can work wonders in lake management and restoration programmes (Novak and Chambers 2014).

5 Conclusion

The present insight into compartmentalization of nutrients and traces in epilimnion, surface sediments and plant components reflects dilution and dissipation of toxic and nontoxic contaminants over a spatiotemporal scale. The fate and behaviour of contaminants in a freshwater body depends on its hydrology, bioaccumulation and sedimentation processes. The retention of toxic materials is avoided by recycling and retention via biogeochemical pathways.

Acknowledgements The authors would like to express sincere appreciation to Quality Control cum Leaf Tissue Analysis Lab, Sheri Kashmir Agricultural University of Science and Technology (SKAUST), Srinagar-191121; Central Instrumentation Facility (CIF), Pondicherry University-605014 and Indian Institute of Technology and Management (IITM), Chennai- 600036 for their helpful attitude in facilitating lab and instrumentation services.

References

Abell, J. M., Ozkundakci, D., & Hamilton, D. P. (2010). Nitrogen and phosphorus limitation of phytoplankton growth in New Zealand lakes: Implications for eutrophication control. *Ecosystems, 13*(7), 966–977.

Abubakr, A., & Kundangar, M. R. D. (2009). Three decades of Dal lake pollution. *Restoration, Ecology, Environment and Conservation, 15*(4), 825–833.

Akin, B. S., Atıcı, T., Katircioglu, H., & Keskin, F. (2010). Investigation of water quality on Gokçekayadam lake using multivariate statistical analysis, in Eskişehir, Turkey. *Environmental Earth Sciences, 63*(6), 1251–1261.

Algesten, G., Sobek, S., Bergstrom, A. K., Agren, A., Tranvik, L. J., & Jansson, M. (2003). Role of lakes for organic carbon cycling in the boreal zone. *Global Change Biology, 10*(1), 141–147.

Allan, I., Vrana, B., Greenwood, R., Mills, G., Knutsson, J., Holmberg, A., & Guigues, N. (2006). Strategic monitoring for the European water framework directive. *Trends in Analytical Chemistry, 25*(7), 704–715.

Allison, S. D., Czimczik, C. I., & Treseder, K. K. (2008). Microbial activity and soil respiration under nitrogen addition in Alaskan boreal forest. *Global Change Biology, 14*(5), 1156–1168.

Alloway, B. J. (1995). *Heavy metals in soils* (2nd ed.). London: Blackie Academic and Professional.

Amari, T., Ghnaya, T., Debez, A., Taamali, M., Youssef, N. B., Lucchini, G., & Abdelly, C. (2014). Comparative Ni tolerance and accumulation potentials between Mesembryanthemumcrystallinum (halophyte) and Brassica juncea: Metal accumulation, nutrient status and photosynthetic activity. *Journal of Plant Physiology, 171*(17), 1634–1644.

Ammar, R., Kazpard, V., Wazne, M., El Samrani, A. G., Amacha, N., Saad, Z., & Chou, L. (2015). Reservoir sediments: A sink or source of chemicals at the surface water-groundwater interface. *Environmental Monitoring and Assessment, 187*(9), 579.

ANZECC/ARMCANZ. (2000). *Australian and New Zealand guidelines for fresh and marine water quality*. Australian and New Zealand Environment and Conservation Council and Agriculture and Resource Management Council of Australia and New Zealand, Canberra 1–103.

APHA. (2005). *APHA, AWWA, & WEF. Standard methods for the examination of water and wastewater*, 21st edn., Washington, DC.

Arshid, S., Wani, A. A., Ganie, A. H., & Khuroo, A. A. (2011). On correct identification, range expansion and management implications of *Myriophyllum aquaticum* in Kashmir Himalaya, India. *Check List, 7*(3), 299–302.

Badar, B., Romshoo, S. A., & Khan, M. A. (2012). Integrating biophysical and socioeconomic information for prioritizing watersheds in a Kashmir Himalayan lake: A remote sensing and GIS approach. *Environmental Monitoring and Assessment, 185*(8), 6419–6445.

Bastami, K. D., Neyestani, M. R., Shemirani, F., Soltani, F., Haghparast, S., & Akbari, A. (2015). Heavy metal pollution assessment in relation to sediment properties in the coastal sediments of the southern Caspian Sea. *Marine Pollution Bulletin, 92*(1), 237–243.

Batley, G. E. (2000). *Implications of the new ANZECC/ARMCANZ water quality guidelines for mining companies*. In: Grundon, NJ & Bell, LC. In Proceedings of the Fourth Australian Workshop on Acid Mine Drainage 221–229.

Beck, M. W., Tomcko, C. M., Valley, R. D., & Staples, D. F. (2014). Analysis of macrophyte indicator variation as a function of sampling, temporal, and stressor effects. *Ecological Indicators, 46*, 323–335.

Belkhiri, L., & Narany, T. S. (2015). Using multivariate statistical analysis, geostatistical techniques and structural equation modeling to identify spatial variability of groundwater quality. *Water Resources Management, 29*(6), 2073–2089.

BerzasNevado, J. J., Rodriguez Martín-Doimeadios, R. C., Guzman Bernardo, F. J., Jimenez Moreno, M., Ortega Tardio, S., Sánchez-Herrera Fornieles, M. M., & Doncel Perez, A. (2009). Integrated pollution evaluation of the Tagus River in Central Spain. *Environmental Monitoring and Assessment, 156*(1), 461–477.

Bierman, P., Lewis, M., Ostendorf, B., & Tanner, J. (2009). A review of methods for analysing spatial and temporal patterns in coastal water quality. *Ecological Indicators, 11*(1), 103–114.

Bornette, G., & Puijalon, S. (2011). Response of aquatic plants to abiotic factors: A review. *Aquatic Sciences, 73*(1), 1–14.

Boyd, C. E., Tucker, C. S., & Somridhivej, B. (2016). Alkalinity and hardness: Critical but elusive concepts in aquaculture. *Journal of the World Aquaculture Society, 47*(1), 6–41.

Bu, H., Meng, W., Zhang, Y., & Wan, J. (2014). Relationships between land use patterns and water quality in the Taizi River basin, China. *Ecological Indicators, 41*, 187–197.

Bunn, S. E., & Davies, P. M. (2000). Biological processes in running waters and their implications for the assessment of ecological integrity. In *Assessing the ecological integrity of running waters* (pp. 61–70). Dordrecht: Springer.

Canavan, R. W., Slomp, C. P., Jourabchi, P., Van Cappellen, P., Laverman, A. M., & Van den Berg, G. A. (2006). Organic matter mineralization in sediment of a coastal freshwater lake and response to salinization. *Geochimicaet Cosmochimica Acta, 70*(11), 2836–2855.

Chambers, P. A., Lacoul, P., Murphy, K. J., & Thomaz, S. M. (2008). Global diversity of aquatic macrophytes in freshwater. *Hydrobiologia, 595*(1), 9–26.

Chandrasekaran, A., Ravisankar, R., Harikrishnan, N., Satapathy, K. K., Prasad, M. V. R., & Kanagasabapathy, K. V. (2015). Multivariate statistical analysis of heavy metal concentration in soils of Yelagiri Hills, Tamilnadu, India–Spectroscopical approach. *Spectrochimica Acta Part A: Molecular and Biomolecular Spectroscopy, 137*, 589–600.

Choi, K. Y., Kim, S. H., Hong, G. H., & Chon, H. T. (2012). Distributions of heavy metals in the sediments of south Korean harbors. *Environmental Geochemistry and Health, 34*(1), 71–82.

Chon, H. S., Ohandja, D. G., & Voulvoulis, N. (2012). The role of sediments as a source of metals in river catchments. *Chemosphere, 88*(10), 1250–1256.

Cook, C. D. (1996). *Aquatic and wetland plants of India: A reference book and identification manual for the vascular plants found in permanent or seasonal fresh water in the subcontinent of India south of the Himalayas* (Vol. 198548214, pp. 1–385). Oxford: Oxford University Press.

Cotner, J. B., Kenning, J., & Scott, J. T. (2009). The microbial role in littoral zone biogeochemical processes: Why Wetzel was right. *Verhandlungen des Internationalen Verein Limnologie, 30*, 981–984.

De Jonge, V. N., Elliott, M., & Orive, E. (2002). Causes, historical development, effects and future challenges of a common environmental problem: Eutrophication. *Hydrobiologia, 475*(1), 1–19.

De Vicente, I., Guerrero, F., & Cruz-Pizarro, L. (2010). Chemical composition of wetland sediments as an integrator of trophic state. *Aquatic Ecosystem Health and Management, 13*(1), 99–103.

Dean, J. R., Eastwood, W. J., Roberts, N., Jones, M. D., Yigitbaşıoglu, H., Allcock, S. L., & Leng, M. J. (2015). Tracking the hydro-climatic signal from lake to sediment: A field study from Central Turkey. *Journal of Hydrology, 529*, 608–621.

Dhote, S. (2007). Role of Macrophytes in improving water quality of an aquatic eco-system. *Journal of Applied Sciences and Environmental Management, 11*(4), 133–135.

Downing, J. A., Cole, J. J., Middelburg, J. J., Striegl, R. G., Duarte, C. M., Kortelainen, P., & Laube, K. A. (2008). Sediment organic carbon burial in agriculturally eutrophic impoundments over the last century. *Global Biogeochemical Cycles, 22*(1), 1–10.

EEA. (2012). *European waters – Assessment of status and pressures*. Copenhagen: European Environment Agency.
El-Otify, A. M. (2015). Evaluation of the physicochemical and chlorophyll-a conditions of a subtropical aquaculture in Lake Nasser area, Egypt. *Journal of Basic and Applied Sciences, 4*(4), 327–337.
Elser, J. J., Andersen, T., Baron, J. S., Bergstrom, A. K., Jansson, M., Kyle, M., & Hessen, D. O. (2009). Shifts in lake N: P stoichiometry and nutrient limitation driven by atmospheric nitrogen deposition. *Science, 326*(5954), 835–837.
Estefan, G., Sommer, R., Ryan, J. (2013). *Methods of soil, plant and water analysis*. International Center for Agricultural Research in the Dry Areas (ICARDA), Beirut, Lebanon.
FAO (Food and agriculture organization) (2010) The wealth of waste: the economics of wastewater use in agriculture. *Water Reports* (35).
Feichtinger, F., Smidt, S., & Klaghofer, E. (2002). Water and nitrate fluxes at a forest site in the North Tyrolean Limestone Alps. *Environmental Science and Pollution Research, 9*(2), 31.
Garbey, C., Murphy, K. J., Thiebaut, G., & Muller, S. (2004). Variation in phosphorus content in aquatic plant tissues offers an efficient tool for determining plant growth strategies along a resource gradient. *Fresh Water Biology, 49*, 1–11.
Garland, J. L., Mackowiak, C. L., & Zabaloy, M. C. (2010). Organic waste amendment effects on soil microbial activity in a corn–rye rotation: Application of a new approach to community-level physiological profiling. *Applied Soil Ecology, 44*(3), 262–269.
Ghosh, S. K. (2005). *Illustrated aquatic and wetland plants in harmony with mankind* (pp. 1–225). Kolkata: Standard Literature.
Goher, M. E., Farhat, H. I., Abdo, M. H., & Salem, S. G. (2014). Metal pollution assessment in the surface sediment of Lake Nasser, Egypt. *The Egyptian Journal of Aquatic Research, 40*(3), 213–224.
Gothberg, A., Greger, M., Holm, K., & Bengtsson, B. E. (2004). Influence of nutrient levels on uptake and effects of mercury, cadmium, and lead in water spinach. *Journal of Environmental Quality, 33*(4), 1247–1255.
Gottschall, N., Boutin, C., Crolla, A., Kinsley, C., & Champagne, P. (2007). The role of plants in the removal of nutrients at a constructed wetland treating agricultural (dairy) wastewater, Ontario, Canada. *Ecological Engineering, 29*(2), 154–163.
Grabas, G. P., Blukacz-Richards, E. A., & Pernanen, S. (2012). Development of a submerged aquatic vegetation community index of biotic integrity for use in Lake Ontario coastal wetlands. *Journal of Great Lakes Research, 38*(2), 243–250.
Greenacre, M., & Primicerio, R. (2013). *Multivariate analysis of ecological data*. Foundation BBVA.
Gudasz, C., Bastviken, D., Steger, K., Premke, K., Sobek, S., & Tranvik, L. J. (2010). Temperature-controlled organic carbon mineralization in lake sediments. *Nature, 466*(7305), 1478–1481.
Guerra-Garcia, J. M., & Garcia-Gomez, J. C. (2005). Assessing pollution levels in sediments of a harbour with two opposing entrances. Environmental implications. *Journal of Environmental Management, 77*(1), 1–11.
Gunn, I. D., O'Hare, M., Carvalho, L., Roy, D. B., Rothery, P., & Darwell, A. M. (2010). Assessing the condition of lake habitats: A test of methods for surveying aquatic macrophyte communities. *Hydrobiologia, 656*(1), 87–97.
Gupta, A., Ronghang, M., Kumar, P., Mehrotra, I., Kumar, S., Grischek, T., & Knoeller, K. (2015). Nitrate contamination of riverbank filtrate at Srinagar, Uttarakhand, India: A case of geogenic mineralization. *Journal of Hydrology, 531*, 626–637.
Gupta, P. K. (2004). *Soil, plant, water and fertilizer analysis*. Bikaner: Agro Botanica, Vyas Nagar.
Gurnell, A. M., Bertoldi, W., & Corenblit, D. (2012). Changing river channels: The roles of hydrological processes, plants and pioneer fluvial landforms in humid temperate, mixed load, gravel bed rivers. *Earth-Science Reviews, 111*(1), 129–141.

Han, Y. M., Du, P. X., Cao, J. J., & Posmentier, E. S. (2006). Multivariate analysis of heavy metal contamination in urban dusts of Xi'an, Central China. *Science of the Total Environment, 355*, 176–186.

Harguinteguy, C. A., Cirelli, A. F., & Pignata, M. L. (2014). Heavy metal accumulation in leaves of aquatic plant Stuckeniafiliformis and its relationship with sediment and water in the Suquíariver (Argentina). *Microchemical Journal, 114*, 111–118.

Hasler, C. T., Butman, D., Jeffrey, J. D., & Suski, C. D. (2016). Freshwater biota and rising pCO2? *Ecology Letters, 19*(1), 98–108.

Havens, K. E., Sharfstein, B., Brady, M. A., East, T. L., Harwell, M. C., Maki, R. P., & Rodusky, A. J. (2004). Recovery of submerged plants from high water stress in a large subtropical lake in Florida, USA. *Aquatic Botany, 78*(1), 67–82.

Hayakawa, A., Ikeda, S., Tsushima, R., Ishikawa, Y., & Hidaka, S. (2015). Spatial and temporal variations in nutrients in water and riverbed sediments at the mouths of rivers that enter Lake Hachiro, a shallow eutrophic lake in Japan. *Catena, 133*, 486–494.

Heathcote, A. J., & Downing, J. A. (2012). Impacts of eutrophication on carbon burial in freshwater lakes in an intensively agricultural landscape. *Ecosystems, 15*(1), 60–70.

Ibanez, S., Talano, M., Ontañon, O., Suman, J., Medina, M. I., Macek, T., & Agostini, E. (2016). Transgenic plants and hairy roots: Exploiting the potential of plant species to remediate contaminants. *New Biotechnology, 33*(5), 625–635.

Iqbal, J., Tirmizi, S. A., & Shah, M. H. (2013). Statistical apportionment and risk assessment of selected metals in sediments from Rawal Lake (Pakistan). *Environmental Monitoring and Assessment, 185*(1), 729–743.

Ismail, Z., Othman, S. Z., Law, K. H., Sulaiman, A. H., & Hashim, R. (2014). Comparative performance of water hyacinth (Eichhorniacrassipes) and water lettuce (Pistastratiotes) in preventing nutrients build-up in municipal wastewater. *CLEAN - Soil Air Water, 43*(4), 521–531.

Jeelani, G., & Shah, A. Q. (2006). Geochemical characteristics of water and sediment from the dal Lake, Kashmir Himalaya: Constraints on weathering and anthropogenic activity. *Environmental Geology, 50*(1), 12–23.

Ji, Y., Zhang, J., Li, R., Pan, B., Zhang, L., & Chen, X. (2015). Distribution and partitioning of heavy metals in sediments of the Xinjiang River in Poyang Lake region, China. *Environmental Progress and Sustainable Energy, 34*(3), 713–723.

Jing, L. D., Xi Wu, C., Tong Liu, J., Guang Wang, H., & Yi Ao, H. (2013). The effects of dredging on nitrogen balance in sediment-water microcosms and implications to dredging projects. *Ecological Engineering, 52*, 167–174.

Jing, L., Liu, X., Bai, S., Wu, C., Ao, H., & Liu, J. (2015). Effects of sediment dredging on internal phosphorus: A comparative field study focused on iron and phosphorus forms in sediments. *Ecological Engineering, 82*, 267–271.

Johnson, J. A., & Newman, R. M. (2011). A comparison of two methods for sampling biomass of aquatic plants. *Journal of Aquatic Plant Management, 49*(1), 1–8.

Kalff, J. (2002). *Limnology: Inland water ecosystems* (Vol. 592). New Jersey: Prentice Hall.

Karanlik, S., Agca, N., & Mehmet, Y. (2011). Spatial distribution of heavy metals content in soils of Amik Plain (Hatay, Turkey). *Environmental Monitoring and Assessment, 173*, 181–191.

Khan, F. A., & Ansari, A. A. (2005). Eutrophication: An ecological vision. *The Botanical Review, 71*(4), 449–482.

Kissoon, L. T., Jacob, D. L., Hanson, M. A., Herwig, B. R., Bowe, S. E., & Otte, M. L. (2013). Macrophytes in shallow lakes: Relationships with water, sediment and watershed characteristics. *Aquatic Botany, 109*, 39–48.

Knoll, L. B., Vanni, M. J., Renwick, W. H., & Kollie, S. (2014). Burial rates and stoichiometry of sedimentary carbon, nitrogen and phosphorus in Midwestern US reservoirs. *Freshwater Biology, 59*(11), 2342–2353.

Lake, P. S., Palmer, M. A., Biro, P., Cole, J., Covich, A. P., Dahm, C., & Verhoeven, J. O. S. (2000). Global change and the biodiversity of freshwater ecosystems: Impacts on linkages between above-sediment and sediment biota all forms of anthropogenic disturbance—Changes in land

use, biogeochemical processes, or biotic addition or loss—Not only damage the biota of freshwater sediments but also disrupt the linkages between above-sediment and sediment-dwelling biota. *Bioscience, 50*(12), 1099–1107.
Lazzarino, J. K., Bachmann, R. W., Hoyer, M. V., & Canfield, D. E. (2009). Carbon dioxide supersaturation in Florida lakes. *Hydrobiologia, 627*(1), 169–180.
Lenzi, M., Gennaro, P., Renzi, M., Persia, E., & Porrello, S. (2012). Spread of Alsidiumcorallinum C. Ag. in a Tyrrhenian eutrophic lagoon dominated by opportunistic macroalgae. *Marine Pollution Bulletin, 64*(12), 2699–2707.
Lin, Y. F., Jing, S. R., Wang, T. W., & Lee, D. Y. (2002). Effects of macrophytes and external carbon sources on nitrate removal from groundwater in constructed wetlands. *Environmental Pollution, 119*(3), 413–420.
Liu, J. L., Liu, J. K., Anderson, J. T., Zhang, R., & Zhang, Z. M. (2014). Potential of aquatic macrophytes and artificial floating island for removing contaminants. *Plant Biosystems – An International Journal Dealing with all Aspects of Plant Biology, 150*(4), 702–709.
Lopes, M. L., Rodrigues, A. M., & Quintino, V. (2014). Ecological effects of contaminated sediments following a decade of no industrial effluents emissions: The sediment quality triad approach. *Marine Pollution Bulletin, 87*(1), 117–130.
Lukacs, B. A., Dévai, G. Y., & Tothmeresz, B. (2009). The function of macrophytes inrelation to environmental variables in eutrophic backwaters and mortlakes. *Phytocoenologia, 39*, 287–293.
Lukacs, B. A., Tothmeresz, B., Borics, G., Varbiróo, G., Juhasz, P., Kiss, B., & Eros, T. (2015). Macrophyte diversity of lakes in the Pannon Ecoregion (Hungary). *Limnologica-Ecology and Management of Inland Waters, 53*, 74–83.
Maanan, M., Saddik, M., Maanan, M., Chaibi, M., Assobhei, O., & Zourarah, B. (2015). Environmental and ecological risk assessment of heavy metals in sediments of Nador lagoon, Morocco. *Ecological Indicators, 48*, 616–626.
MacDonald, D. D., Ingersoll, C. G., & Berger, T. A. (2000). Development and evaluation of consensus-based sediment quality guidelines for freshwater ecosystems. *Archives of Environmental Contamination and Toxicology, 39*(1), 20–31.
Macias, C. G., Schifter, I., Lluch-Cota, D. B., Mendez-Rodriguez, L., & NHernandez-Vazquez, S. (2006). Distribution, enrichment and accumulation of heavy metals in coastal sediments of Salina Cruz Bay, Mexico. *Environmental Monitoring and Assessment, 118*(1), 211–230.
Manap, N., & Voulvoulis, N. (2015). Environmental management for dredging sediments–the requirement of developing nations. *Journal of Environmental Management, 147*, 338–348.
Marberly, S. C. (1996). Diel, episodic and seasonal changes in pH and concentration of inorganic carbon in a productive lake. *Freshwater Biology, 35*, 579–598.
Markich, S. J., Brown, P. L., Batley, G. E., Apte, S. C., & Stauber, J. L. (2001). Incorporating metal speciation and bioavailability into water quality guidelines for protecting aquatic ecosystems. *Australasian Journal of Ecotoxicology, 7*(2), 109–122.
Matusiewicz, H. (2003). *Wet digestion methods* (Vol. 41, pp. 193–233). Amsterdam: Elsevier.
Mazej, Z., & Germ, M. (2009). Trace element accumulation and distribution in four aquatic macrophytes. *Chemosphere, 74*(5), 642–647.
McElarney, Y., Rasmussen, P., Foy, R., & Anderson, N. (2010). Response of aquatic macrophytes in Northern Irish softwater lakes to forestry management; eutrophication and dissolved organic carbon. *Aquatic Botany, 93*(4), 227–236.
MEA (Millennium Ecosystem Assessment). (2005). *Ecosystems and human Well-being: Current state and trends*. Washington, DC: Island Press.
Mediolla, L. L., Domingues, M. C. D., & Sandoval, M. R. G. (2008). Environmental assessment of and active tailings pile in the state of Mexico (Central Mexico). *Research Journal of Environmental and Earth Sciences, 2*, 197–208.
Michard, G., Sarazin, G., Jezequel, D., Alberic, P., & Ogier, S. (2001). Annual budget of chemical elements in a eutrophic lake, Aydatlake (Puy-de-Dôme), France. *Hydrobiologia, 459*(1), 27–46.

Michelan, T. S., Thomaz, S., Mormul, R. P., & Carvalho, P. (2010). Effects of an exotic invasive macrophyte (tropical signalgrass) on native plant community composition, species richness and functional diversity. *Freshwater Biology, 55*(6), 1315–1326.

Min, K., Kang, H., & Lee, D. (2011). Effects of ammonium and nitrate additions on carbon mineralization in wetland soils. *Soil Biology and Biochemistry, 43*(12), 2461–2469.

Moiseenko, T. I., Gashkina, N. A., & Dinu, M. I. (2016). Enrichment of surface water by elements: Effects of air pollution, acidification and eutrophication. *Environmental Processes, 3*(1), 39–58.

Moreno, M., Semprucci, F., Vezzulli, L., Balsamo, M., Fabiano, M., & Albertelli, G. (2011). The use of nematodes in assessing ecological quality status in the Mediterranean coastal ecosystems. *Ecological Indicators, 11*(2), 328–336.

Morford, J. L., Emerson, S. R., Breckel, E. J., & Kim, S. H. (2005). Diagenesis of oxyanions (V, U, Re, and Mo) in porewaters and sediments from a continental margin. *Geochimica et Cosmochimica Acta, 69*, 502–5032.

Muller, B., Meyer, J. S., & Gachter, R. (2016). Alkalinity regulation in calcium carbonate-buffered lakes. *Limnology and Oceanography, 61*(1), 341–352.

Muller, G. (1979). Heavy metals in the sediments of the Rhine: Changes since 1971. *Umschau, 79*(24), 778–783.

Muller, G. (1981). The heavy metal pollution of the sediments of the Neckar and its tributaries: An inventory. *Chemical Zeitung, 105*, 157–164.

National Wetland Atlas: Jammu and Kashmir. (2010). SAC/RESA/AFEG/NWIA/ATLAS/16/2010, Space Applications Centre, ISRO, Ahmedabad, India, 176.

Nilin, J., Moreira, L. B., Aguiar, J. E., Marins, R., de Souza Abessa, D. M., da Cruz Lotufo, T. M., & Costa-Lotufo, L. V. (2013). Sediment quality assessment in a tropical estuary: The case of Ceará River, Northeastern Brazil. *Marine Environmental Research, 91*, 89–96.

Njenga, J. W. (2004). Comparative studies of water chemistry of four tropical lakes in Kenya and India. *Asian Journal of Water Environment and Pollution, 1*(1–2), 87–97.

Nnaji, C. C., & Agunwamba, J. C. (2014). Quality assessment of water receiving effluents from crude oil flow stations in Niger Delta, Nigeria. *Water and Environment Journal, 28*(1), 104–113.

Novak, P. A., & Chambers, J. M. (2014). Investigation of nutrient thresholds to guide restoration and management of two impounded rivers in South-Western Australia. *Ecological Engineering, 68*, 116–123.

Olsen, S., Jeppesen, E., Moss, B., Ozkan, K., Beklioglu, M., Feuchtmayr, H., & Sondergaard, M. (2014). Factors influencing nitrogen processing in lakes: An experimental approach. *Freshwater Biology, 60*(4), 646–662.

Ouma, H., & Mwamburi, J. (2014). Spatial variations in nutrients and other physicochemical variables in the topographically closed Lake Baringo freshwater basin (Kenya). *Lakes and Reservoirs: Research and Management, 19*(1), 11–23.

Pandit AK (2002) Topical evolution of lakes in Kashmir Himalaya. *Natural resources of western Himalaya* (pp. 213–242). Valley Book House, Srinagar.

Pandit, A. K. (1999). *Freshwater ecosystems of the Himalaya.* New York/London: Parthenon Publishing.

Paramasivam, K., Ramasamy, V., & Suresh, G. (2015). Impact of sediment characteristics on the heavy metal concentration and their ecological risk level of surface sediments of Vaigairiver, Tamilnadu, India. *Spectrochimica Acta Part A: Molecular and Biomolecular Spectroscopy, 137*, 397–407.

Parker, J. L., & Bloom, N. S. (2005). Preservation and storage techniques for low-level aqueous mercury speciation. *Science of the Total Environment, 337*(1), 253–263.

Pulido, C., Lucassen, E. C., Pedersen, O. L. E., & Roelofs, J. G. (2011). Influence of quantity and lability of sediment organic matter on the biomass of two isoetids, Littorellauniflora and Echinodorusrepens. *Freshwater Biology, 56*(5), 939–951.

Qingjie, G., Jun, D., Yunchuan, X., Qingfei, W., & Liqiang, Y. (2008). Calculating pollution indices by heavy metals in ecological geochemistry assessment and a case study in parks of Beijing. *Journal of China University of Geosciences, 19*(3), 230–241.

Quilliam, R. S., Van Niekerk, M. A., Chadwick, D. R., Cross, P., Hanley, N., Jones, D. L., & Oliver, D. M. (2015). Can macrophyte harvesting from eutrophic water close the loop on nutrient loss from agricultural land? *Journal of Environmental Management, 152*, 210–217.

Radojevic, M., & Bashkin, V. N. (2006). *Practical environmental analysis* (2nd Edn).The Royal Society of Chemistry, Thomas Graham House, Science Park, Milton Road, Cambridge CB4 0 WF, UK.

Raj, S. M., & Jayaprakash, M. (2007). Distribution and enrichment of trace metals in marine sediments of Bay of Bengal, off Ennore, south-east coast of India. *Environmental Geology, 56*(1), 207–217.

Rashid, I., Romshoo, S. A., Amin, M., Khanday, S. A., & Chauhan, P. (2016). Linking human-biophysical interactions with the trophic status of Dal Lake, Kashmir Himalaya, India. *Limnologica-Ecology and Management of Inland Waters, 62*, 84–96.

Reimann, C., Filzmoser, P., Garrett, R., & Dutter, R. (2008). *Statistical data analysis explained: Applied environmental statistics with R*. Hoboken: Wiley.

Rothwell, J. J., Dise, N. B., Taylor, K. G., Allott, T. E. H., Scholefield, P., Davies, H., & Neal, C. (2010). A spatial and seasonal assessment of river water chemistry across North West England. *Science of the Total Environment, 408*(4), 841–855.

Ryan, J., Estefan, G., & Rashid, A. (2007). *Soil and plant analysis laboratory manual*. Aleppo: International Center for Agricultural Research in the Dry Areas (ICARDA).

Sadasivam, S., & Manikam, A. (2005). *Biochemical methods* (2nd ed.). New Delhi: New Age International.

Sany, S. B. T., Hashim, R., Rezayi, M., Salleh, A., & Safari, O. (2014). A review of strategies to monitor water and sediment quality for a sustainability assessment of marine environment. *Environmental Science and Pollution Research, 21*(2), 813–833.

Sarwar, N., Saifullah Malhi, S. S., Zia, M. H., Naeem, A., Bibi, S., & Farid, G. (2010). Role of mineral nutrition in minimizing cadmium accumulation by plants. *Journal of the Science of Food and Agriculture, 90*, 925–937.

Sass, L. L., Bozek, M. A., Hauxwell, J. A., Wagner, K., & Knight, S. (2010). Response of aquatic macrophytes to human land use perturbations in the watersheds of Wisconsin lakes, USA. *Aquatic Botany, 93*(1), 1–8.

Schaller, J., Vymazal, J., & Brackhage, C. (2013). Retention of resources (metals, metalloids and rare earth elements) by autochthonously/allochthonously dominated wetlands: A review. *Ecological Engineering, 53*, 106–114.

Scheffer, M. (1989). Alternative stable states in eutrophic, shallow freshwater systems: A minimal model. *Hydrobiological Bulletin, 23*(1), 73–83.

Schneider, B., Cunha, E. R., Marchese, M., & Thomaz, S. M. (2015). Explanatory variables associated with diversity and composition of aquatic macrophytes in a large subtropical river floodplain. *Aquatic Botany, 121*, 67–75.

Selig, U., & Schlungbaum, G. (2003). Characterisation and quantification of phosphorus release from profundal bottom sediments in two dimictic lakes during summer stratification. *Journal of Limnology, 62*(2), 151–162.

Shaltout, K. H., Galal, T. M., & El-Komi, T. M. (2014). Biomass, nutrients and nutritive value of PersicariasalicifoliaWilld in the water courses of Nile Delta, Egypt. *RendicontiLincei, 25*(2), 167–179.

Sierszen, M. E., Morrice, J. A., Trebitz, A. S., & Hoffman, J. C. (2012). A review of selected ecosystem services provided by coastal wetlands of the Laurentian Great Lakes. *Aquatic Ecosystem Health & Management, 15*(1), 92–106.

Singh, S. P., & Singh, B. P. (2010). Geothermal evolution of the evaporite-bearing sequences of the Lesser Himalaya, India. *International Journal of Earth Sciences, 99*(1), 101–108.

Sistla, S. A., Appling, A. P., Lewandowska, A. M., Taylor, B. N., & Wolf, A. A. (2015). Stoichiometric flexibility in response to fertilization along gradients of environmental and organismal nutrient richness. *Oikos, 124*(7), 949–959.

Smolders, A. J. P., Lamers, L. P. M., Lucassen, E. C. H. E. T., Van der Velde, G., & Roelofs, J. G. M. (2006). Internal eutrophication: How it works and what to do about it—A review. *Chemistry and Ecology, 22*(2), 93–111.
Sondergaard, M., Johansson, L. S., Lauridsen, T. L., Jorgensen, T. B., Liboriussen, L., & Jeppesen, E. (2010). Submerged macrophytes as indicators of the ecological quality of lakes. *Freshwater Biology, 55*(4), 893–908.
Sondergaard, M., Phillips, G., Hellsten, S., Kolada, A., Ecke, F., Mäemets, H., & Oggioni, A. (2013). Maximum growing depth of submerged macrophytes in European lakes. *Hydrobiologia, 704*(1), 165–177.
Srebotnjak, T., Carr, G., de Sherbinin, A., & Rickwood, C. (2012). A global water quality index and hot-deck imputation of missing data. *Ecological Indicators, 17*, 108–119.
Taylor, S. R., & McLennan, S. M. (1995). The geochemical evolution of the continental crust. *Reviews of Geophysics, 33*(2), 241–265.
Tchobanoglous, G., Burton, F. L., & Stensel, H. D. (2003). *Wastewater engineering: Treatment and reuse*. New York: Metcalf & Eddy Inc, McGraw Hill.
Thompson, J. B., Schultze-Lam, S., Beveridge, T. J., & Des Marais, D. J. (1997). Whiting events: Biogenic origin due to the photosynthetic activity of cyanobacterial picoplankton. *Limnology and Oceanography, 42*(1), 133–141.
Tian, D. L., Xiang, W. H., Yan, W. D., Kang, W. X., Deng, X. W., & Fan, Z. (2007). Biological cycles of mineral elements in a young mixed stand in abandoned mining soils. *Journal of Integrative Plant Biology, 49*(9), 1284–1293.
Tranvik, L. J., Downing, J. A., Cotner, J. B., Loiselle, S. A., Striegl, R. G., Ballatore, T. J., & Kortelainen, P. L. (2009). Lakes and reservoirs as regulators of carbon cycling and climate. *Limnology and Oceanography, 54*(6–2), 2298–2314.
Turner, A., Millward, G. E., & Le Roux, S. M. (2004). Significance of oxides and particulate organic matter in controlling trace metal partitioning in a contaminated estuary. *Marine Chemistry, 88*(3), 179–192.
U. S. Environmental Protection Agency. (1994). *Method 2007: Determination of Metals and Trace Elements in Water and Wastes By Inductively Coupled Plasma – Atomic Emission Spectrometry*, Revision 4.4, EMMC.
Udeigwe, T. K., Teboh, J. M., Eze, P. N., HashemStietiya, M., Kumar, V., Hendrix, J., & Kandakji, T. (2015). Implications of leading crop production practices on environmental quality and human health. *Journal of Environmental Management, 151*, 267–279.
UNEP. (2007). *Global Environment Outlook 4 (GEO-4): environment for development*. United Nations Environment Programme, Nairobi, Kenya.
Urban, N. R., Brezonik, P. L., Baker, L. A., & Sherman, L. A. (2009). Sulfate reduction and diffusion in sediments of Little Rock Lake, Wisconsin. *Limnology and Oceanography, 39*(4), 797–815.
USEPA. (2014). *The assessment and TMDL tracking and implementation system*. National Summary of State Information. http://iaspub.epa.gov/waters10/attains_nation_cy.control#total_assessed_waters
Vorosmarty, C. J., McIntyre, P. B., Gessner, M. O., Dudgeon, D., Prusevich, A., Green, P., Glidden, S., Bunn, S. E., Sullivan, C. A., Liermann, C. R., & Davies, P. M. (2010). Global threats to human water security and river biodiversity. *Nature, 467*, 555–561.
Wagner, B., Lotter, A. F., Nowaczyk, N., Reed, J. M., Schwalb, A., Sulpizio, R., & Zanchetta, G. (2009). A 40,000-year record of environmental change from ancient Lake Ohrid (Albania and Macedonia). *Journal of Paleolimnology, 41*(3), 407–430.
Wali, A., Kawachi, A., Bougi, M. S. M., Dhia, H. B., Isoda, H., Tsujimura, M., & Ksibi, M. (2015). Effects of metal pollution on sediments in a highly saline aquatic ecosystem: Case of the Moknine Continental Sebkha (Eastern Tunisia). *Bulletin of Environmental Contamination and Toxicology, 94*(4), 511–518.

Wang, C. Y., Sample, D. J., Day, S. D., & Grizzard, T. J. (2015). Floating treatment wetland nutrient removal through vegetation harvest and observations from a field study. *Ecological Engineering, 78*, 15–26.

Wei, B., & Yang, L. (2010). A review of heavy metal contaminations in urban soils, urban road dusts and agricultural soils from China. *Microchemical Journal, 94*(2), 99–107.

Wersal, R. M., & Madsen, J. D. (2011). Influences of water column nutrient loading on growth characteristics of the invasive aquatic macrophyte Myriophyllum aquaticum (Vell.)Verdc. *Hydrobiologia, 665*(1), 93–105.

Wetzel RG (2001) *Land–water interfaces: Larger plants*. Limnology, 3rd Edn., Academic Press, San Diego.

Wiik, E., Bennion, H., Sayer, C. D., Davidson, T. A., McGowan, S., Patmore, I. R., & Clarke, S. J. (2015). Ecological sensitivity of marl lakes to nutrient enrichment: Evidence from Hawes water, UK. *Freshwater Biology, 60*(11), 2226–2247.

Wood, K. A., Stillman, R. A., Clarke, R. T., Daunt, F., & O'Hare, M. T. (2012). Understanding plant community responses to combinations of biotic and abiotic factors in different phases of the plant growth cycle. *PLoS One, 7*(11), 49824.

Wright, J. F., Sutcliffe, D. W., & Furse, M. T. (2000). Assessing the biological quality of fresh waters: RIVPACS and other techniques. In *Assessing the biological quality of fresh waters: RIVPACS and other techniques*. Freshwater Biological Association.

Xiangcan, J. (2003). Analysis of eutrophication state and trend for lakes in China. *Journal of Limnology, 62*(1s), 60–66.

Xu, Y., Xie, R., Wang, Y., & Sha, J. (2014). Spatio-temporal variations of water quality in Yuqiao Reservoir Basin, North China. *Frontiers of Environmental Science & Engineering, 9*(4), 649–664.

Yu, S., Yu, G. B., Liu, Y., Li, G. L., Feng, S., Wu, S. C., & Wong, M. H. (2012). Urbanization impairs surface water quality: Eutrophication and metal stress in the Grand Canal of China. *River Research and Applications, 28*(8), 1135–1148.

Yuan, Z., Taoran, S., Yan, Z., & Tao, Y. (2014). Spatial distribution and risk assessment of heavy metals in sediments from a hypertrophic plateau lake Dianchi, China. *Environmental Monitoring and Assessment, 186*(2), 1219–1234.

Zaier, H., Ghnaya, T., Rejeb, K. B., Lakhdar, A., Rejeb, S., & Jemal, F. (2010). Effects of EDTA on phytoextraction of heavy metals (Zn, Mn and Pb) from sludge-amended soil with Brassica napus. *Bioresource Technology, 101*(11), 3978–3983.

Zargar, U. R., Chishti, M. Z., Yousuf, A. R., & Fayaz, A. (2012). Infection level of monogenean gill parasite, Diplozoonkashmirensis (Monogenea, Polyopisthocotylea) in the Crucian Carp, Carassiuscarassius from lake ecosystems of an altered water quality: What factors do have an impact on the Diplozoon infection? *Veterinary Parasitology, 189*(2), 218–226.

Zhang, C., Qiao, Q., Piper, J. D., & Huang, B. (2011). Assessment of heavy metal pollution from a Fe-smelting plant in urban river sediments using environmental magnetic and geochemical methods. *Environmental Pollution, 159*(10), 3057–3070.

Zhang, F., Yan, X., Zeng, C., Zhang, M., Shrestha, S., Devkota, L. P., & Yao, T. (2012). Influence of traffic activity on heavy metal concentrations of roadside farmland soil in mountainous areas. *International Journal of Environmental Research and Public Health, 9*, 1715–1731.

Zheng, Y., Wang, X. C., Ge, Y., Dzakpasu, M., Zhao, Y., & Xiong, J. (2015). Effects of annual harvesting on plants growth and nutrients removal in surface-flow constructed wetlands in northwestern China. *Ecological Engineering, 83*, 268–275.

Zhu, X., Ji, H., Chen, Y., Qiao, M., & Tang, L. (2013). Assessment and sources of heavy metals in surface sediments of Miyun Reservoir, Beijing. *Environmental Monitoring and Assessment, 185*(7), 6049–6062.

Remediation of Pesticides Through Microbial and Phytoremediation Techniques

Sartaj Ahmad Bhat, Humaira Qadri, Guangyu Cui, and Fusheng Li

Abstract Pesticides contamination in the environment presents a real hazard to human beings and other aquatic and terrestrial life. If not controlled, the contamination can lead to serious problems to the environment. In order to keep this contamination at a low level, some sustainable and cost-effective alternatives methods are required. Remediation techniques, such as microbial remediation and phytoremediation are reliable and efficient methods that utilize microbes and plants to eliminate the pesticide residues in the environment. These techniques offer useful and effective alternatives to physical and chemical remediation processes for being economically and ecologically sustainable. This chapter discusses present remediation techniques for the removing of pesticides from the natural environment.

Keywords Microbes · Pesticides · Plants · Pollution

1 Introduction

Pesticides occurrence in the environment has become a major environmental issue. These are the toxic chemical substances used in agriculture for removing crop pests, insects, rodents, unwanted plants, pathogenic fungi etc. (Rani and Dhania 2014). On the other hand, enormous consumption may result in the accumulation of these substances in the freshwater (Surface and ground) as well as soil (Mohany et al. 2011). Pesticide residues in surface water are a major concern as they pose a great threat to aquatic environment including humans. Pesticides, particularly the organochlorines and organophosphates, enter into any freshwater body via agriculture

S. A. Bhat (✉) · F. Li
River Basin Research Center, Gifu University, Gifu, Japan

H. Qadri
Department of Environmental Sciences, School of Sciences, Sri Pratap College Campus, Cluster University of Srinagar, Srinagar, Jammu and Kashmir, India

G. Cui
Graduate School of Engineering, Gifu University, Gifu, Japan

H. Qadri et al. (eds.), *Fresh Water Pollution Dynamics and Remediation*,
https://doi.org/10.1007/978-981-13-8277-2_13

run-off (Karunya and Saranraj 2014).There are various ways by which these chemicals can gain entry into water bodies which include surface run-off, spray equipments, industrial effluent, drift into ponds, lakes, streams and river water and aerial sprays to control water-inhibiting pests (Carter and Heather 1995; Singh and Mandal 2013). Pesticides are categorised into various categories such as insecticides, herbicides, fungicides, acaricides, nematicides, molluscicides, and rodenticides (London and Meyers 1995; Rani et al. 2017). On the basis of chemical composition, pesticides can be categorized into: (i) organochlorines (ii) organophosphates (iii) carbamates and (iv) substituted urea. Amongst all, organochlorines are the most hazardous as well as persistent organic pollutants (POPs) and pose greater risk to the environment. Various authors have reported that low concentrated pesticides exist around the globe (Moschet et al. 2014). Moreover, various remediation techniques might be opted to safeguard the health and environment of human beings because pesticides can exceed the prescribed standard levels. Several physical, chemical and biological methods along with adsorption, oxidation, catalytic degradation, membrane filtration and biological treatment have been developed as treatment technology in the remediation of pesticides (Rodante et al. 1992; Smith et al. 2004; Zinovyev et al. 2005; Lin and Lin 2007; Li et al. 2010; Rani et al. 2017).Usually, biological treatment is most preferable because the physical and chemical treatments are highly expensive. Bioremediation technology which involves the utilization of microbes is widely used to diminish the pesticides. In this chapter, a detailed account about the type of pesticides and the current status of the bioremediation techniques has been presented.

2 Bioremediation of Pesticides

Pesticide occurrence leads to various deleterious effects to environment as well as humans, and therefore the need for remediation techniques arises. Bioremediation technology over the conventional methods (physical and chemical) is most promising technique for cleaning the environment. It is an ecologically sustainable and cost-effective technique which can solve the environmental problems related to water bodies (Sahu 2014). It uses plants and microorganisms for the removal or alteration of pesticides into less toxic substances by processes like degradation and biotransformation. Bioremediation has two main processes, i.e. in situ and ex situ. In situ bioremediation process involves adding of nutrients to the organisms in aquatic bodies degradation. Its capability is dependent on factors like biogeochemical and hydrogeological conditions which control the process of biodegradation (Verardo et al. 2017). Ex situ bioremediation involves the transfer of media from one site to another handling the pollutants at a particular site and treating elsewhere.

3 Strategies for Pesticides Bioremediation

3.1 Microbial Bioremediation

Microbial remediation is an efficient tool to clean up the pesticide contaminated sites. The toxic chemicals/substances get converted into low-level toxic substances by the process microbial remediation process. Many studies related to the bioremediation of pesticides from the environment were carried out using microbial remediation technique (Saez et al. 2014; Kurade et al. 2016; Pan et al. 2017). The main advantages of microbial remediation of pesticides are; it is easy to culture, high microbial population and fast mutation. Under suitable conditions (warm temperature, suitable humidity, good pH, air circulation), the microbial degradation can be quick with complete degradation. The commonly used microbes for the pesticidal bioremediation are *Bacillus* sp., *Mycobacterium* sp., *Pseudomonas* sp., *Pandoraea* sp., *Klebsiella* sp., *Phanerochaete Chrysosporium* (Rani and Dhania 2014; Kumar et al. 2018). The remediation of pesticides using microorganisms is presented in Table 1.

3.1.1 Bacteria in Bioremediation of Pesticides

Remediation by bacteria is a most relevant method of bioremediation due to its rapid growth, economical feasibility and easy operation. Bacterial bioremediation can occur through aerobic and anaerobic states. Diuron an active herbicide or algicide present in the water and soil gets mineralized by *Anthrobacter sulfonivorans*, *Variovorax soli*, and *Advenella* sp. of bacterial consortium (Villaverde et al. 2017). Briceno et al. (2016) studied the removal of diazinon pesticide by making use of

Table 1 Remediation of pesticides using microbes

Microorganism	Pesticide	References
Alcaligenes faecalis	Endosulfan	Kong et al. (2013)
Chlorella vulgaris	Diazinon	Kurade et al. (2016)
Ochrobactrum sp.	DDT	Pan et al. (2017)
Phlebia	Aldrin	Xiao et al. (2011)
Pseudomonas aeruginosa	Chlorpyrifos	Kharabsheh et al. (2017)
Streptomyces consortium	Lindane	Saez et al. (2014)
Arthrobacter protophormiae	p-Nitrophenol	Paul et al. (2006)
Burkholderia cepacia	Carbofuran	Plangklang and Reungsang (2010)
Pseudomonas sp., *Pseudomonas putida, Micrococcus lylae, Pseudomonas aureofaciens,* and *Acetobacter liquefaciens*	Malathion	Goda et al. (2010)
Micrococcus sp.	Phenylurea	Sharma et al. (2010)

single and mixed culture of *Streptomyces* sp. The authors observed that mixed culture exhibited the high removal of diazinon pesticide than single culture. According to Fuentes et al. (2017) *Streptomyces* sp. of bacteria has an important function in the remediation of pesticides due to its mycelial development and enhanced growth rates. The bacterial bioremediation technique is cost efficient than chemical process and can be carried out on-site. Thus, the use of bacteria in bioremediation of pesticides is an efficient technique for various polluted environments.

3.1.2 Fungal Bioremediation

Various fungal strains like *Penicillium* (Peng et al. 2012), *Aspergillus* (Mohamed et al. 2011), *Phanerochaete* (Chrinside et al. 2011) are involved in the remediation of pesticides (Maqbool et al. 2016). According to Maqbool et al. (2016) lengthy mycelial networks of fungi and the capacity to use organic material as a source of growth make them the best method for bioremediation. Ellegaard-Jensen et al. (2014) mineralized the phenyl urea herbicide diuron pesticide using a consortium of fungi-bacteria. The fungal-bacterial consortium contributes to the microbial environment with each other for their development. Fungal bioremediation of pesticides occurs by the discharge of an mixture of extracellular enzymes like laccases, polyphenol oxidases, lignin peroxidases which plays an important role in fungal bioremediation process. The intracellular enzymes, such as reductases, methyltransferases, and cytochrome oxygenase, are also involved in the remediation of organic pollutants. Pesticides like clothianidin get biotransformed by the white rot fungus such as *Phanerochaete sordida* which converts clothianidin into non-toxicmetabolite N-(2-chlorothiazol-5-methyl)-N-methyl urea (TZMU) (Mori et al. 2017). White rot fungus has a presence of extracellular enzymes such as lignin peroxidase, manganese peroxidase, and laccase which indicates that it is a good degrader of pesticides.

3.1.3 Cyanobacterial Bioremediation

Algae convert organic substances into a high economic value new molecule. The species of microalgae is highly recommended for the ecosystems contaminated with the lindane pesticide because it is very harmful for the environment and humans. According to Thengodkar and Sivakami (2010) degradation of pesticides in the presence of microbial enzymes (alkaline phosphatise) secreted by *Spirulina platensis* has the potential to hydrolyze the pesticides like chlorpyrifos to its non-toxic primary metabolite 3, 5, 6-trichloro-2-pyridinol. Kabra et al. (2014) studied the degradation of atrazine by the microalgal species *Chlamydomonas mexicana*. The carbohydrate content gets increased in the algae which proved that *C. mexicana* can evacuate the pesticides at pollutedstreams. The rate of remediation depends on selected algal strain and nature of pollutants which can be influenced by environmental factors such as nutrients, water, pH, salinity, oxygen tension, temperature, light intensity, etc. And physicochemical parameters like molecular chemical structure, weight concentration, toxicity, etc. (Priyadarshani et al. 2011; Varsha et al. 2011). The remediation of pesticides using microalgaeis presented in Table 2.

3.2 Removal of Pesticides Through Phytoremediation

Phytoremediation is a technique that involves growing pesticide/metal tolerant plants having pesticide/metal accumulating potential to remediate the contaminated site. These plants can accumulate, absorb and detoxify chemical substances from the site through their metabolic processes (plant oxidative enzymes). Non-toxic substances can be produced by some phytoremediation mechanisms such as phytostabilization, phytotransformation, phytovolatilization, phytofiltration and phytoextraction (Bhat et al. 2016). Various studies have reported the biodegradation of pollutants in plants. The uptake capacity of pesticides by the phytoextraction mechanism of plants implied that it is a good phytoremediator of pesticide endosulfan (Mitton et al. 2016). The aquatic plants are also able to transform the organic contaminates. In plants, silica is an important element which stimulates the resistance of the plant against the pests, pathogens, salinity and metal toxicity (Romeh 2015a; Romehand Hendawi 2017). Suresh et al. (2005) reported that *Cichorium intybus* and *Brassica juncea* plants are efficient in the degradation of DDT. The hairy root cultures of both the plants improved the uptake and breakdown of DDT. The disadvantage of phytoremediation process is that it has a prolonged remediation period than the microbial remediation. The various plant species which have shown the pesticide phytoremediation potentials are presented in Table 3.

Plants can accrue or metabolize a variety of organic compounds, including, imidacloprid (Byrne and Toscano 2005), triazophos (Cheng et al. 2007), chlorpyrifos (Prasertsup and Naiyanan 2011; Romeh and Hendawi 2013), methyl parathion (Khan et al. 2011) and atrazine (Wang et al. 2012). Aquatic plants like *Eichhornia crassipes*, *Lemna minor* and *Elodea canadensis* have been used in water treatment due to their high photosynthetic activity, high growth rate, easy harvesting and high

Table 2 Remediation of pesticides using microalgae

Pesticide	Microalgae	References
Atrazine	*Chlamydomonas mexicana*	Kabra et al. (2014)
Chlorpyrifos	*Spirulina platensis*	Thengodkar and Sivakami (2010)
DDT, parathion	*Scenedesmus obliquus*	Semple et al. (1999)
Herbicide fluroxypyr	*Chlamydomonas reinhardtii*	Zhang et al. (2011)
Mirex	*Chlorococcum* sp.	Semple et al. (1999)
Naphthalene, DDT	*Dunaliella* sp., *Cylindrotheca* sp.	Biswas et al. (2015)
Organophosphorus and organochlorine	*Synechococcus elongates*, *Microcystes aeruginosa*	Vijayakumar (2012)
Toxaphene, Methoxychlor	Chlorella sp.	Semple et al. (1999)

Table 3 Pesticide remediation potential of various plant species

Pesticide	Plant used	References
Endosulfan	*Solanum lycopersicum, Helianthus, Glycine max, Medicago sativa*	Mitton et al. (2016)
Azoxystrobin	*Plantago major*	Romeh (2015b)
Fenpropathrin	*Spirodela polyrhiza*	Xu et al. (2015)
DDT	*Solanum lycopersicum, Helianthus, Glycine max, Medicago sativa*	Mitton et al. (2014)
Endosulfan sulphate	*Zea mays*	Somtrakoon et al. (2014)
Lindane	*Jatropha curcas* L.	Abhilash et al. (2013)
Cypermethrin	*Pennisetum pedicellatum*	Dubey and Fulekar (2013)
Hexachlorobenzene	*Typha latifolia*	Zhou et al. (2013)
Atrazine	*Acorus calamus*	Wang et al. (2012)
Endosulfan	*Brassica campestris*	Mukherjee and Kumar (2012)
Metalaxyl, trifluralin	*Sambucus nigra, Salix alba*	Warsaw et al. (2012)
Atrazine, diazonin, permethrin	*Leersia oryzoides*	Moore and Locke (2012)
Phoxim	*Allium fistulosum*	Wang et al. (2011)
HCHs	*Withania somnifera*	Abhilash and Nandita (2010)
Dimethoate, malathion	*Nasturtium officinale*	Al-Qurainy and Abdel-Megeed (2009)
Chlordane	*Cucumis sativus*	Gent et al. (2007)
DDE	*Brassica juncea,Brassica napus*	White et al. (2005)
Aldicarb	*Zea mays, Vigna radiate, Vigna unguiculata*	Sun et al. (2004)
Butachlor	*Triticum vulgare*	Yu et al. (2003)
Chlordane	*Spinacia oleracea*	Mattina et al. (2003)

pollutant absorption rates (Abdul Waheed et al. 2014). According to Xia and Ma (2006), The uptake and phytodegradation of pesticides by *Eichhornia crassipes* in water bodies can be used as potential, economical and alternative biological method.. Cost effectiveness, longer storage capacity and minimal usage of chemicals are the other potential benefits from this plant (Li et al. 2011). However, the removal efficiency of *E. crassipes*, *P. strateotes* for pyrethroids has been observed to be significantly higher as compared to that of organochlorine (Riaz et al. 2017).

According to Dosnon-Olette et al. (2010), *Lemna minor* and *Spirodela polyrhiza*, is able to remove dimethomorph as long as its concentration does not become too toxic and inhibit depuration mechanisms. *Lemna minor* sensitivity towards dimethomorph was correlated with their ability to remove it from water. *Lemna minor,* also known as duck weed withstands cold weather (1.7–35 °C) and grows rapidly within a week under optimum pH 6 (Prasertsup and Ariyakanon 2011). It also decontaminates heavy metal and organic pollutants such as pesticides by rhizofiltration (Sasmaz et al. 2017).

A. gramineus has the ability to sorb many OP and OC pesticides (diazinon, fenitrothion, malathion, parathion, dieldrin, HCB) and assists in their removal from the aquatic ecosystems (Buyan et al. 2009). *Plantago major* L. is able to take up cyanophos from water by roots as well as by leaves, so *Plantago major* L. may be used for phytoremediation of water contaminated with cyanophos insecticide (Ahmed 2014).

Acorus calamus L. has been reported to exhibit remarkable phytoremediation potential in terms of biomass growth as well as atrazine removal (Roman et al. 2012). *Azolla caroliniana* and *Lemna gibba* have also been reported to remove atrazine from the water (Guimarães et al. 2011). The rate of removal of two fungicides (dimethomorph and pyrimethanil) from water by five macrophyte species (*L. minor*, *S. polyrhiza*, *C. aquatica*, *C. palustris* and *E. canadensis*) was assessed that *L. minor* and *S. polyrhiza* showed the highest removal efficiency for the two fungicides (Rachel et al. 2009).

4 Conclusions

Remediation technique (including microbial remediation and phytoremediation) is a promising technology that makes use of microbes and plants to return a contaminated site to a safe condition. These techniques are economically viable and ecologically effective to reduce the growing pesticide contaminated sites. Combining both microbial and phytoremediation is an approach that ensures a more efficient clean-up of contaminated sites. However, the success largely depends on the microorganism species involved in the bioremediation process.

References

AbdulWaheed, A., Awang, S., & Sarva, P. (2014). The comparison of phytoremediation abilities of water mimosa and water hyacinth. *ARPN Journal of Science and Technology, 4*(12), 722–731.

Abhilash, P. C., & Nandita, C. (2010). *Withania somnifera* Dunal-mediated dissipation oflindane from simulated soil: Implications for rhizoremediation of contaminated soil. *Journal of Soils and Sediments, 10*, 272–282.

Abhilash, P. C., Singh, B., Srivastava, P., Schaeffer, A., & Singh, N. (2013). Remediation of lindaneby *Jatropha curcas* L.: Utilization of multipurpose species for rhizoremediation. *Biomass and Bioenergy, 51*, 189–193.

Ahmed-Ali, R. (2014). Phytoremediation of cyanophos insecticide by *Plantago major* L. in water. *Journal of Environmental Health Science and Engineering, 12*, 38.

Al-Qurainy, F., & Abdel-Megeed, A. (2009). Phytoremediation and detoxification of two organophosphorous pesticides residues in Riyadh area. *World Applied Sciences Journal, 6*(7), 987–998.

Bhat, S. A., Bhatti, S. S., Singh, J., Sambyal, V., Nagpal, A., & Vig, A. P. (2016). Vermiremediation and phytoremediation: Eco approaches for soil stabilization. *Austin Environmental Sciences, 1*(2), 1006.

Biswas, K., Paul, D., & Sinha, S. N. (2015). Biological agents of bioremediation: A concise review. *Frontiers Environmental Microbiology, 1*, 39–43.

Briceno, G., Schalchli, H., Rubilar, O., Tortella, G. R., Mutis, A., Benimeli, C. S., Palma, G., & Diez, M. C. (2016). Increased diazinon hydrolysis to 2-isopropyl-6-methyl-4-pyrimidinol in liquid medium by as specific Streptomyces mixed culture. *Chemosphere, 156*, 195–203.

Buyan, C., Janjit, I., and Jae, S.R (2009). Phytoremediation of organophosphorus and organochlorine pesticides by Acorus gramineus. *Environmental Engineering Research, 14*(4), 226–236.

Byrne, F. J., & Toscano, N. C. (2005). Uptake and persistence of imidacloprid in grapevines treated by chemigation. *Crop Protection 2006, 25*, 831–834.

Carter, A. D., & Heather, A. I. J. (1995). Pesticides in groundwater. In G. A. Best & A. D. Ruthven (Eds.), *Pesticides – developments, impacts, and controls* (pp. 113–123). London: The Royals Society of Chemistry.

Cheng, S., Jin, X., Huiping, X., Liping, Z., & Zhenbin, W. (2007). Phytoremediation of triazophos by canna indicalinn. in a hydroponic system. *International Journal of Phytoremediation, 9*, 453–463.

Chrinside, A. E., Ritter, W. F., & Radosevich, M. (2011). Biodegradation of aged residues of atrazine and alachlor in a mix-load site soil by fungal enzymes. *Applied and Environmental Soil Science, 2011*, 1–10. https://doi.org/10.1155/2011/658569.

Dosnon-Olette, R., Couderche, M., El Arfaoui, A., Sayen, S., & Eullaffroy, P. (2010). Influence of initial pesticide concentrations and plant population density on dimethomorph toxicity and removal by two duckweed species. *Science of the Total Environment, 408*(10), 2254–2259.

Dubey, K. K., & Fulekar, M. H. (2013). Investigation of potential rhizospheric isolate forcypermethrin degradation. *Biotech, 3*, 33–43.

Ellegaard-Jensen, L., Knudsen, B. E., Johansen, A., Albers, C. N., Aamand, J., & Rosendahl, S. (2014). Fungal–bacterial consortia increase diuron degradation in water-unsaturated systems. *Science of the Total Environment, 466*, 699–705.

Fuentes, M. S., Raimondo, E. E., Amoroso, M. J., & Benimeli, C. S. (2017). Removal of a mixture of pesticides by a *Streptomyces* consortium: Influence of different soil systems. *Chemosphere, 173*, 359–367.

Gent, M. P. N., White, J. C., Parrish, Z., Isleyen, M., Eitzer, B., & Incorvia Mattina, M. J. (2007). Uptake and translocation of p,p′-dichlorodiphenyldichloroethylene supplied in hydroponics solution to Cucurbita. *Environmental Toxicology and Chemistry, 26*(12), 2467–2475.

Goda, S. K., Elsayed, I. M., Khodair, T. A., El-Sayed, W., & Mohamed, M. E. (2010). Screening for and isolation and identification of malathion-degrading bacteria: Cloning and sequencing agene that potentially encodes the malathion-degrading enzyme, carboxylestrase in soil bacteria. *Biodegradation, 21*, 903–913.

Guimarães, F. P., Aguiar, R., Karam, D., Oliveira, J. A., Silva, J. A. A., Santos, C. L., Sant'anna-Santos, B. F., & Lizieri-Santos, C. (2011). Potential of macrophytes for removing atrazine from aqueous solution. *Planta daninha, 29* (no.spe Viçosa), 1137–1147.

Kabra, A. N., Ji, M. K., Choi, J., Kim, J. R., Govindwar, S. P., & Jeon, B. H. (2014). Toxicity of atrazine and itsbioaccumulation and biodegradation in a green microalga, *Chlamydomonas mexicana. Environmental Science and Pollution Research, 21*, 12270–12278.

Karunya, S. K., & Saranraj, P. (2014). Toxic effects of pesticide pollution and its biological control by microorganisms: A review. *Applied Journal of Hygiene, 3*, 1–10.

Khan, N. U., Bhavya, V., Nazeeb, I., & Paddu, K. S. (2011). Phytoremediation using an indigenous crop plant (wheat): The uptake of methyl parathion and metabolism of p-nitrophenol. *Indian Journal of Science and Technology, 4*, 1661–1667.

Kharabsheh, H. A., Han, S., Allen, S., & Chao, S. L. (2017). Metabolism of chlorpyrifos by Pseudomonas aeruginosa increases toxicity in adult zebrafish (Danio rerio). *International Biodeterioration & Biodegradation, 121*, 114–121.

Kong, L., Zhu, S., Zhu, L., Xie, H., Su, K., Yan, T., Wang, J., Wang, J., Wang, F., & Sun, F. (2013). Biodegradation of organochlorine pesticide endosulfan by bacterial strain *Alcaligenes faecalis*JBW4. *Journal of Environmental Sciences, 25*, 2257–2264.

Kumar, P. S., Carolin, C. F., & Varjani, S. J. (2018). Pesticides bioremediation. In S. J. Varjani et al. (Eds.), *Bioremediation: Applications for environmental protection and management. Energy, environment, and sustainability*. Singapore: Springer. https://doi.org/10.1007/978-981-10-7485-1_10.

Kurade, M. B., Kim, J. R., Govindwar, S. P., & Jeon, B. (2016). Insights into microalgae mediated biodegradation of diazinon by *Chlorella vulgaris*: Microalgal tolerance to xenobiotic pollutants and metabolism. *Algal Research, 20*, 126–134.

Li, F. B., Li, X. M., Zhou, S. G., Zhuang, L., Cao, F., Huang, D. Y., Xu, W., Liu, T. X., & Feng, C. H. (2010). Enhanced reductive dechlorination of DDT in an anaerobic system of dissimilatory iron-reducing bacteria and iron oxide. *Environmental Pollution, 158*, 1733–1740.

Li, S., Feng-Ying, Z., Yang, H., & Jian-Cong, N. (2011). Thorough removal of inorganic and organic mercury from aqueous solutions by adsorption on Lemna minor powder. *Journal of Hazardous Materials, 186*(1), 423–429.

Lin, C., & Lin, K. S. (2007). Photocatalytic oxidation of toxic organohalides with TiO2/UV: The effects of humic substances and organic mixtures. *Chemosphere, 66*, 1872–1877.

London, L., & Myers, J. E. (1995). General patterns of agricultural chemical usage in the southern regions of South Africa. *The South African Journal of Science, 91*, 508.

Maqbool, Z., Hussain, S., Imran, M., Mahmood, F., Shahzad, T., Ahmed, Z., Azeem, F., & Muzammil, S. (2016). Perspectives of using fungi as bioresource for bioremediation of pesticides in the environment: A critical review. *Environmental Science and Pollution Research, 23*, 16904–16925.

Mattina, M. J. I., Lannucci-Berger, W., Musante, C., & White, J. C. (2003). Concurrent plant uptake of heavy metals and persistent organic pollutants from soil. *Environmental Pollution, 124*, 375–378.

Mitton, F. M., Miglioranza, K. S. B., Gonzalez, M., Shimabukuro, V. M., & Monserrat, J. M. (2014). Assessment of tolerance and efficiency of crop species in the phytoremediation of DDT polluted soils. *Ecological Engineering, 71*, 501–508.

Mitton, F. M., Gonzalez, M., Monserrat, J. M., & Miglioranza, K. S. B. (2016). Potential use of edible crops in the phytoremediation of endosulfan residues in soil. *Chemosphere, 148*, 300–306.

Mohamed, A. T., El-Hussain, A. A., El-Siddig, M. A., & Osman, A. G. (2011). Degradation of oxyfluorfen herbicide by soil microorganism biodegradation of herbicides. *Biotechnology, 10*, 274–279.

Mohany, M., Badr, G., Refaat, I., & El-Feki, M. (2011). Immunological and histological effects of exposure to imidacloprid insecticide in male albino rats. *African Journal of Pharmacy and Pharmacology, 5*, 2106–2114.

Moore, M. T., & Locke, M. A. (2012). Phytotoxicity of atrazine, S-metolachlor, and permethrin to *Typha latifolia* (Linnaeus) germination and seedling growth. *Bulletin of Environmental Contamination and Toxicology, 89*, 292–295.

Mori, T., Wang, J., Tanaka, Y., Nagai, K., Kawagishi, H., & Hirai, H. (2017). Bioremediation of the neonicotinoid insecticide clothianidin by the white-rot fungus *Phanerochaete sordida*. *Journal of Hazardous Materials, 321*, 586–590.

Moschet, C., Wittmer, I., Simovic, J., Junghans, M., Piazzoli, A., Singer, H., Stamm, C., Leu, C., & Hollender, J. (2014). How a complete pesticide screening changes the assessment of surface water quality. *Environmental Science & Technology, 48*, 5423–5432.

Mukherjee, I., & Kumar, A. (2012). Phytoextraction of endosulfan: A remediation technique. *Bulletin of Environmental Contamination and Toxicology, 88*, 250–254.

Pan, X., Xu, T., Xu, H., Fang, H., & Yu, Y. (2017). Characterization and genome functional analysis of the DDT-degrading bacterium *Ochrobactrum* sp. DDT-2. *Science of the Total Environment, 592*, 593–599.

Paul, D., Pandey, G., Meier, C., van-der-Meer, J. R., & Jain, R. K. (2006). Bacterial community structure of a pesticide-contaminated site and assessment of changes induced in community structure during bioremediation. *FEMS Microbiology Ecology, 57*, 116–127.

Peng, X., Huang, J., Liu, C., Xiang, Z., Zhou, J., & Zhong, G. (2012). Biodegradation of bensulphuron-methyl by a novel *Penicillium pinophilum* strain BP-H-02. *Journal of Hazardous Materials, 213*, 216–221.

Plangklang, P., & Reungsang, A. (2010). Bioaugmentation of carbofuran by *Burkholderia cepacia* pcl 3 in a bioslurry phase sequencing batch reactor. *Process Biochemistry, 45*(2), 230–238.

Prasertsup, P., & Ariyakanon, N. (2011). Removal of Chlorpyrifos by water lettuce (Pistia stratiotes L.) and duckweed (Lemna minor L.). *The International Journal of Phytoremediation, 13*(4), 383–395.

Prasertsup, P., & Naiyanan, A. (2011). Removal of chlorpyrifos by water lettuce (Pistia stratiotes l.) and duckweed (Lemna minor L.). *International Journal of Phytoremediation, 13*, 383–395.

Priyadarshani, I., Sahu, D., & Rath, B. (2011). Microalgal bioremediation: Current practices and perspectives. *Journal of Biochemical Technology, 3*, 299–304.

Rachel, O., Michel, C., & Philippe, E. (2009). Phytoremediation of fungicides by aquatic macrophytes: Toxicity and removal rate. *Ecotoxicology and Environmental Safety, 72*(8), 2096–2101.

Rani, K., & Dhania, G. (2014). Bioremediation and biodegradation of pesticide from contaminated soil and water—A noval approach. *International Journal of Current Microbiology and Applied Sciences, 3*, 23–33.

Rani, M., Shanker, U., & Jassal, V. (2017). Recent strategies for removal and degradation of persistent and toxic organochlorine pesticides using nanoparticles: A review. *Journal of Environmental Management, 190*, 208–222.

Riaz, G., Amtul Bari, T., Shakir, I., Abdullah, Y., Mateen, A., Khan, A. M., Mahfooz, Y., & Baqar, M. (2017). Phytoremediation of organochlorine and pyrethroid pesticides by aquatic macrophytes and algae in freshwater systems. *The International Journal of Phytoremediation, 19*(10), 894–898.

Rodante, F., Marrosu, G., & Catalani, G. (1992). Thermal-analysis and kinetic-study of decomposition processes of some pesticides. *Journal of Thermal Analysis and Calorimetry, 38*, 2669–2682.

Roman, M., Wojciech, B., Paweł, C., Łukasz, Ł., & Łukasz, C. (2012). Phytoremediation potential of three wetland plant species toward atrazine in environmentally relevant concentrations. *Polish Journal of Environmental Studies, 21*(3), 697–702.

Romeh, A. A. (2015a). Enhancing agents for phytoremediation of soil contaminated by cyanophos. *Ecotoxicology and Environmental Safety, 117*, 124–131.

Romeh, A. A. (2015b). Evaluation of the phytoremediation potential of three plant species for azoxystrobin-contaminated soil. *International Journal for Environmental Science and Technology, 12*, 3509–3518.

Romeh, A. A., & Hendawi, M. Y. (2013). Chlorpyrifos insecticide uptake by plantain from polluted water and soil. *Environmental Chemistry Letters, 11*, 163–170.

Romeh, A. A., & Hendawi, M. Y. (2017). Biochemical interactions between *Glycine max* L. silicondi-oxide (SiO2) and plant growth-promoting bacteria (PGPR) for improving phytoremediation of soil contaminated with fenamiphos and its degradation products. *Pesticide Biochemistry and Physiology, 142*, 32–43. https://doi.org/10.1016/j.pestbp.2017.01.001.

Saez, J. M., Alvareza, A., Benimeli, C. S., & Amoroso, M. J. (2014). Enhanced lindane removal from soil slurry by immobilized *Streptomyces* consortium. *International Biodeterioration & Biodegradation, 93*, 63–69.

Sahu, O. (2014). Reduction of organic and inorganic pollutant from waste water by algae. *International Letters of Natural Sciences, 8*, 1–8.

Sasmaz, M., Obek, E., & Sasmaz, A. (2017). The accumulation of La, Ce and Y by Lemna minor and Lemna gibba in the Keban gallery water, Elazig Turkey. *Water Environment Journal, 13*(4), 383–395.

Semple, K. T., Ronald, B. C., & Stefan, S. (1999). Biodegradation of aromatic compounds by microalgae. Mini review. *FEMS Microbiology Letters, 170*, 291–300.

Sharma, P., Chopra, A., Cameotra, S. S., & Suri, C. R. (2010). Efficient biotransformation of herbicide diuron by bacterial strain *Micrococcus* sp. PS-1. *Biodegradation, 21*, 979–987.

Singh, B., & Mandal, K. (2013). Environmental impact of pesticides belonging to newer chemistry. In A. K. Dhawan, B. Singh, M. Brar-Bhullar, & R. Arora (Eds.), *Integrated pest management* (pp. 152–190). Jodhpur: Scientific Publishers.

Smith, E., Smith, J., Naidu, R., & Juhasz, A. L. (2004). Desorption of DDT from a contaminated soil using cosolvent and surfactant washing in batch experiments. *Water, Air, and Soil Pollution, 151*, 71–86.

Somtrakoon, K., Kruatrachue, M., & Lee, H. (2014). Phytoremediation of endosulfansulfate-contaminated soil by single and mixed plant cultivations. *Water, Air, and Soil Pollution, 225*, 1886.

Sun, H., Xu, J., Yang, S., Liu, G., & Dai, S. (2004). Plant uptake of aldicarb from contaminated soil and its enhanced degradation in the rhizosphere. *Chemosphere, 54*, 569–574.

Suresh, B., Sherkane, P., Kale, S., Eapen, S., & Ravishankar, G. (2005). Uptake and degradation of DDT by hairy root cultures of *Cichorium intybus* and *Brassica juncea*. *Chemosphere, 61*, 1288–1292.

Thengodkar, R. R., & Sivakami, S. (2010). Degradation of Chlorpyrifos by an alkaline phosphatase from the cyanobacterium Spirulina platensis. *Biodegradation, 21*, 637–644.

Varsha, Y. M., Naga, D. C. H., & Chenna, S. (2011). An emphasis on xenobiotic degradation in environmental cleanup, review article. *Journal of Bioremediation & Biodegradation, 4*, 1–10.

Verardo, E., Atteia, O., & Prommer, H. (2017). Elucidating the fate of a mixed toluene, DHM, methanol, and i-propanol plume during in situ bioremediation. *Journal of Contaminant Hydrology, 201*, 6–18. https://doi.org/10.1016/j.jconhyd.2017.04.002.

Vijayakumar, S. (2012). Potential applications of cyanobacteria in industrial effluents—A review. *Journal of Bioremediation & Biodegradation, 3*, 1–6.

Villaverde, J., Rubio-Bellido, M., Merchan, F., & Morillo, E. (2017). Bioremediation of diuron contaminated soils by a novel degrading microbial consortium. *Journal of Environmental Management, 188*, 379–386.

Wang, F. Y., Tong, R. J., Shi, Z. Y., Xu, X. F., & He, X. H. (2011). Inoculations with arbuscular mycorrhizal fungi increase vegetable yields and decrease phoxim concentrations in carrot and green onion and their soils. *PLoS One, 6*(2), e16949.

Wang, Q., Zhang, W. X., Li, C., & Xiao, B. (2012). Phytoremediation of atrazine by three emergent hydrophytes in a hydroponic system. *Water Science and Technology, 66*(6), 1282–1288.

Warsaw, A. L., Fernandez, R. T., Kort, D. R., Cregg, B. M., Rowe, B., & Vandervoort, C. (2012). Remediation of metalaxyl, trifluralin, and nitrate from nursery runoff using container-grown woody ornamentals and phytoremediation areas. *Ecological Engineering, 47*, 254–263.

White, J. C., Parrish, Z., Isleyen, M., Gent, M., Iannucci-Berger, W., Eitzer, B., & Mattina, M. (2005). Uptake of weathered p,p′-DDE by plant species effective at accumulating soil elements. *Microchemical Journal, 81*, 148–155.

Xia, H., & Ma, X. (2006). Phytoremediation of ethion by water hyacinth (Eichhornia crassipes) from water. *Bioresource Technology, 97*(8), 1050–1054.

Xiao, P., Mori, T., Kamei, I., Kiyota, H., Takagi, K., & Kondo, R. (2011). Novel metabolic pathways of organochlorine pesticides dieldrin and aldrin by the white rot fungi of the genus *Phlebia*. *Chemosphere, 85*, 218–224.

Xu, X. J., Sun, J. Q., Nie, Y., & Wu, X. L. (2015). *Spirodela polyrhiza* stimulates the growth of its endophytes but differentially increases their fenpropathrin-degradation capabilities. *Chemosphere, 125*, 33–40.

Yu, Y. L., Chen, Y. X., Luo, Y. M., Pan, X. D., He, Y. F., & Wong, M. H. (2003). Rapid degradation of butachlor in wheat rhizosphere soil. *Chemosphere, 50*, 771–774.

Zhang, S., Qiu, C. B., Zhou, Y., Jin, Z. P., & Yang, H. (2011). Bioaccumulation and degradation of pesticide fluroxypyr are associated with toxic tolerance in green alga Chlamydomonas reinhardtii. *Ecotoxicology, 20*, 337–347.

Zhou, Y., Tigane, T., Li, X., Truu, M., Truu, J., & Mander, U. (2013). Hexachlorobenzene dechlorination in constructed wetland mecocosms. *Water Research, 47*, 102–110.

Zinovyev, S. S., Shinkova, N. A., Perosa, A., & Tundo, P. (2005). Liquid phase hydrodechlorination of dieldrin and DDT over Pd/C and Raney-Ni. *Applied Catalysis B: Environmental, 55*, 39–48.

Trends in Phytomanagement of Aquatic Ecosystems and Evaluation of Factors Affecting Removal of Inorganic Pollutants from Water Bodies

Abdul Barey Shah, Rana Pratap Singh, and U. N. Rai

Abstract The deterioration of water quality due to the increasing unsustainable developmental activities like production processes carried at high energy inputs, discharge of untreated municipal/industrial wastewater coupled with runoff from agricultural fields led to build up of toxic inorganic contaminants including heavy metals and Reactive Nitrogenous Species (RNS) into the water bodies. Intake of water contaminated with heavy metals and nitrogenous ions (nitrate, nitrite and ammonium) by humans and other life forms may lead to disruption of various metabolic activities, leading to cardiovascular, neurological, renal disorders. Different technologies and methods are being employed to remediate these pollutants from water. Phytoremediation is an economical, ecofriendly and aesthetically pleasing technology that makes the use of plant systems to remove and/or detoxify pollutants from the environment. The efficiency of the decontamination or remediation function of aquatic macrophytes depends on several factors like water physico-chemistry, plant physiology, plant genotype, sediment geochemistry and nature of contaminant or pollutant. Also water remediation by macrophytes can be significantly improved by appropriate selection of plant species which is built on the type of substances to be removed, the topography of the area, microclimate, hydrological conditions, accumulation capacities of the plant species etc. This write-up provides some insights in phytoremediation of inorganic pollutants and factors affecting their removal.

Keywords Constructed wetlands · Heavy metals · Phytoremediation · Water treatment · Floating wetlands

A. B. Shah (✉) · R. P. Singh
Department of Environmental Science, Babasaheb Bhimrao Ambedkar (Central) University, Lucknow, India

U. N. Rai
Plant Ecology and Environmental Science Division, CSIR-NBRI, Lucknow, India

H. Qadri et al. (eds.), *Fresh Water Pollution Dynamics and Remediation*,
https://doi.org/10.1007/978-981-13-8277-2_14

1 Introduction

Since Middle Ages due to human activities, natural aquatic ecosystems, in particular the freshwater systems and estuaries, are not only getting polluted to a large extent, but are also sentenced equally to continuing pollution due to toxic metal deposition in sediment environments. Today, water availability both in terms of quality and quantity is a problem all over the world. Mainly water stressed developing countries are experiencing the worse of this problem due to increasing world population and industrialization (EPA 2001; Adewuyi et al. 2014; Agca et al. 2014). So far quantity is concerned, many countries have conflicts and others undergone mutual/multilateral agreements for water sharing and distribution (Rai 2012; Campbell et al. 2016). The enforcement of several water regulations and strategies worldwide coupled with the need for long term environmental sustainability has demanded the need for various stringent regulations for drinking water supply and wastewater discharge. (Arias et al. 2011; Onur et al. 2014). These methods include reverse osmosis, nano-filtration, chemical precipitation, ion exchange, disinfection, adsorption by activated carbons and sewage treatment plant (STPs) processes. Most of these methods require high energy are expensive, and don't completely remove the heavy metals and excessive nutrients (Mojiri 2012), but merely transfer the pollutants from the wastewater to residual sludge being disposed off by land filling (Tolu and Atoke 2012) whereby the pollutants ultimately make their way to fresh water supplies. Therefore, these procedures do not provide acceptable solution to pollution problem. From the economic and ecological viewpoint, the requirement of an unconventional eco-friendly and cost-effective techniques are recommended for the cleanup of hazardous waste site by conventional technologies is expected to cost the United States alone $400 billion (Salt et al. 1995; Rousseau et al. 2004). Methods using living wetland plants to remove metals from water appear to be an alternative. Plants that have a high metal bioaccumulation capacity and a good tolerance to high metal concentrations over long periods of time are necessary (Rai et al. 2013). Over the past decade, phytoremediation has become an efficient and effective technique of environment cleaning because of different plants capability to accumulate the pollutants at levels which are thousands times higher than background concentrations. The plants are able to remediate many toxic elements from reservoirs, sequestering the pollutants from waters as food elements, fundamentally through their root system using the products of decomposition for their growth (Ali et al. 2013). Water plants are reported to accumulate Mn, Zn, Cu, Fe, Pb, Cd, Cr, As and Ni, and other elements and substances. The plants are more tolerant to the inorganic contaminants as they have differential capacity to assimilate or scavenge degradable and non-biodegradable inorganic contaminants. These potentials of plants have emerged as a major area of phytotechnological studies and they have been studied for the phytoremediation potential for removal of toxic contaminants from contaminated soil and water (Rai 2010; Vymazal and Kropfelova 2011; Rawat et al. 2012; Bauddh and Singh 2012; Rai et al. 2015).

2 Contamination of Water Bodies with Inorganic Pollutants

In surface water systems, heavy metals originate from both natural and anthropogenic sources. Currently, metal pollution due to anthropogenic sources exceeds the natural inputs. Metal pollution mainly results because of fossil fuels burning, mining and smelting of metalliferous ores, sewage, pesticides, fertilizers and municipal wastes (Abou-Elela et al. 2013). The development of the nations requires the production of energy. However, energy-production technology and environmental pollution are intimately linked with each other. Energy-intensive processes and chlor-alkali industries for the manufacture of agrochemicals deteriorate the water quality of lakes and reservoirs due to the discharge of various pollutants; especially a range of heavy metals (Rai 2012). Heavy metals constitute a heterogeneous group of elements; having a specific gravity greater than 4.0 and a relatively high density (approximately 5 g/cm^3) as their common characteristics (Clijsters et al. 1999). The heavy metal load from domestic wastewater and sewage alone signifies that this will be a continuing problem for science and humankind. Water in lakes, rivers and other important aquatic systems can get heavily polluted, owing to the volume, flow and the proximity to point sources of pollution. Toxic metal pollution of water streams and groundwater is a major environmental and health problem that still needs an effective and affordable technological resolution.

3 Effect of Toxic Heavy Metals and Other Inorganics on Different Life Forms

From a common biological as well as plant physiological viewpoint, essential and non-essential heavy metals are recognized. Living beings need traces of some heavy metals, that include copper, molybdenum, iron, manganese, strontium, cobalt, nickel, vanadium and zinc, and these are known as essential heavy metals. Non-essential heavy metals of serious concern in the environment include chromium, nickel, mercury, cadmium, arsenic, silver and lead, pose great risks to human health (Kennish 1992; Muchuweti et al. 2006; Rashed 2010; Kumar et al. 2013). Essential heavy metals play vital roles as constituents of metallo-proteins, as cofactors in enzymatic catalysis, and in a wide array of other cellular processes. At supra-optimal concentration however, they become phytotoxic, induce leaf chlorosis, and reduce growth. The hazard of heavy metal contaminants in water lies in two features of their impact. Firstly, the greater ability of heavy metals to persist in natural ecosystems for a prolonged phase. Secondly, their ability to get accumulated at successive trophic levels in biological chain, thereby causing chronic and acute diseases. For example, cadmium and zinc can damage brain, heart and kidney and also lead to acute gastrointestinal and respiratory problems (Lokhande et al. 2011). The application of industrial and domestic effluents on agricultural lands, which may contain high concentrations of heavy metals, is a common practice in some parts of the world. These toxic metals, upon getting concentrated in plant tissues have serious effects on the plant themselves and also pose a threat to man and animals. Episodes of the metal pollution such as Minamata Episode due to Methyl mercury, Itai- itai or Ouch Ouch due to Cadmium have taken toll on human populations. Table 1; lists some of the toxic heavy metals and nutrients with their harmful effects. (Adopted from; Cardwell et al. 2002; Bellos and Sawidis 2005; Basile et al. 2012; Rawat et al. 2012; Rosli and Yahya 2012; Iqbal et al. 2013 Wang and Yu. 2014).

Another potential concern for toxic heavy metals is their transfer and accumulation in the bodies of animals or human beings through the food chains, which significantly damages DNA and have carcinogenic effects by their mutagenic ability (Haloi and Sarma 2012). Examples include Cu, Cd, and Cr, have been associated with different health effects extending from dermatitis to development of different

Table 1 Effect of toxic heavy metals reported in aquatic ecosystem on human health and plants

Heavy metal	Effect on humans	Effect on plants
Cadmium	Damage to brain, gastrointestinal and respiratory problems, kidney and liver damage	Retarded plant growth, seed germination and lipid content, induction of the production of phytochelatins
Arsenic	Cutaneous and visceral malignancies, black foot disease, severe vomiting, Diarrhoea	Biochemical dysfunction at cellular level, damage to proteins and lipids.
Lead	Kidney damage, heart ailments, reproductive problems, bone weakness	Reduction in chlorophyll production, decreased plant growth; increases superoxide dismutase
Mercury	Foetal brain damage, damage to kidney, lungs, heart, neurological problems	Retarded uptake of water, photosynthetic activity, accumulation of phenol and proline, decreased antioxidant enzymes
Chromium	Haemolysis, renal and liver failure, allergies, dermatitis, Foetal deaths, lung cancers	Decreases enzyme activity and plant growth; produces membrane damage, chlorosis and root Damage
Copper	Gastrointestinal distress, liver or kidney damage (long term exposure)	Inhibition of photosynthesis, lessening of plant growth and reproductive process; decrease in thylakoid surface area
Iron	Increased pulse rates and respiration, hypertension, drowsiness, congestion of blood vessels	
Zinc	Vomiting, renal damage, cramps	Reduces seed germination; increase in the growth of plants and ATP/chlorophyll ratio
Nickel		Reduces seed germination, dry mass accumulation, protein production, chlorophylls and Enzymes; increases free amino acids
Manganese	Growth retardation, fever, sexual impotence, muscles fatigue, eye blindness.	Brown spots on mature leaves, intervienal chlorosis and necrosis, deformation of young leaves and growth retardation

cancers in body (Agarwal et al. 2007). In addition, the occurrence of some metals in the environment as radioactive isotopes (*e.g.*, U238, Cs137, Pt239, and Sr90), are potential threats to health (Fawzy et al. 2012).

4 Conventional Remediation Technologies

Since the quality of drinking water supply is affected by the presence of heavy metal pollution and wastewater discharge, thereby affecting humans, animals and plants, efforts have been made during the last two decades to lessen pollution causes and remediate polluted water assets. Though various technological advancements have been made to remove pollutants from water but certain drawbacks and limitations are associated with them. Table 2 lists the advantages and disadvantages of conventional methods of water treatment (Carty et al. 2008; Rai and Tripathi 2009; Osorio et al. 2011; Saeed and Sun 2012).

5 Removal, Uptake and Stabilization of Inorganic Pollutants from Water by Different Aquatic Macrophytes

Phytoremediation, the use of plants to remove pollutants from the environment, is a growing field of research in environmental studies because of the advantages of its environmental friendliness, cost effectiveness and the possibility of harvesting the plants for the extraction of absorbed contaminants such as metals that cannot be easily biodegraded for recycling (Maine et al. 2004; Skinner et al. 2007; Malik 2007). Over the last two decades, phytoremediation has become progressively

Table 2 Conventional methods of water treatment

Methods	Advantages	Disadvantages
Chemical precipitation	Convenient, self-operation, low maintenance, low capital cost.	Replenishment of chemicals, requirement of extra coagulation and flocculation, toxic sludge generation
Coagulation-flocculation	Settlement of suspended solids in less time, enhanced sludge deposition	Disposal of sludge and its associated costs
Ion exchange	Time efficient, no sludge production, high metal removal efficiency, better performance in acidic pH range	Less suitable as few metals are not exchangeable through ion exchange resins, high capital cost
Reverse osmosis	Greater ionic species removal, can also operate at high temperatures, reduces the concentration of dissolved organic compounds	Expensive to procure and operate, elevated pressure makes the technique costly and sensitive to operating conditions
Nanofiltration	Operates at low pressures than reverse osmosis	Costly, membrane fouling

known technique for removal of contaminants from water, wastewater and shallow soil environments. Further, the technique is solar-driven, aesthetically pleasing, passive in nature and useful for remediation of shallow plumes with low to medium levels of contamination (EPA 2001; Wang et al. 2011). Wetland ecosystems act as natural filters and have been effectively used for the treating and removing toxic chemicals from wastewater through absorption by specific plants. Certain wetlands are being engineered and have been designed and constructed to utilize natural processes involving wetland vegetation, soil, and the associated microbial assemblages to assist in treating wastewaters. They are designed to take advantage of many of the same processes that occur in natural wetlands, but do so within a more controlled environment. They are referred to as constructed wetlands. The use of constructed wetland system is a reasonable option for treating contaminated water by simulating natural wetlands, owing to lower cost, fewer operation and maintenance requirements, and little reliance on energy inputs (Zhang et al. 2012). Constructed wetland (CW) is a highly efficient and biogeochemical system for the treatment of contaminated waters arising from different sources such as industrial, mining, domestic and highway sectors and also offers an ecofriendly alternative for traditional wastewater treatment systems. Constructed wetlands have been used for a variety of purposes, from rehabilitating areas where wetlands were previously located, to serving very specific functions such as wastewater treatment. This system has been found to be able to remove various pollutants and nutrients from wastewater (Vymazal 2007; Bindu et al. 2008) and has also been successfully used to treat wastewater with high concentrations of nutrients (Ghosh and Singh 2005; Tee et al. 2012). Much interest has been focused on constructed wetlands for removing toxic metals from wastewater and drinking water sources in recent years (Maine et al. 2004; Hadad et al. 2006; Jayaweera et al. 2008). The CW system consists of natural pathways of aquatic macrophytes that not only amass pollutants directly into their tissues but also act as catalysts for purification processes usually taking place in below ground part of plants called as rhizosphere. In CW, interactions in the substratum of plants remove most of the metals from contaminated water (Liu et al. 2007). The permanent or temporarily anoxic condition in wetland soil helps to create an environment for immobilization of heavy metals in the highly reduced sulfite or metallic form and plants may play an important role in metal removal through filtration, adsorption, cation exchange, and root-induced chemical changes in the rhizosphere (Liu et al. 2007; Kadlec and Wallace 2009). Numerous factors including pH of water and sediment, mobilization and uptake from the soil, compartmentalization and sequestration within the root, efficiency of xylem loading and transport (transfer factors), distribution between metal sinks in the aerial parts, sequestration and storage in leaf cells as well as the plant growth and transpiration rates can also effect the remediation processes of the contaminated sites (Hadadet al. 2006; Soda et al. 2012). Although, emergent macrophytes are mostly used for constructed wetlands but the design of the systems in terms of media as well as the flow regime varies. The most common CW systems are constructed with a horizontal subsurface flow (HF constructed wetlands), though vertical flow (VF constructed wetlands) systems are becoming more prevalent (Vymazal and Kropfelova 2011). Among the different types of CWs, Horizontal Sub-surface Flow Constructed Wetlands (HSSFCWs) are most widely used and became low-impact alternatives to more conventional wastewater treatment processes. In a typical HSSFCW, wastewater is maintained at a constant depth and flows horizontally below the surface of the bed and has been proven to be efficient in removing pollutants, organic matter and pathogens. Sewage treatment efficiency of Angular Horizontal Subsurface Flow Constructed wetland using *Colocasiae sculenta* has been reported by Chavan and Dhulap (2012) and it was observed that reduction in electrical conductivity (EC) by 23.68%, total suspended solids (TSS) by 46.15%, total dissolved solids (TDS) by 50.08%, total solids (TS) by 49.34%, biological oxygen demand (BOD) by 54.30%, chemical oxygen demand (COD) by 58.69%, nitrates (NO_3) by 59.48%, phosphates (PO_4) by 46.99% and sulfates (SO_4) by 39.32% against treatment of sewage in the control bed in which EC was reduced by 11.62%, TSS by 27.90%, TDS by 32.66%, TS by 29.94%, BOD by 31.26%, COD by 39.81%, NO3 by 23.93, PO4 by 20.89 and SO4 by 16.48% respectively. Thus constructed wetlands can also be used for removing organic pollution load of wastewater.

Aquatic macrophytes, which play important roles in aquatic ecosystems, have shown great potential to sequester selected heavy metals and nutrients through their root systems and by uptake through their plant bodies. It has been reported that these plants can accumulate heavy metals 100, 000 times greater than in the associated water (Mishra and Mishra 2008). Therefore, they have been used for heavy metal and nutrient removal from a variety of sources (Rai 2010, 2012). Phytoremediation exploits plant's distinctive biological mechanisms for the elimination of pollutants from the environment for human benefit. The phytoremediation technique relies upon the following processes:

5.1 *Phytodegradation*

Phytodegradation primarily removes the organic contaminants from the environment by employing internal and external metabolic pathways carried out by plants. It includes the application of plants to uptake, store and degrade pollutants within their tissue whereby plants metabolize and destroy contaminants (Ghosh and Singh, 2005). During this process plants take-up metal contaminants directly from the water/soil or release root exudates that help in the degradation of pollutants via co-metabolism in the rhizosphere.

5.2 *Phytoextraction*

Primarily this method is used for waste laden metals whereby plant roots absorb, translocate and store these metals along with other nutrients and water. Metal compounds that have been effectively phytoextracted included, cadmium, arsenic, nickel, zinc, copper and chromium. This process is known to occur either continuously (natural) using hyper-accumulators or induced through the addition of chelates such as EDTA to increase the bioavailability of metals (Utmazian and Wenzel 2006). This has been also realized that phytoextraction can be used for the retrieval of precious metals such as gold, silver, platinum and palladium, which represents the vast possibilities of this phytoremediation technique with regards to mining.

5.3 *Rhizofiltration*

Usually aquatic plants perform this process. The hyperaccumulating aquatic plants adsorb and absorb pollutants from aquatic environments i.e., water and wastewater (Rahman and Hasegawa 2011). A plant suitable for rhizofiltration process can remove toxic metals from water column over a prolonged period with its rapid-growth root system. Numerous plant species are recognized to be highly effective in confiscating toxic metals such as, Cr, Cu, Cd, Ni, Zn and Pb from polluted water (EPA 2001; Rai 2012).

5.4 *Phytovolatilization*

This phytoremediation technique is the plants ability to take up toxic metals from the growth medium and then transform and volatilize them through the leaves into atmosphere. There is a transformation of pollutants within the plant body, as the water travels along the plant's vascular system from the roots to the leaves, whereby the contaminants evaporate or volatilize into the air surrounding the plant. Some of these contaminants can pass through the plants to the leaves and volatilize into the atmosphere at comparatively low concentrations (Ghosh and Singh 2005).

5.5 *Phytostabilization*

Phytostabilization, also referred to as in-place inactivation, is primarily used for the remediation of soil, sediment and sludge (EPA 2001). The process of phytostabilizationis dependent upon the tolerance ability of a plant to a pollutant. It involves the application of plant roots to minimize metal mobility and bioavailability in the soil. During the process, the contaminants may be precipitated in the rhizosphere, absorbed and accumulated by roots or adsorbed onto the root system of plants. Thus preventing the mobility of the contaminants, thereby reducing their entry in the food chain (Jadia and Fulekar 2009). For the removal of heavy metal from aquatic ecosystems some of the common aquatic macrophytes used are listed in Table 3. (From Hadad et al. 2006; Padmavathiamma and Li 2007; Liu et al. 2007;

Table 3 Some common heavy metal accumulating aquatic macrophytes

Aquatic macrophytes	Heavy metal accumulation
Azolla filiculoids	Cr, Ni, Zn, Fe, Pb, As, Hg, Cd
Azolla pinnata	Cd, Cu, Zn, Hg
Ceratophyllum demersum	Cu, Cr, Pb, Hg, Fe, Mn, Zn, Ni
Eichhornia crassipes	Cd, Pb, Cu, As, Ni, Cr, Zn, Hg, Co, Al
Hydrilla verticillata	Cu, Hg, Fe, Ni, Pb
Lemna spp.	Pb, Mn, Cu, Cd, Cr, Hg, Ni, Fe
Mentha aquatic	Cd, Zn, Cu, Fe, Hg
Nymphaea alba	Cr, Cd, Pb, Ni, Zn, Mn, Fe, Co
Phragmitis australis	Fe, Mn, Zn, Cu
Potamogeton crispus	Cu, Pb, Mn, Fe, Cd
Salvinia spp.	Cu, Fe, Ni, Zn
Spirodela polyrrhiza	As, Hg
Typha domingensis	Fe, Mn, Zn, Al, Ni
Wolfia globosa	As

Rai and Tripathi 2009; Rahman and Hasegawa 2011; Hegazy et al. 2011; Singh et al. 2012; Mojiri 2012; Tolu and Atoke 2012; Souza et al. 2013; Shah et al. 2015.)

6 Factors Affecting the Removal Rates of Inorganic Pollutants from Water Using Aquatic Macrophytes

Bioavailability of metals in water for accumulation is influenced by various factors such as pH, temperature, redox potential, chemical speciation, seasonal changes, sediment type, salinity and organic matter (Shuping et al. 2011). Both root and shoot tissues of aquatic marcophytes have been demonstrated for metal accumulation and translocation under natural ecosystems (Shah et al. 2015). Generally, metal levels were much higher in the roots as compared to shoot of the plants, which are in line with the reports of other researchers (Rieumont et al. 2007). The inhibition of metal translocation to shoot parts of plants be ascribed to formation of complex compounds with -COOH group thus leading to low mobility of metals from root to shoot (Cradwell et al. 2002). However at certain sites, it was observed that plants accumulated metals in higher proportions in shoots than root and differential metal accumulation by different plant species may be for compartmentalization and translocation in the vascular units of plants (Wu et al. 2011). The competence of phytoremediation differs significantly between species as different routes and mechanisms of ion uptake are operative in each species, which are based on their morphological, physiological, genetic and anatomical characteristics (Rehman and Hasegawa 2011). Differences in the bioaccumulation rates and translocation factors among the two plant species reflect the abundance of the metals studied and the intrinsic abilities of the plants to sequester the metals (Anning et al. 2013).

The inorganic nutrients in water i.e., nitrite, nitrate, ammonium and phosphate degrade the quality of water and deplete the dissolved oxygen present in water. Nitrite being a natural constituent of the nitrogen cycle in ecosystems, and its existence in the environment is a serious problem due to its well-known toxicity to animals (Sinha and Nag 2011). Its removal from water is necessary in order to reduce its harm to humans and animals, as they cannot assimilate nitrite like bacteria and plants (Alonso and Camargo 2009). Our study pertaining to the changes in the inorganic nutrients reveals that there is a significant decrease in the concentration of inorganic pollutants from the treatment system. Similar results pertaining to the current study were reported by Rawat et al. (2012). The potential rate of nutrient uptake by plant is limited by its net productivity (growth rate) and the concentration of nutrients in the plant tissues (Vymazal 2007). Reduction in the concentration of nitrate could be due to the increased plant uptake rather than microbial denitrification. Bindu et al. (2008) reported the removal of nitrate might be due to uptake by the plant roots. Denitrification is believed to be the major pathway for ammonia removal in the constructed wetland (Rai et al. 2013). The mechanism of the phosphorus removal is reported to occur by complexation, precipitation, sorption and assimilation into microbial and plant biomass. The key role of wetland plants for the removal of phosphate is through direct uptake and provision of favourable conditions for microorganisms that use phosphorus as a nutrient (Mbuligwe 2004). Nutrient removals by plants have been found to account for 15–80% nitrogen and 24–80%phosphorus (Rawat et al. 2012). Differential nutrient uptake by plant species can be accounted to nutrient loading rates, specific abilities of different plants and climates (Vymazal 2007).

7 Floating Wetland Beds: An Innovative and Effective Treatment for Water Remediation

Floating Wetlands are somewhat a novel and innovative alternative of treatment wetlands and pond technology that offer great potential for treatment of polluted waters (Ladislas et al. 2015). Floating Treatment Wetlands (FTWs) are an evolving variant of constructed wetland technology which comprises of emergent wetland plants developing hydroponically on structures. The floating islands develop an efficient plant of hungry microbes, waiting for nutrients to consume, whose byproduct is food for other organisms in the complex mesh of a wetland (Chaudhuri et al. 2014). This technique is accurate sustainability, which appears a far cry from the many plotted and dysfunctional water cleansing structures that are common today. For storm-water holding ponds the floating islands have impressive potential to purify nutrient rich water and create multidimensional habitat at the same time (Maltais-Landry et al. 2009).

National Institute of Water & Atmospheric Research (NIWA) at New Zealand evaluated the Application of Floating Wetlands for Enhanced Storm-water Treatment and stated "In existing systems, the FTW (floating treatment wetland) may become a low cost option to upgrade existing stormwater ponds for removal of fine particles and associated metals." It depicts that further research is mandatory to identify key treatment processes in floating wetland systems. Primary studies are revealing that cattail (*Typha latifolia*) is not the only potential wetland hero for water purification but also, there is the carbon sequestering abilities of the microbes within these systems (Tanner and Headley 2011).

The floating islands are a "stacked function tool" for water management. Municipalities with a floating wetland in the form of retention pond near the commercial sector of town can afford the following functions at very low costs: clean-up of water by the removal of soluble nutrients like nitrogen and phosphorous, providing habitat for flora and fauna, in a way that is visible and aesthetically pleasing; creating a new riparian edge that can be planted with indigenous wetland plants; mobilizing the local community to get involved in planting the islands; creating a completely new form of environmental stewardship; sequestering atmospheric carbon dioxide and other greenhouse gases; offer municipal water managers another tool for water treatment that is cost effective, ecofriendly, unremarkable and attractive (Hubbard 2010). Artificially created floating wetlands have been used with varying success for a number of applications to date, such as water quality improvement, habitat enhancement and aesthetic purposes in ornamental ponds. In terms of water quality improvement, the main applications of FTWs reported to date have been for the treatment of: stormwater, combined stormwater-sewer, overflow sewage. Previous studies have shown that floating wetlands can be effective systems for removing the dissolved metals present in runoff. However, such processing methods are designed for implementation in situ, directly on the surface of existing retention basins (Sukias et al. 2010). Other studies have reported on systems floating on the surface of runoff retention basins (airport, residential areas) or lakes or rivers. Although, floating wetlands are increasingly used to treat a variety of types of wastewater, the construction of the system remains mainly empirical and a research effort is needed to define design parameters more precisely. Most studies of floating treatment systems have analyzed Chemical Oxygen Demand (COD), Biochemical Oxygen Demand (BOD), nutrients (N and P) and suspended solids (TSS). To our knowledge, the role played in the decontamination process by the microbial biofilm that develops on the surface of roots suspended in the water column has been little studied. Previous research has shown that *Juncus* and *Carex*are able to grow and accumulate Ni, Cd, and Zn under hydroponic conditions in the laboratory (Ladislas et al. 2013). The plant roots flourish through the floating mat and into the water below. As well as storing nutrients directly from the water column (rather than the bottom sediments), the roots develop a large surface area for adsorption and biofilm attachment (Tanner and Headley 2011). Because FTWs can stand deep and changing water levels, they can be used in conditions where use of conventional surface-flow wetlands with bottom-rooted emergent aquatic macrophytes would be unacceptable. As such, FTWs assimilate the nutrient attenuation proficiencies of wetlands with the elasticity of deeper pond systems, and so increase the variety of conditions where wetland ecotechnologies can be applied for water quality improvement (Headley and Tanner 2008).

8 Conclusions

The minimization of environmental and health impacts for the presence of inorganic pollutants (heavy metals and nutrients) in aquatic systems necessitates the application of diverse treatment practices. This has necessitated the need for cost-effective, viable and ecofriendly technologies for safe drinking water supply and wastewater discharge viz-a-viz its treatment, that preserves precious natural resources and biological lives. Phytoremediation is one new approach that offers more ecological benefits and a cost-effective alternative. Though, a cheaper method, it requires technical strategy, expert project designers and with field experiences that choose the proper species and cultivators for particular toxic pollutants (metals, nutrients organics) and regions. Phytoremediation technology needs more attention in various areas such as gene manipulation, harvesting and recycling tools. A multidisciplinary research work that incorporates the developments of natural sciences, environmental engineers and policy makers is indispensable for bigger attainment of green technologies as a potential tool for inorganic pollutants removal and their management in aquatic ecosystems.

References

Abou-Elela, S. I., Golinielli, G., Abou-Taleb, E. M., & Hellal, M. S. (2013). Municipal wastewater treatment in horizontal and vertical flows constructed wetlands. *Ecological Engineering, 61*, 460–468.

Adewuyi, G. O., Etchie, A. T., & Etchie TO. (2014). Health risk assessment of exposure to metals in a Nigerian water supply. *Human and Ecological Risk Assessment: An International Journal, 20*, 29–44.

Agarwal, R., Kumar, R., & Behari, J. R. (2007). Mercury and lead content fish species from the river Gomti, Lucknow, India, as biomarker of contamination. *Bulletin of Environment Contamination and Toxicology, 78*, 118–122.

Agca, N., Karanlık, S., & Odemiş, B. (2014). Assessment of ammonium, nitrate, phosphate, and heavy metal pollution in groundwater from Amik Plain, southern Turkey. *Environment Monitoring Assessment*. https://doi.org/10.1007/s10661-014-3829-z.

Ali, H., Khan, E., & Sajad, M. A. (2013). Phytoremediation of heavy metals-concepts and applications. *Chemosphere, 9*, 869–881.

Alonso, A., & Camargo, J. A. (2009). Effects of pulse duration and post-exposure period on the nitrite toxicity to a freshwater amphipod. *Ecotoxicology and Environmental Safety, 72*(7), 2005–2008.

Anning, A. K., Korsah, P. E., & Addo-Fordjour, P. (2013). Phytoremediation of wastewater with *Limnocharis flava, Thalia geniculata* and *Typha latifolia* in constructed wetlands. *International Journal of Phytoremediation, 15*(5), 452–464.

Arias, M. F. C., Bru, L. V., Rico, D. P., & Galvan, P. V. (2011). Kinetic behaviour of sodium and boron in brackish water membranes. *Journal of Membrane Science, 368*, 86–94.

Basile, A., Sorbo, S., Conte, B., Cobianchi, R. C., Trinchella, F., Capasso, C., & Carginale, V. (2012). Toxicity, accumulation, and removal of heavy metals by three aquatic macrophytes. *International Journal of Phytoremediation, 14*(4), 374–387.

Bauddh, K., & Singh, R. P. (2012). Cadmium tolerance and its phytoremediation by two oil yielding plants *Ricinus communis* (L.) and *Brassica juncea* (L.) from the contaminated soil. *International Journal of Phytoremediation, 14*(8), 772–785.

Bellos, D., & Sawidis, T. (2005). Chemical pollution monitoring of the River Pinios (Thessalia—Greece). *Journal of Environmental Management, 76*, 282–292.

Bindu, T., Sylas, V. P., Mahesh, M., Rakesh, P. S., & Ramasamy, E. V. (2008). Pollutant removal from domestic wastewater with Taro (*Colocasia esculenta*) planted in a subsurface flow system. *Ecological Engineering, 33*(1), 68–82.

Campbell, C., Greenberg, R., Mankikar, D., & Ross, R. D. (2016). A case study of environmental injustice: The failure in Flint. *International Journal of Environmental Research and Public Health, 13*, 951. https://doi.org/10.3390/ijerph13100951.

Cardwell, A., Hawker, D., & Greenway, M. (2002). Metal accumulation in aquatic macrophytes from Southeast Queensland, Australia. *Chemosphere, 48*, 653–663.

Carty, A., Scholz, M., Heal, K., Gouriveau, F., & Mustafa, A. (2008). The universal design, operation and maintenance guidelines for farm constructed wetlands (FCW) in temperate climates. *Bioresource Technology, 99*, 6780–6792.

Chaudhuri, D., Majumder, A., Misra, A. K., & Bandyopadhyay, K. (2014). Cadmium removal by *Lemna minor* and *Spirodela polyrhiza*. *International Journal of Phytoremediation, 16*(11), 1119–1132.

Chavan, B. L., & Dhulap, V. P. (2012). Sewage treatment with constructed wetland using *Panicum maximum* forage grass. *Journal of Environmental Science and Water Resources, 1*(9), 223–230.

Clijsters, H., Cuypers, A., & Vangronsveld, J. (1999). Physiological responses to heavy metals in higher plants: Defence against oxidative stress. *Zeitschrift für Naturforschung C, 54*, 730–734.

EPA. (2001). *A Citizen's guide to phytoremediation.*

Fawzy, M. A., El-sayed Badr, N., El-Khatib, A., & Abo-El-Kassem, A. (2012). Heavy metal biomonitoring and phytoremediation potentialities of aquatic macrophytes in River Nile. *Environment Monitoring Assessment, 184*, 1753–1771.

Ghosh, M., & Singh, S. P. (2005). A review on phytoremediation of heavy metals and utilization of it's by products. *Asian Journal of Energy and Environment, 6*(4), 18.

Hadad, H. R., Maine, M. A., & Bonetto, C. A. (2006). Macrophytes growth in a pilot-scale constructed wetland for industrial wastewater treatment. *Chemosphere, 63*, 1744–1753.

Haloi, N., & Sarma, H. P. (2012). Heavy metal contaminations in the groundwater of Brahmaputra flood plain: An assessment of water quality in Barpeta District, Assam (India). *Environmental Monitoring and Assessment, 184*(10), 6229–6237.

Headley, T. R., & Tanner, C. C. (2008). *Floating wetlands for stormwater treatment: Removal of copper, zinc and fine particulates* (Technical Report 2008–030), Auckland Regional Council Auckland, NZ, 38p.

Hegazy, A. K., Abdel-Ghani, N. T., & El-Chaghaby, G. A. (2011). Phytoremediation of industrial wastewater potentiality by *Typha domingensis*. *International Journal of Environmental Science & Technology, 8*(3), 639–648.

Hubbard, R. K. (2010). Floating vegetated mats for improving surface water quality. In V. Shah (Ed.), *Emerging environmental technologies* (pp. 211–244). New York: Springer.

Iqbal, H., Ishfaq, M., Ullah, M., & Ahmad, W. (2013). Physico-chemical analysis of drinking water in district Kohat, Khyberpakhtunkhwa, Pakistan. *International Journal of Basic Medical Sciences and Pharmacy, 3*, 2.

Jadia, C. D., & Fulekar, M. H. (2009). Phytoremediation of heavy metals: Recent techniques. *African Journal of Biotechnology, 8*(6), 921–928.

Jayaweera, M. W., Kasturiarachchi, J. C., Kularatne, R. K., & Wijeyekoon, S. L. (2008). Removal of aluminium by constructed wetlands with water hyacinth (*Eichhornia crassipes* (Mart.) Solms) grown under different nutritional conditions. *Journal of Environmental Science and Health Part A, 42*(2), 185–193.

Kadlec, R. H., & Wallace, S. D. (2009). *Treatment wetlands* (2nd ed.). Boca Raton: Taylor and Francis group. ISBN: 9781-56670-526-4.

Kennish, L. (1992). Toxicity of heavy metals: Effects of Cr and Se on humans health. *Journal of Indian Public Health Education, India, 2*, 36–64.

Kumar, N., Mallick, S., Yadava, R. N., Singh, A. P., & Sinha, S. (2013). Co-application of selenite and phosphate reduces arsenite uptake in hydroponically grown rice seedlings: Toxicity and defence mechanism. *Ecotoxicology and Environmental Safety, 911*, 71–179.

Ladislas, S., Gérente, C., Chazarenc, F., Brisson, J., & Andrès, Y. (2013). Performances of two macrophytes species in floating treatment wetlands for cadmium, nickel, and zinc removal from urban stormwater runoff. *Water, Air, and Soil Pollution, 2*(224), 1408.

Ladislas, S., Gérente, C., Chazarenc, F. J., & Brisson, A. Y. (2015). Floating treatment wetlands for heavy metal removal in highway stormwater ponds. *Ecological Engineering, 80*, 85–91.

Liu, J., Dong, Y., Xu, H., Wang, D., & Xu, J. (2007). Accumulation of Cd, Pb and Zn by 19 wetland plant species in constructed wetland. *Journal of Hazardous Materials, 147*(3), 947–953.

Lokhande, R. S., Singare, P. U., & Pimple, D. S. (2011). Toxicity study of heavy metals pollutants in waste water effluent samples collected from Taloja industrial estate of Mumbai, India. *Resources and Environment, 1*(1), 13–19.

Maine, M. A., Suñé, N. L., & Lagger, S. C. (2004). Chromium bioaccumulation: Comparison of the capacity of two floating aquatic macrophytes. *Water Research, 38*(6), 1494–1501.

Malik, A. (2007). Environmental challenge vis-a-vis opportunity: The case of water hyacinth. *Environment International, 33*(1), 122–138.

Maltais-Landry, G., Maranger, R., Brisson, J., & Chazarenc, F. (2009). Nitrogen transformations and retention in planted and artificially aerated constructed wetlands. *Water Research, 43*(2), 535–545.

Mbuligwe, S. E. (2004). Comparative effectiveness of engineered wetland systems in the treatment of anaerobically pre-treated domestic wastewater. *Ecological Engineering, 23*(4), 269–284.

Mishra, S. S., & Mishra, A. (2008). Assessment of physico-chemical properties and heavy metal concentration in Gomti river. *Research in Environment and Life Sciences, 1*(2), 55–58.

Mojiri, A. (2012). Phytoremediation of heavy metals from municipal wastewater by *Typha domingensis*. *African Journal of Microbiology Research, 6*(3), 643–647.

Muchuweti, M., Birkett, J. W., Chinyanga, E., Zvauya, R., Scrimshaw, M. D., & Lister, J. N. (2006). Heavy metal content of vegetables irrigated with mixtures of wastewater and sewage sludge in Zimbabwe: Implication for human health. *Agric. Ecosyts. Environ., 112*, 41–48.

Onur, C. T., Cengiz, T., Harun, B., & Anil, Y. (2014). Constructed wetlands as green tools for management of boron mine wastewater. *International Journal of Phytoremediation, 16*, 537–553.

Osorio, A. C., Villafane, P., Caballero, V., & Manzano, Y. (2011). Efficiency of mesocosm scale constructed wetland systems for treatment of sanitary wastewater under tropical conditions. *Water, Air, and Soil Pollution, 220*, 161–171.

Padmavathiamma, P. K., & Li, L. Y. (2007). Phytoremediation technology: Hyperaccumulation of metals in plants. *Water, Air, & Soil Pollution, 184*, 105–126.

Rahman, M. A., & Hasegawa, H. (2011). Aquatic arsenic: Phytoremediation using floating macrophytes. *Chemosphere, 83*, 633–646.

Rai, P. K. (2010). Seasonal monitoring of heavy metals and physicochemical characteristics in a lentic ecosystem of subtropical industrial region, India. *Environmental Monitoring and Assessment, 165*(1–4), 407–433.

Rai, P. K. (2012). An eco-sustainable green approach for heavy metals management: Two case studies of developing industrial region. *Environmental Monitoring and Assessment, 184*(1), 421–448.

Rai, P. K., & Tripathi, B. D. (2009). Comparative assessment of *Azollapinnata* and *Vallisneria spiralis* in Hg removal from GB Pant Sagar of Singrauli Industrial region, India. *Environmental Monitoring and Assessment, 148*(1–4), 75–84.

Rai, U. N., Tripathi, R. D., Singh, N. K., Upadhyay, A. K., Dwivedi, S., Shukla, M. K., & Nautiyal, C. S. (2013). Constructed wetland as an ecotechnological tool for pollution treatment for conservation of Ganga river. *Bioresource Technology, 148*, 535–541.

Rai, U. N., Upadhyay, A. K., Singh, N. K., Dwivedi, S., & Tripathi, R. D. (2015). Seasonal applicability of horizontal sub-surface flow constructed wetland for trace elements and nutrient removal from urban wastes to conserve Ganga River water quality at Haridwar, India. *Ecological Engineering, 81*, 115–122.

Rashed, M. N. (2010). Monitoring of contaminated toxic and heavy metals from mine tailings through age accumulation in soil and some wild plants at Southeast Egypt. *Journal of Hazardous Materials, 178*, 739–746.

Rawat, S. K., Singh, R. K., & Singh, R. P. (2012). Remediation of nitrite contamination in ground and surface waters using aquatic macrophytes. *Journal of Environmental Biology, 33*, 51–56.

Rieumont, S. O., Lima, L., De la Rosa, D., Graham, D. W., Columbie, I., Santana, J. L., & Sanchez, M. J. (2007). Water hyacinths (*Eichhornia crassipes*) as indicators of heavy metal impact of a large landfill on the Almendares river near Havana, Cuba. *Bulletin of Environment Contamination and Toxicology, 79*, 583–587.

Rosli, N. R. M., & Yahya, K. (2012). *Assessment of nutrients and sediment loading in a tropical river system in Malaysia*. International conference on environment, chemistry and biology IPCBEE, 49.

Rousseau, D. P. L., Vanrolleghem, P. A., & De Pauw, N. (2004). Constructed wetlands in Flanders: A performance analysis. *Ecological Engineering, 23*, 151–163.

Saeed, T., & Sun, G. (2012). A review on nitrogen and organics removal mechanisms in subsurface flow constructed wetlands: Dependency on environmental parameters, operating conditions and supporting media. *Journal of Environmental Management, 112*, 429–448.

Salt, D. E., Blaylock, M., Nanda-Kumar, P. B. A., Dushenkov, V., Ensly, B. D., Chet, I., & Raskin, I. (1995). Phytoremediation: A novel strategy for the removal of toxic elements from the environment using plants. *Bio-Technology, 13*, 468–475.

Shah, A. B., Rai, U. N., & Singh, R. P. (2015). Correlations between some hazardous inorganic pollutants in the Gomti River and their accumulation in selected macrophytes under aquatic ecosystem. *Bulletin of Environmental Contamination and Toxicology, 94*, 783–790.

Shuping, L. S., Snyman, R. G., Odendaa, P. J., & Ndakidemi, P. A. (2011). Accumulation and distribution of metals in *Bolboschoenus maritimus* (Cyperaceae), from a South African River. *Water Air Soil Pollution, 216*, 319–328.

Singh, D., Gupta, R., & Tiwari, A. (2012). Potential of duckweed (*Lemna minor*) for removal of lead from wastewater by phytoremediation. *Journal of Pharmacy Research, 5*(3).

Sinha, S. N., & Nag, P. K. (2011). Air pollution from solid fuels. In J. O. Nriagu (Ed.), *Encyclopedia of environmental health* (Vol. 1, p. 46).

Skinner, K., Wright, N., & Porter-Goff, E. (2007). Mercury uptake and accumulation by four species of aquatic plants. *Environmental Pollution, 145*(1), 234–237.

Soda, S., Hamad, T., Yamaoka, Y., Ike, M., Nakazato, H., Saeki, Y., Kasamatsu, T., & Sakurai, Y. (2012). Constructed wetlands for advanced treatment of wastewater with a complex matrix from a metal-processing plant: Bioconcentration and translocation factors of various metals in Acorus gramineus and Cyperus alternifolius. *Ecological Engineering, 39*, 63–70.

Souza, F. A., Dziedzic, M., Cubas, S. A., & Maranho, L. T. (2013). Restoration of polluted waters by phytoremediation using *Myriophyllum aquaticum* (Vell.)Verdc., Haloragaceae. *Journal of Environmental Management, 120*, 5–9.

Sukias, J. P. S., Yates, C. R., & Tanner, C. C. (2010). *Assessment of floating treatment wetlands for remediation of eutrophic lake waters – Maero Stream (Lake Rotoehu)*. NIWA client report for environment bay of plenty, HAM2010-104, NIWA, Hamilton Dec 2010.

Tanner, C. C., & Headley, T. R. (2011). Components of floating emergent macrophyte treatment wetlands influencing removal of stormwater pollutants. *Ecological Engineering, 37*, 474–486.

Tee, H. C., Lim, P. E., Seng, C. E., & Nawi, M. A. (2012). Newly developed baffled subsurface flow constructed wetland for the enhancement of nitrogen removal. *Bioresource Technology, 104*, 235–242.

Tolu, O. A., & Atoke, O. O. (2012). Achieving environmental sustainability in wastewater treatment by phytoremediation with water hyacinth (*Eichhornia crassipes*). *Journal of Sustainable Development*, 5–7.

Utmazian, M. N. D. S, & Wenzel, W. W. (2006). *Phytoextraction of metal polluted soils in Latin America*. Environmental Applications of Poplar and Willow Working Party, pp 18–20.

Vymazal, J. (2007). Removal of nutrients in various types of constructed wetlands. *Science of the Total Environment, 380*(1), 48–65.

Vymazal, J., & Kropfelova, L. (2011). A three-stage experimental constructed wetland for treatment of domestic sewage: First 2 years of operation. *Ecological Engineering, 37*(1), 90–98.

Wang, H., & Yu, X. (2014). A review of the protection of sources of drinking water in China. *Natural Resources Forum*. https://doi.org/10.1111/1477-8947.12036.

Wang, H., Zhang, H., & Cai, G. (2011). An application of phytoremediation to river pollution remediation. *Procedia Environmental Sciences, 10*, 1904–1907.

Wu, H., Zhang, J., Li, P., Zhang, J., Xie, H., & Zhang, B. (2011). Nutrient removal in constructed microcosm wetlands for treating polluted river water in northern China. *Ecological Engineering, 37*, 560–568.

Zhang, D. Q., Gersberg, R. M., Hua, T., Zhu, J., Tuan, N. A., & Tan, S. K. (2012). Pharmaceutical removal in tropical sub-surface flow constructed wetlands at varying hydraulic rates. *Chemosphere, 87*, 273–277.

Urban Pond Ecosystems: Preservation and Management Through Phytoremediation

Mohd. Muzamil Bhat, R. N. Shukla, and M. Yunus

Abstract Life is impossible without freshwater as it is significantly important for all living creatures on earth. The natural built-up of nutrients in freshwater bodies is an extremely time-consuming process but human interferences have enhanced the rate of contaminating the pond and lake ecosystems with N and P plenty of times than natural cause. Large quantities of untreated effluents from industrial and domestic sector are directly discharged into adjacent recipient freshwater environs (ponds and lakes), which manifolds the concentration of concerned nutrients into these freshwater ecosystems. There is no limit pertaining to the treatment techniques availability, but they are either insufficient or least effective for removing the nuisance contaminates from the wastewaters. Besides, these techniques have plenty of environment related issues, in other words conventional remediation techniques pose threats to the freshwater environs and required high energy and cost for establishment. Employing naturally growing plants in disturbed aquatic environs has been observed a viable technique to clean up the nuisance nutrients and toxic pollutants. Phytoremediants scavenge the harmful substances (nutrients and heavy metals) from disturbed surface waters have recently been explored as substitute to conventional methods.

Keywords Water bodies · Phytoremediation · Restoration · Contamination · *Ludwigia repens* · *Eichhornia crassipes*

M. M. Bhat (✉)
Faculty- Environmental Science, School of Sciences, Islamic University of Science and Technology, Pulwama, Jammu and Kashmir, India

R. N. Shukla
School of Environmental Biology, Awadesh Pratap Singh University, Rewa, Madhya Pradesh, India

M. Yunus
Mohammad Ali Jauhar University, Rampur, Uttar Pradesh, India

H. Qadri et al. (eds.), *Fresh Water Pollution Dynamics and Remediation*,
https://doi.org/10.1007/978-981-13-8277-2_15

1 Introduction

Serious mismanagement has been observed in urban water sector. Fresh water sources of urban areas like rivers, lakes and ponds and in many cases even groundwater has been polluted and depleted. The rainfall is generally seen as a bane rather than boon as it brings floods because the drainage systems are seriously ill designed or mismanaged. Lack of provision of adequate minimum water for vast proportions of poorer segments on the one hand and wasteful use without paying even cost prices by more prosperous segments on the other hand is typical picture of most urban areas. The urban and peri-urban waterbodies (ponds) are being appreciated as significant water resources now. Due to increasing population in urban and peri-urban centres, more stress is on water management of city administration. Proper management of the ponds is now necessary for best water quality. Urban India depends heavily upon various types of water bodies to meet its daily requirement of water (Wagh et al. 2008). There is greater demand for water now due to increasing population, which also implies huge pressure on these water bodies. At the same time, due to the requirement of more land for residential and other urban expansion, the water bodies are being filled up. So, in spite of their over-riding importance, the very existence of urban ponds is critically endangered. This is a highly worrying fact as the role of the ponds in urban and peri-urban milieu is multifaceted. It only just is not a pool of water as this pool, may not be considered has a very significant role of social, ecological and civic importance. The major use is no doubt bathing. A large number of people use these surface water sources for bathing, cleaning and other requirements. Barring drinking, ponds are the only source of water for all requirements for a large number of people working in markets, small factories, living in slums or in poor housing conditions. Fish cultivation is the major productive activity related to these ponds. Most of these water bodies are used for pisciculture (Ray and Majumdar 2005). Environmentally these water bodies serve the purpose of an open space in the crowded urban localities. In many cases the banks of these water bodies are the only spaces for the development of greenery. Local natural life (aquatic, avifauna and terrestrial) sustains around these water bodies. Ponds and the surroundings are one of the most important protectors of biodiversity. Social roles are equally important. Water bodies have generally been associated with different aspects of Indian cultural and religious practices. In urban places, these water bodies also act as a centre of local social and cultural activities. There are clubs and temples by the side of these water bodies. Often there is immersion of idols and fairs are organized on the fields next to these water bodies. Pond water bodies serve as receptors for rainwater inflows and help in maintaining local ground water levels. Thus urban water bodies are a special component in water use management, to which little attention has been paid.

Water is still a critical issue in India, though enjoying a relatively moderate average rainfall. There are disputes between states and countries in sharing of water resources, resulting in violent riots. Most city authorities can not able to supply required minimum water to its citizens. Even in areas with high rainfall, water scarcity has become a problem during non-rainy months. Urbanization and large-scale industrialization implied stress on most of the fresh water resources. The development of new environmental problems as a result of this has given rise to new ideas in the field of monitoring and assessment of aquatic ecosystems. The overall condition or health of aquatic ecosystems is determined by the interaction of all its physical, chemical and biological components, which make up its ecosystem. With the onset of industrialization and urbanization, the world's oceans and other freshwater resources have been increasingly contaminated with sewage, agricultural chemicals, oils, heavy metals, radioactive materials, detergents and many other products of the human settlements. As the earth's population continues to increase rapidly, the growing human need for freshwater is leading to a global water resource crisis. There is a growing consensus that if current trends continue, water scarcity and deteriorating water quality will become critical factors limiting the future economic development, the increase of food production, the provision of basic health and hygiene services to millions of disadvantaged people in the developing countries. Due to the mismanagement of natural resources, the world is heading towards a freshwater crisis which is evident in many parts of the world, varying in scale and intensity (Ramachandra and Solanki 2007). UNESCO's World Water Development Report (WWDR 2003) from its World Water Assessment Program indicates that in the next 20 years, the quantity of water available to everyone is predicted to decrease by 30%. Forty percent of the world's inhabitants currently have insufficient freshwater for minimal hygiene. More than 2.2 million people died in 2000 from diseases related to the consumption of contaminated water or drought. To accommodate the various human needs water resources have been grossly mismanaged resulting in declined water quality and considerable loss of water resources. The various impacts due to anthropogenic activities need to be assessed at regular intervals for its restoration and conservation. Recognizing the importance of water resources to the planet's future, the United Nations General Assembly proclaimed the year 2003 as the 'International Year of Freshwater'. To understand the importance of water, the implications of its mismanagement and to facilitate strategies to manage, restore and conserve this fast degrading natural resource, it becomes essential to understand its ecological status and certain processes associated with water.

In this context, the conservation and restoration of this precious resource is gaining importance and calls for integrated management approaches. Since the beginning of the industrial revolution, increasing human population, economic activities as well as shortcomings in their management have resulted in more pollutants being introduced into watercourses. An increasing number of surface water bodies have come under serious threat of degradation. The global freshwater resources are under increasing pressure (GWP Technical Advisory Committee 2000). The anthropogenic impact on aquatic ecosystems has become a crucial topic of increasing concern. These problems have led to the adoption of an integrated approach to the management of water resources, which is called Integrated Water Resources Management (IWRM). Keeping in view the ever increasing problems of water, there is a need to study the status of urban pond water bodies with the aim to device sustainable method for their remediation.

2 Distribution of Water on Earth and Threats to Water Quality

The water resource of earth is not equally distributed. Out of the total water found on earth surface 97.47% is present in oceans and seas while as freshwater resources merely account 2.53% of which 69.6% is stored in continental ice caps, 30.1% in aquifers and 0.26 as surface water in the form of lakes and rivers. Lentic systems occupy only 0.007% of the worlds freshwater. Most of the freshwater available on earth is difficult to use and is tied up in glaciers and deep groundwater (Ramachandra et al. 2002).

2.1 Threats to Water Resources

As is true of all organisms, our very existence depends on water. Water makes up 60–70% of all living matter. However, in the present era, this valuable resource is not only being over-exploited but is also being seriously degraded due to various anthropogenic activities and has become a global issue. Therefore only less than 1% of all water on earth is available for human consumption. According to the United Nations Environment Program (UNEP), 25% of the world's population may soon suffer from chronic water shortages. The major threats to water resources are from point sources (industrial effluents, sewage etc.) and from non-point sources (agriculture, urban etc.). Point source pollution originates from detectable and/or distinct sources that discharge directly into the water system through a definite outlet such as ditches, drainpipes, channels and effluent outlets releasing organic loads, heavy metals and toxic nutrients. These sources of pollution are comparatively easy to monitor and control. Whilst non-point source pollution enters the water system by a multitude of pathways, such as urban runoff, agricultural runoff, animal and human waste, atmospheric deposition, seepage, groundwater flow and river course modification (Pegram and Gorgens 2001). These diffuse sources are becoming gradually more important and arise over a wider area and are often difficult to monitor and control than point sources of pollution (Yang et al. 2014). Dumping of solid wastes, acid precipitation, thermal pollution, chemical spills, mine drainage etc., also deteriorates the quality of water by changing its physical, chemical and biological properties and then systematically disturbs the delicate food web. Worldwide, wetlands are threatened by excessive amounts of anthropogenic nutrients and metals through industrial wastewater and agricultural runoff (Mitsch and Gooselink 2007). Heavy metals pose a serious threat to humans due to their persistent toxic nature and bioaccumulation in the food web (Zhang et al. 2009; Meitei and Prasad 2013; Singh et al. 2013; Zhang et al. 2014; Asefi and Zamani-Ahamadmahmoodi 2015; Bortey-Sam et al. 2015). Understanding the implications of each of these threats requires aquatic system characterizations involving detailed ecological understanding.

2.2 *Freshwater Quality*

Contamination of water bodies is a serious problem faced by areas under water stress such as Rajasthan. It is well established that domestic sewage and industrial effluents falling into natural water bodies change the water quality and lead to eutrophication. Characteristics of water bodies influence the water quality individually and in combination with different pollutants, thus influencing the biota (Srivastava et al. 2003; Hoo et al. 2004; Smitha et al. 2007). The accumulation of metals and organic pollutants in vital organs of fishes result in long term toxic effects (Gupta and Srivastava 2006; Kumar and Riyazuddin 2006; Tilak et al. 2007; Karthikeyan et al. 2007; Singh et al. 2008). Structural and functional abnormalities are induced in different organs of fishes by these metals and organic pollutants (Dorval et al. 2003; Gupta and Srivastava 2006). With increasing industrialization, urbanization and growth of population, India's environment has become fragile and has been causing concern (Mohapatra and Singh 1999). Urbanization has direct impact on water bodies as the settlement takes place around the vicinity of water bodies and due to lack of space people have tendency to encroach upon the lake (Pavendan et al. 2011). Organic enrichment of the lake through floral offerings, idol immersion and decomposition of aquatic weeds are also the significant causes of its eutrophication. Several studies have been conducted to understand the physical and chemical properties of lakes, ponds and reservoirs such as the Halai Reservoir, Kolovoi Lake, Kalyani reservoirs, Salim Ali Lake, Dahikhura reservoir, and wetlands in urban Coimbatore in India (Gowd and Kotaiah 2000; Thorat and Sultana 2000; Mohanraj et al. 2000; Shastri and Pendse 2001). In such studies the characteristics of water bodies were taken into consideration with reference to physical, chemical and biological properties. Gupta et al. (2001) have used only chemical characteristics of water bodies of Udaipur in their observations. Srivastava et al.(2003) studied the physicochemical properties of various water bodies in and around Jaipur. Bhat et al. (2009, 2012a, b) investigated the water quality status of some urban ponds of Lucknow, U.P., India and concluded that water quality of the city is polluted as the results are above the permissible limits. The city sewage discharge, agriculture and urban runoff and continuous dumping of waste materials especially sanitary waste are affecting the water quality of these urban water bodies. There is considerable need for better understanding of these small impoundments so that they can be managed effectively. In reflection to global concern, water pollution has become one of the major issues faced by the world because most of the rivers have been contaminated (Chan 2012; Kusin et al. 2016a). Off-site pollution from industrial areas and chemical use for agricultural purposes are suggested to be the main sources of toxic contamination in the environment (Fulazzaky et al. 2010; Biswas and Tortajada 2011; Chan 2012; Othman et al. 2012; Al Badaii et al. 2013; Kusin et al. 2016a). The presence of heavy metals such as Cu, Zn, Pb, Mn and Fe in aquatic environments could bring adverse effects on human health, aquatic life as well as environment (Kusin et al. 2016b).

2.3 Lentic Pollution from Catchment Area

Surface water bodies typically get enriched with nutrients however; this natural nutrient enrichment process is normally very slow. Human interference with the lake environment will have a predominant effect on the ageing and accelerates cultural eutrophication. Some of the potential sources for the degradation of water quality are, commercial development in cities, massive quantities of sewage, tremendous increase in population, unsustainable agricultural activities and organic pollution into the inland surface water bodies like ponds, lakes, rivers and streams have enormously increased the concentration of several nutrients like phosphates, sulphates, nitrates, etc. Land use activities have a prominent effect on the physicochemical properties of water bodies like ponds, streams, rivers, lakes and shallow aquifers. Rainfall clears the air and land surfaces like the roof tops, parking lots, agricultural lands, etc. But simultaneously it washes away some materials like sediments, animal wastes, fertilizers, toxic substances, mine discharges, etc. and transports it to the nearby water resources. Bhat et al. (2013) assessed the pollution sources in water bodies of Lucknow city and reported that city sewage discharge, agriculture and settlement runoff and dumping of municipal and commercial wastes are potentially affecting the chemistry of these urban water resources. There is an utmost need for proper understanding of pollution dynamics in both lentic and lotic water bodies in order to manage them efficiently.

2.3.1 Road Runoff

Land use in the watershed or catchment area play a very crucial role in maintaining the quality of water bodies. Urban runoff from roads and highways have been identified as a vital source of many pollutants and contaminants such as nutrients, salts, metals, organic acids, polyaromatic hydrocarbons (PAHs) and other toxic persistent organic compounds (Mahvi and Mardani 2005; Dhananjayan et al. 2012; Bhat et al. 2013). Storm runoff from roads, highways, parking slots and moter workshops have provided evidences of toxicity in many bioassay experiments (Polkowska et al. 2001). However, a significant variation in contamination of runoff has also been recorded with respect to season, last rainfall, amount of precipitation, duration etc. Many cases have reported the phenomenon of first-flush effect (Deletic and Maksimovic 1998) and therefore episodic toxicity due to road run-off can be expected.

Urban catchment area is composed of 20% by roads, but the road runoff contributes approximately 50% of the total solids and almost 30% of hydrocarbon compounds diverted directly without any treatment to receiving water bodies (Sriyaraj and Shutes 2001). Suspended and settleable solids, heavy metals, hydrocarbons and salts are the major pollutants present in urban runoff with the major sources as roads and vehicle wear (Mungur et al. 1995). Urban runoff poses a great threat to hydrologic and pedologic quality of natural inland water bodies

(Mitsch and Gosselink 1993). Hydrologic change has a direct effect on sediment properties like nutrient availability, soil salinity and pH. Heavy metals such as lead (Pb), zinc (Zn), copper (Cu) and cadmium (Cd) are bioaccumulative pollutants which pose a serious threat to aquatic ecology particularly when short summer storms shower follow a prolonged dry period as the pollutants and contaminants have already got accumulated along the road surface, in verges and in the drainage system. Although heavy rainfall can have less detrimental effect on aquatic ecosystems due to greater dilution effect (Mungur et al. 1995). The detrimental or toxic effect of heavy metals on aquatic organisms depends on multiple factors but the nature of metal and its form is utmost important. Free metal ions, metal adsorbed on organic and inorganic complexes and metal bound to organic and inorganic particulate matter have their different tendencies to affect aquatic organisms. Water chemistry such as pH, EC, hardness and the organic matter content potentially play a great role in the final effect of metal pollutants and their speciation (Depledge et al. 1993).

2.3.2 Run-off from Agricultural Sources

The agricultural industry is one of the most potential contributors to net primary productivity through nutrient runoff, sediment load, pesticides, herbicides, pollutants and contaminants (USEPA 1998). Agricultural and horticultural crops require more elevated use of chemical pesticides and nutrients than natural forest cover and grasslands. Agricultural operations such as ploughing and tillage make soil particles prone to erosion during rainfall. Furthermore, the land under settlements, lawns and gardens in more intensively managed which finally results in even more augmentation of contaminants and pollutants. Urban settlements encourage the process of runoff and decrease the down ground infiltration and percolation as most of land is concrete in urban areas. Higher solid matter and nutrients from agricultural areas and settlements are transported to receiving water resources. Thus, unsustainable urban sprawl coupled with indiscriminate use of agrochemicals makes it a very challenging to conserve the existing water resources. Biophysical simulation methods and their application in limnology is gaining more acceptance in scientific platforms as these methods are trying to locate the potential sources of nutrients and other contaminates in the receiving water bodies (Marzen et al. 2000; Bhuyan et al. 2001). Several models based on geospatial and spatial data helps in accurately locating the point and non-point sources of pollutants. Runoff from urban settlements and pavements has been confirmed as the potential factor for surface water quality deterioration and encourages the entry of sediments, nutrients, heavy metals, oils, hydrocarbons, pesticides etc. (Novotny 1999; Schreiber et al. 2001; Lazzarotto et al. 2005). Environmental engineering techniques like buffer zones, ponds, tanks, wetlands and riparian zones have been found very effective in controlling physical pollutants from runoff so have been presented as treatment trains for urban areas (Yin and Mao 2002).

The quality of water in natural inland waters is related to geomorphology, climate and land-use in the catchment (drainage basin). The size and slope of the catchment, precipitation, wind, temperature, erosion, vegetation and soil structure all play a role in the catchment water quality (Schindler 1997). Land management of the catchment for agriculture, forestry, horticulture, conservation, industry and urban areas influences the quality of water that enters the aquatic system (Johnes et al. 1996). Agricultural practices such as land clearance, irrigation, drainage, pesticide use, soil enrichment and animal waste will have consequences for the quality and quantity of water in the rivers and lakes in the catchment (Elliott and Sorrell 2002). Strong linkages between vegetation cover in the catchment and water chemistry have been shown for relatively unproductive lakes in the United Kingdom (Maberly et al. 2003) and implied for lakes in some regions of North America (Lougheed et al. 2001).

Heavy metal contamination of lake and reservoir sediments derives from both atmospheric and catchment inputs. The former is generally diffuse and represents short, medium and long distance atmospheric transport (Nriagu 1979; Renburg 1986; Foster and Dearing 1987). Catchment inputs usually consist of point sources, including mine waste and accidental spillage, industrial and sewage works discharges, and leakage from land-fill sites (Furstner and Wittmann 1979; Christenson and Chien 1981; Dearing et al. 1981; Hakanson and Jansson 1983). Diffuse catchment inputs may come from urban storm runoff, particularly when the storm sewer capacity is exceeded by high runoff. Interpretation of lake sediment heavy metal content is made difficult, however, when several sources interact. For example, lakes and reservoirs in heavily industrialized regions will receive both a direct and an indirect atmospheric input: the latter deriving from eroded catchment soils which may either be enriched with atmospherically derived heavy metals, or diluted by sources depleted of metals, such as channel bank erosion (Foster and Dearing 1987).

2.3.3 Urban Settlement Run-off

Waste water runoff from domestic areas, industrial sectors, streets and buildings constitute one of the most potential sources of area pollution (Carpenter et al. 1998). Water quality of a particular source is depicted by its hydrology, flow, discharge and the various activities which take place in its catchment area (Sonneman et al. 2001). The overall pollution load including sediments, nutrients and other has a very strong positive correlation with the urban development, watershed activities and existing drainage system (Brown et al. 2005). Although some of the studies have revealed that macro invertebrates and fishes also provide us information about the quality of water (Fitzpatrick et al. 2004; Gray 2004; Park and Shin 2007). The fact is that these water bodies which are situated very close to urban areas must be having under continuous monitoring of limnologists in order to conserve their quality and ecology (Walsh 2000).

Most of the point sources are treating their waste water to some extent and there arises a need to deal with the non-point sources accordingly as most of pollution load comes from diffused sources like sedimentation, flooding, temperature rise, dissolved oxygen depletion, eutrophication etc. (Marsalek 1998; Choe et al. 2002). So there arises an increased concern about storm runoff management, so as to control their effect on receiving water bodies fully or partially. Such runoff sewage management is sometimes also referred as best management practices (BMPs). Diffused runoff is composed of are many pollutants which include suspended solids, persistent organic matter, chemical fertiliser residues, hydrocarbons, pathogenic bacteria and heavy metals (ASCE 2002). Poly aromatic hydrocarbons (PAHs) and benzo(a) pyrene have been confirmed as potential carcinogens, their concentration has also been reported by many scientists (Yamane et al. 1993; Asada and Ohgaki 1996). Several recent investigations have reported pathogenic microbes and heavy metals in urban storm water runoff (Mallin et al. 2000; Noble et al. 2000; Lipp et al. 2001; Choe et al. 2002; Farm 2002). The total suspended solids, biochemical oxygen demand, chemical oxygen demand are mostly used to estimate storm water runoff quality (Shinya et al. 2003). Contaminants from storm runoff are highly site-specific due to wide variations in atmospheric dust fall, mean daily traffic movement and land use practice in surrounding areas (Farm 2002).

The process of first flush from storm runoff has been developed mainly due to the fact that first portion of urban runoff is highly polluted. Difficulties arose as the first flush phenomenon was defined in different ways. To examine the first flush of runoff, researchers mostly use curves of cumulative fraction of total pollutant load against the fraction of total cumulative runoff volume for a particular event. Geiger (1987) defined a first flush as occurring when such curves have an initial slope greater than 45% and used the point of maximum divergence from the 45° slope to quantify the first flush volume. Gupta and Saul (1996) used a very similar definition, as did Ashley et al. 1992. French researchers (Saget et al. 1995) put forth a very unique definition of the process as it is the first flush volume which washes 80% of the pollution load and contributes to just 30% of the overall volume. Other experts (Vorreiter and Hickey 1994) have described the process in terms of the pollution load in the first 25% of the overall event volume. After an assessment of 13 separate urban catchment areas, Lee et al. 2002 has suggested that a first flush exists at time *t* if the dimensionless cumulative pollutant mass exceeds the corresponding runoff volume. It must be noted that a more common way for defining the phenomenon is the approach where a fraction of total pollution load is compared with a fraction of total runoff volume, both calculated at the same point which was chosen somewhere in the first part of the runoff cumulative curve. Saget et al. (1995) have selected the point at 30% of the runoff and Vorreiter and Hickey (1994) have chosen 25% point.

Storm water pollution is the change in water quality of a urban water resource by runoff from urban settlements. The storm water runoff leads to hydrological and topological changes in the watershed and finally deteriorates the existing inland water bodies (Joliffe 1995). Unsustainable urban sprawl and concrete pavements increases surface runoff that is discharged more quickly into the receiving water systems (Anon 1981; O'Loughlin 1994). Many of the water bodies in urban areas get polluted due to the phenomenon that pollutants are washed off from land by storms. The polluted storm water from urban areas can completely change the health of water bodies, negatively affect the aquatic ecology, recreation problem, aesthetic loss and cause algal bloom (Settacharnwit et al. 2003). Contributions to water quality impairment from point sources such as industrial effluent and effluent from sewage treatment plants (STP) can be equal to or even greater than that of storm water (Cordery 1976). Pollutant loads discharged from point sources are relatively easy to quantify because the entry point to waterways is fixed and the rate of flow and concentrations are generally known and available (Griffin et al. 1980). On the other hand, storm water pollution has vice versa problem. In addition, runoff load is more temporally and spatially variable than point source because they only occur when catchment encourages runoff after rainfall (Novotny 1994). Furthermore, diffused pollution such as storm water runoff is extremely difficult to monitor and control.

2.3.4 General Point and Diffused Sources

Each and every domestic source may not be a potential source but when all the existing sources like household, agricultural, commercial, landscaping, settlement runoff join together then they exhibit a cumulative effect in the water bodies and initiate many detrimental processes like eutrophication, algal bloom etc. Sediments, lubricants, heavy metals and organic matter load from storm runoff are the potential sources of siltation in lentic and lotic water sources.

2.4 Existing Practices of Treatment

Industrial effluent treatment, exchange resins, chemical precipitation and electrodialysis are the widely used methods worldwide. The overall treatment process is recommended based on various physico chemical and biological parameters and accordingly the methods available like coagulation (flocculation), sedimentation, flotation, ionic exchange, reverse osmosis, extraction, microfiltration, adsorption, etc. is adopted. However, the researchers are trying to switch over to other simple, eco-friendly and cost effective techniques for sewage treatment as these techniques are costly and lead to contamination of water bodies (Galiulin 1994; Salt et al. 1995; Shrivastava and Rao 1997).

3 Phytoremediation: Alternative Eco-Friendly Technology

The eco-friendly and most feasible technique to deal with the aquatic pollution particularly heavy metal contamination is phytoremediation. This is a novel and most powerful concept that makes use of aquatic macrophytes to extract, remove, sequester, detoxify heavy metals or pollutants from both contaminated sites and aquatic environs (Memon et al. 2001). "Phytoremediation basically refers to the use of plants and associated soil microbes to reduce the concentrations or toxic effects of contaminants in the environments" (Greipsson 2011).

3.1 Phytoremediation Approach

Natural vegetation plays a crucial role in accumulating, immobilizing, removing and transforming various toxic pollutants such as heavy metals and pesticides. Plants act as filters for many pollutants in soil and water. Phytoremediation is a promising tool to stabilise the contaminants and pollutants in water and soil environments (USEPA 1999, 2000; Raskin and Ensley 2000). The term "phytoremediation" has been coined in 1991. Its potential dealing with the pollutants is encouraging but requires some more research to be more effective and applicable in future (Table 1).

Phytoremediation can be categorised in many further sub processes like phytoextraction, phytotransformation, phytostabilization, phytodegradation and rhizofiltration.

- *Phytoextraction* or *phytoaccumulation* is the process plantsin which they accumulate or store the target pollutants into their various tissues like roots, shoots and leaves. A wide area can be dealt with this process and later the accumulated pollutants can be managed and recycled effectively.
- *Phytotransformation* or *phytodegradation* This process encourages the transformation of a toxic pollutant to other less toxic pollutant or completely trnsformes into other neutrall product which is biologically sterile product. Hexavalent chromium can be converted to trivalent chromium, which is relatively less mobile and non-carcinogenic.

Table 1 Various phytoremediation removal mechanism processes in different ecosystems (Vidali 2001)

Process	Removal mechanism	Type of ecosystem
Phytoextraction	Direct uptake of toxic metals into the plant tissues and clean-up of polluted sources	Soils
Phytotransformation	By the process of uptake and metabolic activities, the pollutant gets transformed from its toxic form to less toxic form	Surface and groundwater
Phytostabilization	Pollutant gets stabilised in plant tissue through adsorption and absorption	Soil, groundwater, mine drianage
Phytodegradation	Degradation in rhizosphere by microbial activity	Soils, ground water in rhizosphere
Rhizofiltration	Accumulation of toxic metals in root system	Lentic ecosystems
Phytovolatilization	By way evapotranspiration (hydrocarbons, Se and Hg)	Underground water and soils

- *Phytostabilization* is a process in which the contaminants and pollutants get adsorbed on the roots to form a stable mass and may not enter back into the system to pose detrimental effects.
- *Phytodegradation* or *rhizodegradation* is the process in which plants degrade the pollutants in the rhizosphere by means of proteins, enzymes, soil organisms etc. This process makes benefit out of symbiotic relationship between plants and microbes. Plants provides nutritional support to microbes and in turn microbes provide a healthier soil environment.
- *Rhizofiltration* is a simple process in which plants roots uptake the pollutants from soil and water and utilise them in their various metabolic process and thus render target wetland or soil pollution free.

The summary of various phytoremediation techniques along with the mechanism in different mediums is listed in Table 3. Phytoremediation is a very eco-friendly and feasible technique applied mainly to the fields pollutants whose concentration is very less and require more effort which treated chemically. Although this technology has a limitation that this is time consuming process and requires more difficulty in establishing plants at some sites with high toxic levels. Phytoextraction is the main process employed for removal of heavy metals and metalloids from polluted sites (Milic et al. 2012). The term phytotechnology encourages the application of science, technology and engineering to analyse the environmental problems and deal with the problems with plants. This concept involves a broader understanding of the role of plants and significant role within both ecological and economical systems (Mangkoedihardjo 2007).

Phytoremediation is actually the ability of a plant to stabilise a pollutant and can degrade, remove and neutralise by its own metabolic processes. Metals, pesticides, solvents, crude oil, leachate etc. can be degraded by the plants. Potential plants have been identified as sunflower, ragweed, cabbage, geranium etc. (Lasat 2002).

The plants have been identified as very powerful and aesthetically pleasing tool for remediation tool than conventional treatment technologies (Nyer and Gatliff 1996). Phytoremediation is a biological methodin which plant beds are employed to treat nutrients in wastewater which is cost-effective, eco-friendly and very efficient for the control of eutrophication nuisance (Yeh et al. 2015; Roley et al. 2016). The processes are involved in phytoremediation technologies include the nutrient removal from water and simultaneously encourage the microbial growth which intern improves the pollutant removal and stabilisation (Brix 1997; Ma et al. 2016; Wu et al. 2016). Furthermore, the plants enter in competition with the algal polpulation for light and nutrients; thereby discourage the algal bloom in eutrophic water. Large network of plant roots can act as sieves, trapping algae, and other suspended particles in dirty water (Bu and Xu 2013; Qin et al. 2016).

Phytoremediation by Aquatic Macrophytes

Macrophytes are aquatic plants that grow in or around water bodies and can be emergent, submerged or floating. The term aquatic macrophytes refers to the macro forms of aquatic vegetation and includes macroalgae (*e.g.,* the alga *Cladophora*, stoneworts like *Chara*, few species of pteridophytes (mosses, ferns) adapted to aquatic habitat and angiosperms (Wetzel 1975). Macrophytes are valuable to lakes because they provide food and shelter to the aquatic fauna. The oxygen produced by them helps in overall lake functioning. Aquatic macrophytes have been witnessed to have a remarkable potential for heavy metal remediation (Das et al. 2014). *Pistia stratiotes* is a widely used phytoremediant for the wetland systems (Prajapati et al. 2012). For the removal of low cadmium (Cd) levels from water *Limnocharis flava* is a viable option as it has a high potential for bio-concentration, translocation, higher relative growth rate and is easy to culture (Abhilash et al. 2009).

The macrophytes have been categorised by Arber (1920) and Sculthorpe (1967) as:

A. **Aquatic macrophytes attached to substratum**

(i) Emergent Macrophytes: rhizomatous or cormous perennials occuring on aerial or submerged soil at a point where water table is about 0.5 m below the soil. (*e.g., Glyceria, Eleocharis, Typha and Phragmites*).
(ii) Floating leaved Macrophytes: primarily angiosperms occurring on submerged sediments with water depths around 0.5–3 m; submerged leaves are floating with flexble petioles, reproductive parts are floating or aerial, (*e.g.,* the waterlilies *Nuphar* and *Nymphea, Brasenia, Potamogeton natans*).
(iii) Submerged Macrophytes: few pteridophytes (*e.g., Isoetes*), numerous mosses and charophytes (stonewort algae *Chara*, *Nitella*) and many angiosperms.

B. **Free floating macrophytes**

A diverse group of free floating plants that have hairy roots not attached to the substratum; ranging from long plants with rosettes of aerial and floating leaves and well developed submerged roots (*e.g.,Eichhornia, Pistia stratiotes, Trapa, Hydrocharis*) to minute surface floating or submerged plants with few or no roots (*e.g., Lemnaceae, Azolla, Salvinia*); reproductive organs are floating and aerial (*e.g.,* aquatic *Utricularia*) but rarely submerged (*e.g.,Ceratophyllum*).

According to the ecological classification by B.A. Fedchenko (Lukina and Smirnova 1988), all non-terrestrial macrophytes are sub-categorised into five groups depending on the relation of their vegetative organs to air, water, and ground:

1. Amphibious plants;
2. Plants rooted to the bottom of a water body with leaves emerging at the water surface;
3. Rooted plants with vegetative organs submerged in water;
4. Plants floating at the water surface without connection to the bottom; and
5. Completely submerged unrooted plants.

The peculiarities of the accumulation of heavy metals in the plant organs are of importance for the screening of these macrophytic groups to identify the plants effectively accumulating heavy metals. Macrophytes are considered as important component of the aquatic ecosystem not only as food source for aquatic invertebrates, but also act as an efficient accumulator of heavy metals (Chung and Jeng 1974). They are unchangeable biological filters and play an important role in the maintenance of aquatic ecosystem. Aquatic macrophytes are taxonomically closely related to terrestrial plants, but are aquatic phanerogams, which live in a completely different environment. Their characteristics to accumulate metals make them an interesting research objects for testing and modeling ecological theories on evolution and plant succession, as well as on nutrient and metal cycling (Forstner and Whittman 1979). Therefore, it is very important to understand the functions of macrophytes in aquatic ecosystem. The water, sediments and plants in wetlands receiving urban runoff contain higher levels of heavy metals than wetlands not receiving urban runoff. Large aquatic plants are known to accumulate heavy metals in their tissues. Macrophytes take up heavy metals mainly through the root, although uptake through the leaves may also be of significance. As the macrophytes die and decay, the accumulated metals in the decaying macrophytes can increase in the concentration of heavy metals in the sediments. Aquatic plants often grow more vigorously where nutrient loading is high. They are capable of removing water soluble substances from solution and temporarily immobilize them within the system (HO 1988). Bioavailability and bioaccumulation of heavy metals in aquatic ecosystems is gaining tremendous significance globally. Several of the submerged, emergent and free-floating aquatic macrophytes are known to accumulate and bioconcentrate heavy metals (Chow et al. 1976). Aquatic macrophytes take up metals from the water, producing an internal concentration several fold greater than their surroundings. Many of the aquatic macrophytes are found to be the potential scavengers of heavy metals from water and wetlands (Gulati et al. 1979).

Only recently has the value of metal accumulating plants for environmental remediation been fully realized, giving birth to a new cleanup technology termed as phytoremediation (Lasat 2002). This involves the use of plants to reduce or eliminate environmental hazards resulting from accumulation of toxic chemicals and other hazardous wastes. The effectiveness of this technique depends on the capability of the selected plants to grow and accumulate metals under the specific conditions of the site being remediated (Kulli et al. 1999). The ideal plant species to remediate a heavy metal contaminated site would be a rapidly growing, high biomass crop with an extensive root system that can both tolerate and accumulate the contaminants of interest. Therefore, a judicious selection of plant species to be used in remediating metal contaminated waters must be made so that the effectiveness of phytoremediation can be maximized.Several reviews are concerned with this problem. From the generalization by Zolotukhina and Gavrilenko (1989), it follows that the biological availability of a metal is not always directly related to its total content in water, but is determined by the ratio between the ionic and fixed forms, which depends on pH and the presence of organic and inorganic compounds. The accumulation of metals by macrophytes is more intensive in the first hours and during the first day; then, the uptake rate decreases. The time required for saturation is directly related to the metal concentration in water, and the saturation degree is inversely related to it. The uptake rate varies for different chemical elements at similar concentrations. The metal accumulation by plants can be affected by the increased concentrations of other elements. It was also found that plants completely submerged in water accumulated more metals than floating and partially submerged plants. This confirms the role of the contact area between the plant and the aqueous environment. Metals can enter plants directly from the water and by plant's contact with suspended particles. This process involves two stages: the adsorption on the surface and the absorption followed by the metal fixation in plant tissues. The main mechanism for the fixation of metals is the formation of complexes via the addition of ions to the functional groups of organic compounds (carboxyl, amino, imino, hydroxyl, sulpha-hydryl and keto groups). Some analogous assertions were advanced by Eichenberger (1993). It was found that the intake of metals across the cellular membranes of plants depended on their concentration gradient, and the rate of this process was generally proportional to the concentration of free ions rather than to the total metal concentration. It was also found that free-floating macrophytes were characterized by the high absorption of heavy metals. However, the lifecycle of these plants is generally very short; after their death and decomposition, chemical elements return in part to the water or are accumulated in bottom sediments. The absorption of metals by rooted macrophytes is controlled by their content in riverbed sediments, and the adsorption of unrooted macrophytes is determined by their concentration in the water.Aquatic phytoremediation with floating aquatic plants (FAP) for nutrient removal has a large potential, especially in tropical and subtropical regions of the world. Few species such as *Eichhornia crassipes* and various duckweed such as *Lemna minor* and *Spirodela polyrrhiza* have been intensively studied mainly for domestic wastewater treatment (Gijzen and Veenstra 2000; El-Gendy et al. 2005). Nutrient removal from high strength organic wastewater such as swine lagoon effluent (Bergmann et al. 2000a, b) and pig waste anaerobic effluents (Hernández et al. 1997) has also been performed with duckweed. Suelee et al. (2017) investigated the phytoremediation potential of Vetiver Grass(*Vetiveria zizanioides*) where the findings have shown significant implication for treatment of metal-contaminated water. There are some factors which are to be considered for specifying a plant for a particular remediation (Olguin et al. 2003) which include (1) its seasonal efficiency for removing nutrients or pollutants from the wastewater; (2) its productivity under the particular climatic conditions; (3) its capacity to outgrow other aquatic macrophytes in the same environment; (4) the cost of harvesting; (5) the possible use of the harvested biomass. Aquatic macrophytes are effective in purifying waste water and in removing heavy metals (Brix and Schierup 1989; Rai et al. 1995). Some aquatic plants like water hyacinth have been used for waste water treatment (Abbasi and Nipaney 1985). Similarly, *Eichornia, Pistia* and *Salvinia* are known to scavenge inorganic and organic compounds from waste waters (Boyd 1969). *Ipomea aquatica* showed good Cr (VI) uptake ability in waste water effluent (Bhat et al. 2005). Duckweeds are also significant nutrient removers (Nihan and Elmaca 2007). Remediation potential of *Eichhornia crassipes* (water hyacinth), *Pistia stratiotes* (water lettuce) and *Salvinia molesta* (water fern) in the treatment process of textile wastewater was supported by Wickramasinghe and Jayawardana (2018). Several workers have approved the potential for uptake and concentration of heavy metals by aquatic plants (Tremp and Kohler 1995). The final magnified concentrations of heavy metals in plants used in remediation has lead to plant toxicity and tolerance studies (Ernst et al. 1992; van Steveninck et al. 1992); the role of the aquatic plants in the biochemical cycles (Jackson et al. 1994); their use as biological filters (Brix and Schierup 1989; Dunbabin and Bowmer 1992; Ellis et al. 1994) and their use as biomonitors (Mal et al. 2002). The metal uptake kinetics has been studied under laboratory conditions (Brune et al. 1994) which implies that the extent of metal adsorption and its distribution in plants has important consequences on the uptake, residence time and release of the metal (Ellis et al. 1994). Removal efficiency of some common macrophytes with reference to their metal preferences is depicted in Table 2. For metal removal, aquatic macrophytes are competing with other secondary treatments, being the principal mechanism for metal uptake adsorption through roots (Denny and Wilkins 1987).

Therefore, these macrophytes have been used for heavy metal removal from a variety of sources (Miretzky et al. 2004; Hassan et al. 2007; Mishra and Tripathi 2008). In the process of phytoremediation pollutants are collected by plant roots and either decomposed to less harmful forms or accumulated in the plant tissues. The ability of aquatic plants to accumulate toxic metals from water and their physiological impact is well documented (Rai et al. 1995; Rai 2007; Rai and Tripathi 2007a, b). Many free floating, emergent and submerged species have been identified as potential accumulators of heavy metals (Rai 2007). Such plants could be utilized for the amelioration of water quality and for reducing the pollution load in water bodies. Concomitantly with accumulation toxic metals also cause a high level of phytotoxicity in plants as a result of which several physiological and biochemical changes take place in the plant system. These changes are due to the interaction of heavy metals with the sulpha hydryl groups of the enzymes (Van and Clijsters 1990). Macrophytes are considered important components of aquatic systems not only for their role as major food sources for aquatic invertebrates, but also due to their ability to accumulate heavy metals (Chung and Jeng 1974). Many aquatic macrophytes are found to be potential scavengers of heavy metals from the aquatic environment and are being used in wastewater treatment systems (Vardanyan and Ingole 2006). The major benefits of aquatic macrophyte based treatment systems over conventional

Table 2 Removal of metal ions by some common aquatic macrophytes

Plant	Toxic metals	Accumulation (%)	References
Eichhornia crassipes	Fe and Cd	78–82	Sahu et al. (2007), Schneider et al. (1999), and Prakash et al. (1987)
Azolla spp.	Hg	90–92	Kamal et al. (2004)
Ceratophyllum demersum	Zn and Pb	70–82	Keskinkan et al. (2004)
Ipomoea aquatic	Hg	88–92	Gotheberg et al. (2002)
Lemna spp.	Pb	85–93	Gazi and Steven (1999)
Ludwigia repens	Hg	95–98	Pilon-smith and Pilon (2002)
Pistia stratiotes	Cr, Cd and Hg	80–92	Maine et al. (2001)
Potamogeton spp.	Zn and Pb	68–74	Schneider et al. (1999)
Salvinia herzogii	Fe and Cr	74	Maine et al. (2004)

methods include that they are natural systems, have low operating costs, have low energy requirements, and are easy to regenerate (Weerasinghe et al. 2008). Studies have been reported on the accumulation of a variety of metals, such as Pb, Cd, Hg, Cr, and Cu, by several macrophytes, such as *Hydrilla verticillata, Potomogeton pectinatus, Vallisneria spiralis* (Jana and Chaudhari 1982), *Salvinia molesta, Azolla pinnata, Marsilea minuta* (Abbasi and Nipaney 1984; Gupta and Devi 1995), and *Lemna sp*. (Wang et al. 1997). Heavy metals including both essential and non-essential elements have a particular significance in ecotoxicology, since they are highly persistent and all have the potential to be toxic to living organisms (Storelli et al. 2005).Studies on heavy metals in rivers, lakes, fish and sediments (Pote et al. 2008 and Praveena et al. 2008) have been a major environmental focus especially during the last decade. Heavy metals such as copper, iron, chromium and nickel are essential metals since they play an important role in biological systems, whereas cadmium and lead are non-essential metals, as they are toxic, even in trace amounts (Fernandes et al. 2008). For the normal metabolism of the fish, the essential metals must be taken up from water, food or sediment (Kreuzig 2005). These essential metals can also produce toxic effects when the metal intake is excessively elevated (Tüzen 2003). Uptake of metals by terrestrial plants has been studied by several workers, some examples being Van Aardt and Erdmann (2004), Ghaderian et al. (2007) and Wu et al. (2007).

4 Phytoremediation Through Constructed Wetlands

Constructed wetlands are artificial wastewater treatment systems consisting of shallow (usually less than 1 m deep) ponds or channels which have been planted with aquatic plants, and which rely upon natural microbial, biological, physical and chemical processes to treat wastewater (Ismail et al. 1996; (Sooknah and Wilkie 2004; Deaver et al. 2005; Nahlik and Mitsch 2006). They are being constructed worldwide, are designed and operated for wastewater treatment at the secondary and tertiary level (Kadlec and Knight 1996; Gopal 1999). Several physical, chemical and biological processes are involved in the transformation and consumption of organic matter and plant nutrients within the wetland. The most important functions of the macrophytes in the treatment of wastewater relate to physical effects which they induce therein (Brix 1997). For example, wetlands involve settling of suspended particulate matter, which is the prime cause for reduction of BOD levels in the treated wastewaters. The macrophytes provide good conditions for physical filtration and a large surface area for attached microbial growth and activity (Brix 1997). Over the years, the use of artificially constructed wetlands for wastewater treatment has been increasing considerably (Moshiri 1993). The general practice provides evidence that wetlands remove contaminating nutrients and solids from the wastewater. The problem is to maximize their efficiency at the lowest possible cost. Some floating aquatic macrophytes are used in constructed wetlands, mainly in tropical countries (El-Sayed 1999; Singhal and Rai 2003), due to their capacity to

Table 3 Pollutant reduction in constructed wetlands (Brix 1997)

Pollutant	Percentage reduction (%)
Suspended solids	78–82
P	40–44
N	40–48
Organic constituents (litter)	69–74

absorb and store large quantities of nutrients, and their rapid growth rate (Raskin et al. (1994; Ran et al. 2004). Duckweed and water velvet have been shown to accumulate metals such as Fe and Cu by up to 78 times the concentrations in the wastewater (Jain et al. 1989). Pinto et al. (1987) demonstrated that water hyacinth would remove silver from industrial wastewater for subsequent recovery with high efficiency in a fairly short time. The accumulation of some other heavy metals and trace elements in many species of wetland plants has also been demonstrated (Falbo and Weaks 1990; Zayed et al. 1998; Zhu et al. 1999). Percent reduction of different pollutants through constructed wetlands is listed in Table 3.

5 Current Research Areas in Phytoremediation

Phytoremediation projects have been successfully implemented in the developed nations for the cleanup of metal polluted/contaminated soil/water including the restoration of degraded mines. Extensive diversity of native and non-native plants has been used in this strategy. Researchers are examining the exact mechanisms surrounding metal transport in plants, and why some plants can absorb and tolerate high amounts of toxic metals while others cannot. Identified genes are being cloned, and certain plants are being genetically modified to tolerate metal contamination. In some cases, plants are being genetically modified with bacterial genes. For example, researchers at the University of Georgia genetically modified yellow poplar trees with a gene from mercury-resistant bacteria. In the lab, the trees thrived in mercury-heavy soils. The researchers are further studying the trees in a greenhouse setting. Scientists are also studying biodiversity at metal-contaminated sites across Canada. This work may determine whether seed banks can be established from these wild species, and whether they can be successfully grown in greenhouses. This could eventually lead to the creation of an inventory of plants that could be used at other metal-contaminated sites(Ali et al. 2013).

6 Future Perspectives of Phytoremediation

Vetiver grass (*Vetiveria zizanioides,* or *Chrysopogon zizainoides*), a *super-absorbent* and deep rooted perennial grass, could be used for landfill rehabilitation, erosion and leachate control in particular (Truong and Stone 1996). Because of the high tolerance of this grass species to high acidity, alkalinity, heavy metal levels, it is also recommended for the rehabilitation of mining areas (Truong 1999). Therefore, phytoremediation is not merely applicable for water sanitation, but can also be also involved more to water conservation.The harvested plants from the application of phytoremediation, however, need additional measure in a completed treatment cycle. Particular cultivation treatment, such as cutting and harvesting of the plants are needed for maintaining the pollutant removal efficiency. These nutrient accumulating plants may be used for composting and for energy production. In many cases, the harvested plants are dried and incinerated. The heat released from the incineration is used for energy generation. However, plants from remediated sites, which are polluted by hazardous waste, are generally suspected to contain significant amount of hazardous pollutants and need careful disposal measures (Kramer 2005; Abhilash et al. 2012). Furthermore, fast-growing and high-biomass producing aquatic plants could be used for both phytoremediation and energy production (Abhilash et al. 2012).

7 Conclusions

The contamination removal technologies existing so called conventional methods, have been found creating multiple problems to the environment and are nowadays considered unsafe for any type of ecosystem. Plant based viable technology has gained wide range of acceptance as for as efficiency, health of environment, cost and energy is concerned. Phytoremediants not only remove the nutrients efficiently from the disturbed aquatic environs but have great capacity to clean up the toxic metals. Furthermore, modified aquatic plants can further add to the efficiency of remediating the pollutants from disturbed aquatic ecosystems. Therefore, priority should be given to utilise genetically modified aquatic plants for the restoration of disturbed aquatic environs.

References

Abbasi, S. A., & Nipaney, P. C. (1984). The catalytic effect of copper (II), zinc (II) and nickel (II) on the anaerobic digestion of *Salvinia molesta* (Mitchell). In F. A. Curtis (Ed.), *Energy development: New forms, renewable, conservation* (pp. 237–247). Pergamon: Oxford.

Abbasi, S. A., & Nipaney, P. C. (1985). Waste water treatment using aquatic plants. Survivability and growth of *Salvinia molesta* (Mitchell) over water treated with zinc II and subsequent utilization of the harvested weed for energy (biogas) production. *Resource Conservation, 12*, 47–55.

Abhilash, P. C., Pandey, V. C., Srivastava, P., Rakesh, P. S., Chandran, S., Singh, N., & Thomas, A. P. (2009). Phytofiltration of cadmium from water by *Limnocharis flava* (L.) Buchenaugrown in free-floating culture system. *Journal of Hazardous Materials, 170*(2–3), 791–797.

Abhilash, P. C., Powell, J. R., Singh, H. B., & Singh, B. K. (2012). Plant-microbe interactions: Novelapplications for exploitation inmultipurpose remediation technologies. *Trends in Biotechnology, 30*, 416–420.

Al-Badaii, F., Othman, M. S., & Gasim, M. B. (2013). Water quality assessment of the Semenyih River, Selangor, Malaysia. *Journal of Chemistry, 2013*, 1–10. Article ID 871056. https://doi.org/10.1155/2013/871056.

Ali, H., Khan, E., & Sajad, M. A. (2013). Phytoremediation of heavy metals-concepts and applications. *Chemosphere, 91*, 869–881.

American Society of Civil Engineers (ASCE), Environmental Protection Agency (EPA). (2002). *Urban stormwater BMP performance monitoring. A guidance manual for meeting the national stormwater BMP database requirements* (pp. 395–400). Washington DC.

Anon. (1981). *Characterisation of pollution in urban stormwater runoff* (Technical Paper No. 60, p. 8). Canberra: Australian Water Resources Council and Department of National Development and Energy.

Arber, A. (1920). *A study of aquatic angiosperms* (p. 436). Cambridge: Cambridge University Press.

Asada, M., & Ohgaki, S. (1996). Benzo(a) pyrene in urban runoff at the beginning of rainfall. *Journal of Japan Society on Water Environment, 19*, 904–909.

Asefi, M., & Zamani-Ahamadmahmoodi, R. (2015). Mercury concentrationsand health risk assessment for two fish species, Barbus grypus and Barbus luteus from the maroon river,Khuzestan province, Iran. *Environmental Monitoring and Assessment, 187*, 653.

Ashley, R. M., Wotherspoon, D. J. J., Coghlan, B. P., & McGregor, I. (1992). The erosion and movement of sediments and associated pollutants in combined drainage systems. *Water Science and Technology, 25*, 101–114.

Bergmann, B. A., Cheng, J., Classen, J., & Stomp, A. M. (2000a). Nutrient removal from swine lagoon effluent by duckweed. *Transactions of the American Society of Agricultural Engineers, 43*, 263–269.

Bergmann, B. A., Cheng, J., Classen, J., & Stomp, A. M. (2000b). In-vitro selection of duckweed geographical isolates for potential use in swine lagoon effluent renovation. *Bioresource Technology, 73*, 13–20.

Bhat, S. C., Goswami, S., & Ghosh, U. C. (2005). Removal of trace chromium (VI) from contaminated water: Biosorption by *Ipomea aquatica*. *Journal of Environmental Science and Engineering, 47*, 316–321.

Bhat, M. M., Yazdani, T., Narain, K., Yunus, M., & Shukla, R. N. (2009). Water quality status of some urban ponds of Lucknow, Uttar Pradesh. *Journal of Wetlands Ecology, 2*, 67–73.

Bhat, M. M., Narain, K., Andrabi, S. Z., Shukla, R. N., & Yunus, M. (2012a). Assessment of heavy metal pollution in urban pond ecosystems. *Universal Journal of Environmental Research and Technology, 2*(4), 225–232.

Bhat, M. M., Narain, K., Ahmad, A., Shukla, R. N., & Yunus, M. (2012b). Seasonal variations of physico-chemical characteristics in several ponds of Lucknow city affected by urban drainage. *Advances in Environmental Biology, 6*(10), 2654–2662.

Bhat, M. M., Narain, K., Shukla, R. N., & Yunus, M. (2013). Apportionment of pollution loads arising from catchments in pond water bodies. *Advances in Applied Science Research, 4*(4), 436–441.

Bhuyan, S. J., Marzen, L. J., Koelliker, J. K., Harrington, J. A., & Barnes, P. L. (2001). Assessment of runoff and sediment yield using remote sensing, GIS and AGNPS. *Journal of Soil and Water Conservation, 57*, 351–364.

Biswas, A. K., & Tortajada, C. (2011). Water quality management: An introductory framework. *International Journal of Water Resources Development, 27*(1), 5–11.

Bortey-Sam, N., Nakayama, S. M. M., Ikenaka, Y., Akoto, O., Baidoo, E., Mizukawa, H., & Ishizuka, M. (2015). Healthrisk assessment of heavy metals and metalloids in drinkingwater from communities near gold mines in Tarkwa, Ghana. *Environmental Monitoring and Assessment, 187*, 397.

Boyd, C. E. (1969). Vascular aquatic plants for mineral nutrient removal from polluted waters. *Economic Botany, 24*, 95–103.

Brix, H. (1997). Do macrophytes play a role in constructed treatment wetlands? *Water Science and Technology, 35*, 11–17.

Brix, H., & Schierup, H. H. (1989). The use of aquatic macrophytes in water pollution control. *Ambio, 18*, 100–197.

Brown, L. R., Gray, R. H., Hugies, R. M., & Meador, M. R. (2005). Introduction to effects of urbanization on stream ecosystems. *American Fisheries Society Symposium, 47*, 1–8.

Brune, A., Urbach, W., & Dietz, K. (1994). Compartmentation and transport of zinc in barley primary leaves as basic mechanisms involved in zinc tolerance. *Plant, Cell & Environment, 17*, 153–162.

Bu, F. P., & Xu, X. Y. (2013). Planted floating bed performance intreatmentofeutrophicriverwater. *Environmental Monitoring and Assessment, 185*, 9651–9662.

Carpenter, S. R., Caraco, N. F., Correll, D. L., Howarth, R. W., Sharpley, A. N., & Smith, V. H. (1998). Nonpoint pollution of surface waters with phosphorus and nitrogen. *Ecological Applications, 83*, 559–568.

Chan, N. W. (2012). Managing urban rivers and water quality in Malaysia for sustainable water resources. *Water Resources Development, 28*(2), 343–354.

Choe, J. S., Bang, K. W., & Lee, J. H. (2002). Characterization of surface runoff in urban areas. *Water Science and Technology, 45*, 249–254.

Chow, T. J., Snyder, C. B., Snyder, H. G., & Earl, J. L. (1976). Lead content of some marine organisms. *Journal of Environmental Science and Health, Part A, 11*, 33–44.

Christensen, E. R., & Chien, N. K. (1981). Fluxes of arsenic, lead, zinc and cadmium to Green Bay and Lake Michigan sediments. *Environmental Science and Technology, 15*, 553–558.

Chung, I. H., & Jeng, S. S. (1974). Heavy metal pollution of Ta-Tu River. *Bulletin of the Institute of Zoology, Academy of Science, 13*, 69–73.

Cordery, I. (1976). Some effects of urbanization on streams. *Civil Engineering Transactions, Institution of Engineers, Australia, 18*, 7–11.

Das, S., Goswami, S., & Talukdar, A. D. (2014). A study on cadmium phytoremediation potential of water lettuce, Pistia stratiotes L. *Bulletin of Environmental Contamination and Toxicology, 92*(2), 169–174.

Dearing, J. A., Elner, J. J., & Happey-Wood, C. M. (1981). Recent sediment flux and erosional processes in a Welsh upland lake-catchment based on magnetic susceptibility measurements. *Quaternary Research, 16*, 356–372.

Deaver, E., Moore, M. T., Cooper, C. M., & Knight, S. S. (2005). Efficiency of three aquatic macrophytes in mitigating nutrient run-off. *International Journal of Ecology and Environmental Sciences, 31*, 1–7.

Deletic, A., & Maksimovic, C. T. (1998). Evaluation of water quality factors in storm water runoff from paved areas. *Journal of Environmental Engineering, 124*, 869–878.

Denny, H., & Wilkins, D. (1987). Zinc tolerance in Betula spp. II. Microanalytical studies of zinc uptake into root tissues. *The New Phytologist, 106*, 525–534.

Depledge, M. H., Weeks, J. M., & Bjerregaard, P. (1993). Heavy metals. In P. Calow (Ed.), *Handbook of ecotoxicology* (Vol. 2, pp. 79–99). Oxford: Blackwell.

Dhananjayan, V., Muralidharan, S., & Peter, V. R. (2012). Occurrence and distribution of polycyclic aromatic hydrocarbons in water and sediment collected along the Harbour Line, Mumbai, India. *International Journal of Oceanography, 2012*, 1–7. https://doi.org/10.1155/2012/403615.

Dorval, J., Leblond, V. S., & Hentela, A. (2003). Oxidative stress and loss of cortisol secretion in adrenocortical cells of rainbow trout (*Oncorhynchus mykiss*) exposed in vitro to endosulfan, an organochlorine pesticide. *Aquatic Toxicology, 63*, 229–241.

Dunbabin, J. S., & Bowmer, K. H. (1992). Potential use of constructed wetlands for treatment of industrial wastewaters containing metals. *Science of the Total Environment, 111*, 151–168.

Eichenberger, E. (1993). Correlation between necessity and toxicity of metals in water ecosystems. In *Some problems of metal ions toxicity* (pp. 62–87). Moscow: Mir Publishing House.

El-Gendy, A. S., Biswas, N., & Bewtra, J. K. (2005). A floating aquatic system employing water hyacinth for municipal landfill leachate treatment: Effect of leachate characteristics on the plant growth. *Journal of Environmental Engineering and Science, 4*, 227–240.

Elliott, S., & Sorrell, B. (2002). *Lake Manager's handbook, land–water interactions* (pp. 61–62). Wellington: Ministry for the Environment.

Ellis, J., Shutes, R., Revitt, D., & Zhang, T. (1994). Use of macrophytes for pollution treatment in urban wetlands. *Resources, Conservation and Recycling, 11*, 1–12.

EL-Sayed, A. F. M. (1999). Alternative dietary protein sources for farmed tilapia, *Oreochromis* spp. *Aquaculture, 179*, 149–168.

Ernst, W., Verkleij, J., & Schat, H. (1992). Metal tolerance in plants. *Acta Botanica Neerlandica, 41*, 229–248.

Falbo, M. B., & Weaks, T. E. (1990). A comparison of *Eichhornia crassipes* (Pontederiaceae) and *Sphagnum quinquefarium* (Sphagnaceae) in treatment of acid mine water. *Economic Botany, 44*, 40–49.

Farm, C. (2002). Evaluation of the accumulation of sediment and heavy metals in a storm-water detention basin. *Water Science and Technology, 45*, 105–112.

Fernandes, C., Fontaínhas-Fernandes, A., Cabral, D., & Salgado, M. A. (2008). Heavy metals in water, sediment and tissues of *Liza saliens* from Esmoriz–Paramos lagoon, Portugal. *Environmental Monitoring and Assessment, 136*, 267–275.

Fitzpatrick, F. A., Harris, M. A., Arnold, T. L., & Richards, K. D. (2004). Urbanization influences on aquatic communities in North-Eastern Illinois streams. *Journal of the American Water Resources Association, 40*, 461–475.

Forstner, U., & Whittman, G. T. W. (Eds.). (1979). *Metal pollution in the aquatic environment* (p. 486). Berlin/Heidelberg/New York: Springer.

Foster, I. D. L., & Dearing, J. A. (1987). Quantification of long term trends in atmospheric pollution and agricultural eutrophication: A lake-watershed approach. *IAHS Publication, 168*, 173–189.

Fulazzaky, M. A., Seong, T. W., & Masrin, M. I. M. (2010). Assessment of water quality status for the Selangor River in Malaysia. *Water, Air, and Soil Pollution, 205*, 63–77.

Galiulin, R. V. (1994). Inventory and recultivation of the agro landscape soil cover contaminated by different chemical substances. *Agrochemistry, 7*(8), 132–143.

Gazi, N. W. R., & Steven, P. K. S. (1999). Bioremoval of lead from water using *Lemna minor*. *Bioresource Technology, 70*, 225–230.

Geiger WF (1987) Flushing effects in combined drainage systems. In: *Proceedings of the 4th International Conference on Urban Storm Drainage*, Lausanne, Switzerland, pp. 46.

Ghaderian, S. M., Mohttadi, A., Rahiminejad, M. R., & Baker, A. J. M. (2007). Nickel and other metal uptake and accumulation by species of Alyssum (Brassicaceae) from the ultramafics of Iran. *Environmental Pollution, 145*, 293–298.

Gijzen, H. J., & Veenstra, S. (2000). Duckweed-based wastewater treatment for rational resource recovery and reuse. In E. J. Olguin, G. Sánchez, & E. Hérnandez (Eds.), *Environmental biotechnology and cleaner bioprocesses* (pp. 83–100). London: Taylor and Francis.

Gopal, B. (1999). Natural and constructed wetlands for wastewater treatment: Potentials and problems. *Water Science and Technology, 40*(3), 27–35.

Gotheberg, A., Greger, M., & Bengtsson, B. E. (2002). Accumulation of heavy metals in water spinach (*Ipomoea aquatica*) cultivated in the Bangkok region, Thailand. *Environmental Toxicology and Chemistry, 21*, 1934–1939.

Gowd, S. S., & Kotaiah, B. (2000). Seasonal variation of water quality in a tropical Kalyani reservoir, near Tirupati. *Indian Journal of Environmental Protection, 20*, 452–455.

Gray, L. (2004). Changes in water quality and macroinvertebrate communities resulting from urban stormflows in the Provo River, Utah, USA. *Hydrobiologia, 518*, 33–46.

Greipsson, S. (2011). Phytoremediation. *Nature Education Knowledge, 2*, 7.

Griffin, Jr., Gizzard, D. M., Randall, T. J., Hessle, C. W., & Hartigan, J. P. (1980). Analysis of nonpoint pollution export from small catchments. *Journal WPCF, 52*, 780–790.

Gulati, K. L., Nagpaul, K. K., & Bukhari, S. S. (1979). Uranium, boron, nitrogen, phosphorus and potassium in leaves of mangroves, Mahasagar. *Bulletin of the National Institute of Oceanography, 12*, 183–186.

Gupta, M., & Devi, S. (1995). Uptake and toxicity of cadmium in aquatic ferns. *Journal of Environmental Biology, 16*, 131–136.

Gupta, K., & Saul, A. J. (1996). Specific relationships for the first flush load in combined drainage systems. *Water Research, 30*, 1244–1252.

Gupta, P., & Srivastava, N. (2006). Effects of sub-lethal concentrations of zinc on histological changes and bioaccumulation of zinc by kidney of fish *Channa punctatus* (Bloch). *Journal of Environmental Biology, 27*, 211–215.

Gupta, S. C., Rathore, G. S., & Mathur, G. C. D. (2001). Hydro-chemistry of Udaipur lakes. *Indian Journal of Environmental Health, 43*, 38–44.

GWP Technical Advisory Committee. (2000). *TAC background paper no. 4* (pp. 1–15). Stockholm: Global Water Partnership. Integrated Water Resources Management.

Hakanson, L., & Jansson, M. (1983). *Principles of lake sedimentology* (Vol. 3, p. 16). Berlin: Springer.

Hassan, S. H., Talat, M., & Rai, S. (2007). Sorption of cadmium and zinc from aqueous solutions by water hyacinth (*Eichhornia crassipes*). *Bioresource Technology, 98*, 918–928.

Hernández, E., Olguín, E. J., Trujillo, S., & Vivanco, J. (1997). Recycling and treatment of anaerobic effluents from pig waste using Lemna sp. under temperate climatic conditions. In D. L. Wise (Ed.), *Global environmental biotechnology* (pp. 293–304). Amsterdam: Elsevier.

HO, Y. B. (1988). Metal levels in three intertidal macroalgae in Hong Kong waters. *Aquatic Botany, 29*, 367–372.

Hoo, L. S., Samat, A., & Othman, M. R. (2004). The level of selected Heavy metals (Cd, Cu, Fe, Pb, Mn and Zn) at residential area nearby Labu river system Riverbank, Malaysia. *Research Journal of Chemistry and Environment, 8*, 24–29.

Ismail, A. S., Abael-Sabour, R. M., & Radwan. (1996). Water hyacinth as an indicator for heavy metal pollution in different selected sites and water bodies around greater Cairo Egypt. *Journal of Soil Science, 36*, 343–354.

Jackson, L., Rasmussen, J., & Kalff, J. (1994). A mass balance analysis of trace metals in two weedbeds. *Water, Air, and Soil Pollution, 75*, 107–119.

Jain, S. K., Vasudevan, P., & Jha, N. K. (1989). Removal of some heavy metals from polluted water by aquatic plants: Studies on duckweed and water velvet. *Biological Wastes, 28*, 115–126.

Jana, S., & Chaudhari, M. A. (1982). Senescence in submerged aquatic angiosperms: Effects of heavy metal. *The New Phytologist, 90*, 477–484.

Johnes, P. J., Moss, B., & Phillips, G. (1996). The determination of total nitrogen and total phosphorus concentrations in freshwaters from land-use, stocking headage and population data: Testing of a model for use in conservation and water quality management. *Freshwater Biology, 36*, 451–473.

Joliffe, I. B. (1995). Hydrological cycle and impacts of urbanization. In M. G. Sharpin (Ed.), *Environmental aspects of urban drainage, Concord Function Centre, Sydney, 22 August* (pp. 1–20). Sydney: Stormwater Industry Association.

Kadlec, R. H., & Knight, R. L. (1996). Treatment wetlands. In *Lewis Publishers*. Boca Raton: Florida.

Kamal, M., Ghalya, A. E., Mahmouda, N., & Cote, R. (2004). Phytoaccumulation of heavy metals by aquatic plants. *Environment International, 29*, 1029–1039.

Karthikeyan, S., Palaniappan, P. L. R. M., & Sabhanayakam, S. (2007). Influence of pH and water hardness upon nickel accumulation in edible fish *Cirrhinus mrigala*. *Journal of Environmental Biology, 28*, 489–492.

Keskinkan, O., Goksu, M. Z. L., Basibuyuk, M., & Forster, C. F. (2004). Heavy metal adsorption properties of a submerged aquatic plant (*Ceratophyllum demersum*). *Bioresource Technology, 92*, 197–200.

Kramer, U. (2005). Phytoremediation: Novel approaches to cleaning up polluted soils. *Current Opinion in Biotechnology, 16*, 133–141.

Kreuzig, R. (2005). *Phytoremediation: Potential of plants to clean-up polluted soils*. Braunschweig University of Technology, Institute of Ecological Chemistry and Waste Analysis.

Kulli, B., Balmer, M., Krebs, R., Lothenbach, B., Geiger, G., & Schulin, R. (1999). The influence of nitrillotriacetate on heavy metal uptake of lettuce and ryegrass. *Journal of Environmental Quality, 28*, 1699–1705.

Kumar, A. R., & Riyazuddin, P. (2006). Chemical speciation of trace metals in aquatic environment – An overview. *Research Journal of Chemistry and Environment, 10*, 93–103.

Kusin, F. M., Muhammad, S. N., Zahar, M. S. M., & Madzin, Z. (2016a). Integrated river basin management: Incorporating the use of abandoned mining pool and implication on waterquality status. *Desalination and Water Treatment, 57*(60), 29126–29136.

Kusin, F. M., Zahar, M. S. M., Muhammad, S. N., Mohamad, N. D., Zin, Z. M., & Sharif, S. M. (2016b). Hybrid off-river augmentation system as an alternative raw water resource: The hydrogeochemistry of abandoned mining ponds. *Environment and Earth Science, 75*(3), 1–15.

Lasat, M. M. (2002). Phytoextraction of toxic metals: A review of biological mechanisms. *Journal of Environmental Quality, 31*, 109–120.

Lazzarotto, P., Prasuhn, V., Butscher, E., Crespi, C., Fluhler, H., & Stamm, C. (2005). Phosphorus export dynamics from two Swiss grassland catchments. *Journal of Hydrology, 304*, 139–150.

Lee, J. H., Bang, K. W., Choe, J. S., Yu, M. J., & Ketchum, L. H. (2002). First flush analysis of urban storm runoff. *The Science of the Total Environment, 293*, 163–175.

Lipp, E. K., Kurz, R., Vincent, R., Rodriguez-Palacios, C., Farrah, S. R., & Rose, J. B. (2001). The effects of seasonal variability and weather on microbial faecal pollution and enteric pathogens in a subtropical estuary. *Estuaries, 24*, 491–497.

Lougheed, V. L., Crosbie, B., & Chow-Fraser, P. (2001). Primary determinants of macrophyte community structure in 62 marshes across the Great Lakes basin: Latitude, land use, and water quality effects. *Canadian Journal of Fisheries and Aquatic Sciences, 58*, 1603–1612.

Lukina, L. F., & Smirnova, N. N. (1988). *Physiology of higher aquatic plants* (p. 188). Kiev: Naukova Dumka Publishing House.

Ma, L., He, F., Huang, T., Zhou, Q. H., Zhang, Y., & Wu, Z. B. (2016). Nitrogen and phosphorus transformations and balancein a pond-ditch circulation system for rural polluted water treatment. *Ecological Engineering, 94*, 117–126.

Maberly, S. C., King, L., Gibson, C. E., May, L., Jones, R. I., Dent, M. M., & Crawford, J. (2003). Linking nutrient limitation and water chemistry in upland lakes to catchment characteristics. *Hydrobiologia, 506–509*, 83–91.

Mahvi, A. H., & Mardani, G. (2005). Determination of Phenanthrene in urban runoff of Tehran, Capital of Iran. *Iranian Journal of Environmental Health Science and Engineering, 2*(2), 5–11.

Maine, M., Duarte, M., & Sune, N. (2001). Cadmium uptake by floating macrophytes. *Water Research, 35*, 2629–2634.

Maine, M. A., Sune, N. L., & Lagger, S. C. (2004). Chromium bioaccumulation: Comparison of the capacity of two floating aquatic macrophytes. *Water Research, 38*, 1494–1501.

Mal, T., Adorjan, P., & Corbett, A. (2002). Effect of copper on growth of an aquatic macrophyte, Elodea canadensis. *Environmental Pollution, 120*, 307–311.

Mallin, M., Williams, K., Esham, E., & Lowe, R. (2000). Effect of human development on bacteriological water quality in coastal watershed. *Ecological Applications, 10*, 1047–1056.

Mangkoedihardjo, S. (2007). Phytotechnology integrity in environmental sanitation for sustainable development. *Journal of Applied Sciences Research, 3*(10), 1037–1044.

Marsalek, J. (1998). Challenges in urban drainage. In J. Marsalek, C. Maksimovic, E. Zeman, & R. Price (Eds.), *Hydroinformatics tools for planning, design, operation and rehabilitation of sewer systems* (pp. 1–23). Dordech/Boston/London: Kluwer Academic Publishers.

Marzen, L. J., Bhuyan, S. J., Harrington, J. A., Koelliker, J. K., Frees, L. D., & Volkman, C. G. (2000). Water quality modeling in the red rock creek watershed, Kansas. *Proceedings of Applied Geography Conferences, 23*, 175–182.

Meitei, M. D., & Prasad, M. N. V. (2013). Lead (II) and cadmium (II) biosorption. *Chemical Engineer, 1*(3), 200–207.

Memon, A. R., Aktoprakligül, D., ÖZdemür, A., & Vertii, A. (2001). Heavy metal accumulation and detoxification mechanisms in plants. *Turkish Journal of Botany, 25*, 111–121.

Milic, D., Lukovic, J., Ninkov, J., Zeremski-Skoric, T., Zoric, L., Vasin, J., & Milic, S. (2012). Heavy metal content inhalophytic plants frominland and maritime salineareas. *Central European Journal of Biology, 7*, 307–317.

Miretzkey, P., Saralegui, A., & Fernandez Cirelli, A. (2004). Aquatic macrophytes potential for the simultaneous removal of heavy metals (Buenos Aires, Argentina). *Chemosphere, 57*(8), 997–1005.

Mishra, V. K., & Tripathi, B. D. (2008). Concurrent removal and accumulation of heavy metals by the three aquatic macrophytes. *Bioresource Technology, 99*, 7091–7097.

Mitsch, W. J., & Gooselink, J. G. (2007). *Wetlands* (4th ed., pp. 1–34). New York: Wiley.

Mitsch, W. J., & Gosselink, J. G. (1993). *Wetlands* (2nd ed., p. 539). New York: Van Nostrand-Reinhold.

Mohanraj, R., Sathishkumar, M., Azeez, P. A., & Sivakumar, R. (2000). Pollution status of wetlands in urban Coimbatore, Tamilnadu, India. *Bulletin of Environmental Contamination and Toxicology, 64*, 638–643.

Mohapatra, U. K., & Singh, B. C. (1999). Trace metals in drinking water from different sources in old capital city of Cuttak. *Indian Journal of Environmental Health, 41*, 115–120.

Moshiri, G. A. (1993). *Constructed wetlands for water quality improvement*. Boca Raton, FL: Lewis Publishers, CRC Press.

Mungur, A. S., Shutes, R. B. E., Revitt, D. M., & House, M. A. (1995). An assessment of metal removal from highway runoff by a natural wetland. *Water Science and Technology, 32*, 169–175.

Nahlik, A. M., & Mitsch, W. J. (2006). Tropical treatment wetlands dominated by free-floating macrophytes for water quality improvement in Costa Rica. *Ecological Engineering, 28*, 246–257.

Nihan, O., & Elmaca, A. (2007). Performance of duckweed (*Lemna minor* L.) on different types of waste water treatment. *Journal of Environmental Biology, 27*, 307–314.

Noble, R. T., Dorsey, J., Leecaster, M. K., Reid, D., Schiff, K. C., & Weisberg, S. B. (2000). A regional survey of the microbiological water quality along Southern California bight shoreline. *Environmental Monitoring and Assessment, 64*, 435–447.

Novotny, V. (1994). *Diverse solutions for diffuse pollution*, vol. 1, 24–31. Water Quality International, International Association for Water Quality.

Novotny, V. (1999). Integrating diffuse pollution control and water body restoration into watershed management. *Journal of the American Water Resources Association, 35*, 717–727.

Nriagu, J. O. (1979). Global inventory of natural and anthropogenic emissions of trace metals to the atmosphere. *Nature, 279*, 409–411.

Nyer, E. K., & Gatliff, E. G. (1996). Phytoremediation. *Groundwater Monitoring and Remediation, 16*, 58–62.

O'Loughlin, G. (1994). Pollution prevention and politics – The recent experience in Sydney. *Water Science and Technology, 30*, 13–22.

Olguin, E. J., Rodriguez, D., Sanchez, G., Hernandez, E., & Ramirez, M. E. (2003). Productivity, protein content and nutrient removal from anaerobic effluents of coffee wastewater in Salvinia minima ponds under subtropical conditions. *Acta Biotechnologica, 23*, 259–270.

Othman, F., Eldin, M. E. A., & Mohamed, I. (2012). Trend analysis of a tropical urban river water quality in Malaysia. *Journal of Environmental Monitoring, 14*, 3164–3173.

Park, K. S., & Shin, H. W. (2007). Studies on phyto and zooplankton composition and its relation to fish productivity in a west coast fish pond ecosystem. *Journal of Environmental Biology, 28*, 415–422.

Pavendan, P., Anbu selvan, S., & Sebastian rajasekaran, C. (2011). Physico Chemical and microbial assessment of drinking water from different water sources of Tiruchirappalli District, South India. *European Journal of Experimental Biology, 1*(1), 183–189.

Pegram, G. C., Görgens, A. H. M. (2001). *A guide to non-point source assessment* (Water Research Commission Report TT142/01, pp 115).
Pilon-Smiths, E., & Pilon, M. (2002). Phytoremediation of metals using transgenic plants. *Critical Reviews in Plant Sciences, 21*, 439–456.
Pinto, C. L., Caconia, A., & Souza, M. (1987). Utilization of water hyacinth for removal and recovery of silver from industrial wastewater. In D. Athie (Ed.), *The use of macrophytes in water pollution controls* (Water Sci Technol) (Vol. 19, pp. 89–102).
Polkowska, ò., Grynkiewicz, M., Zabiegala, B., & NamieÑnik, J. (2001). Levels of pollutants in runoff water from roads with high traffic intensity in the City of Gda½sk, Poland. *Polish Journal of Environmental Studies, 10*, 351–363.
Pote, J., Haller, L., Loizeau, J. L., Bravo, A. G., Sastre, V., & Wildi, W. (2008). Effects of a sewage treatment plant outlet pipe extension on the distribution of contaminants in the sediments of the bay of Vidy, Lake Geneva, Switzerland. *Bioresource Technology, 99*, 7122–7131.
Prajapati, S. K., Meravi, N., & Singh, S. (2012). Phytoremediation of Chromium and Cobalt using Pistia stratiotes: A sustainable approach. *Proceedings of the International Academy of Ecology and Environmental Sciences, 2*(2), 136–138.
Prakash, O., Mehroira, I., & Kumar, P. (1987). Removal of cadmium from water by water hyacinth. *Journal of Environmental Engineering, 113*, 352–365.
Praveena, S. M., Radojevic, M., Abdullah, M. H., & Aris, A. Z. (2008). Application of sediment quality guidelines in the assessment of mangrove surface sediment in Mengkabong lagoon, Sabah, Malaysia. *Iran Journal of Environmental Health Science and Engineering, 5*(1), 35–42.
Qin, H. J., Zhang, Z. Y., & Liu, M. H. (2016). Site test of phytoremediation of an open pond contaminated with domestic sewage using water hyacinth and water lettuce. *Ecological Engineering, 95*, 753–762.
Rai, P. K. (2007). Wastewater management through biomass of *Azolla pinnata*: An ecosustainable approach. *Ambio: A Journal on Human Environment, 36*, 426–428.
Rai, P. K., & Tripathi, B. D. (2007a). Heavy metals adsorption characteristics of free floating aquatic macrophyte Spirodela poyrhhiza. *Journal of Environmental Research and Development, 5*, 656–660.
Rai, P. K., & Tripathi, B. D. (2007b). Heavy metals removal using nuisance blue green alga Microcystis in continuous culture experiment. *Environmental Sciences, 4*, 53–59.
Rai, U. N., Sinha, S., Tripathi, R. D., & Chandra, P. (1995). Wastewater treatability potential of some aquatic macrophytes: Removal of heavy metals. *Ecological Engineering, 157*, 1–8.
Ramachandra TV, Solanki M (2007) Ecological assessment of lentic water bodies of Bangalore. ENVIS Technical Report, Vol. 25, pp. 8–9.
Ramachandra, T. V., Kiran, R., & Ahalya, N. (2002). *Status, conservation and management of wetlands* (p. 3). New Delhi: Allied Publishers (P) Limited.
Ran, N., Agami, M., & Oron, G. (2004). A pilot study of constructed wetlands using duckweed (*Lemna gibba* L.) for treatment of domestic primary effluent in Israel. *Water Research, 38*, 2241–2248.
Raskin, I., & Ensley, B. D. (2000). *Phytoremediation of toxic metals: Using plants to clean up the environment* (pp. 193–229). New York: Wiley.
Raskin, I., Nanda Kumar, P. B. A., Dushenkov, V., & Salt, D. E. (1994). Bioconcentration of heavy metals by plants. *Current Opinions Biotechnology, 5*, 285–290.
Ray, M. K., & Majumdar, S. (2005). Evaluating economic sustainability of urban and peri-urban waterbodies, a case study from Kolkata Ponds. In N. Sengupta & J. Badyopadhyay (Eds.), *Biodiversity and quality of life* (pp. 135–146). New Delhi: Macmillan.
Renburg, I. (1986). Concentration and annual accumulation of heavy metals in lake sediments: Their significance in studies of the history of heavy metal pollution. *Hydrobiologia, 143*, 379–385.
Roley, S. S., Tank, J. L., Tyndall, J. C., & Witter, J. D. (2016). How cost-effective are cover crops, wetlands, and two-stage ditches for nitrogen removal in the Mississippi River Basin? *Water Resources and Economics, 15*, 43–56.

Saget, A., Chebbo, G., & Desbordes, M. (1995). Urban discharges during wet weather: What volumes have to be treated? *Water Science and Technology, 32*, 225–232.
Sahu, R. K., Naraian, R., & Chandra, V. (2007). Accumulation of metals in naturally grown weeds (aquatic macrophytes) grown on an industrial effluent channel. *Clean, 35*, 261–265.
Salt, D. E., Blaylock, M. N. P., Kumar, B. A., Dushenkov, V., Chet, I., & Raskin, I. (1995). Phytoremediation: A novel strategy for the removal of toxic metals from the environment using plants. *Biotechnology, 13*, 468–474.
Schindler, D. W. (1997). Widespread effects of climatic warming on freshwater ecosystems in North America. *Hydrological Processes, 11*, 1043–1067.
Schneider, I. A. H., Rubio, J., & Smith, R. W. (1999). Effect of some mining chemicals on biosorption of Cu (II) by the nonliving biomass of the freshwater macrophyte Potamogeton lucens. *Minerals Engineering, 12*, 255–260.
Schreiber, J. D., Rebich, R. A., & Cooper, C. M. (2001). Dynamics of diffuse pollution from US southern catchements. *Water Research, 35*, 2534–2542.
Sculthorpe, C. D. (1967). *The biology of aquatic vascular plants* (p. 610). New York: St. Martin's Press.
Settacharnwit, S., Buckney, R. T., & Lim, R. P. (2003). The nutrient status of Nong Han, a shallow tropical lake in North-Eastern Thailand: Spatial and temporal variations. *Lakes & Reservoirs: Research and Management, 8*, 189–200.
Shastri, Y., & Pendse, D. C. (2001). Hydro biological study of Dahikhura reservoir. *Journal of Environmental Biology, 22*, 67–70.
Shinya, M., Tsuruho, K., Konishi, T., & Ishikawa, M. (2003). Evaluation of factors influencing diffuse pollutant loads in urban highway runoff. *Water Science and Technology, 47*, 227–232.
Shrivastava, S., & Rao, K. S. (1997). Observation on the utility of integrated aquatic macrophyte base system for mercury toxicity removal. *Bulletin of Environmental Contamination Toxicology, 59*, 777–782.
Singh, D., Nath, K., Trivedi, S. P., & Sharma, Y. K. (2008). Impact of copper on haematological profile of freshwater fish, *Channa punctatus*. *Journal of Environmental Biology, 29*, 253–257.
Singh, N. K. S., Devi, C. B., Sudarshan, M., Meetei, N. S., Singh, T. B., & Singh, N. R. (2013). Influence of Nambul river on the water quality of fresh water in Loktak lake. *International Journal of Water Resources and Environmental Engineering, 5*(6), 321–327.
Singhal, V., & Rai, J. P. N. (2003). Biogas production from water hyacinth and channel grass used for phytoremediation of industrial effluents. *Bioresource Technology, 86*, 221–225.
Smitha, P. G., Byrappa, K., & Ramaswamy, S. N. (2007). Physico-chemical characteristics of water samples of Bantwal Taluk, South-Western Karnataka, India. *Journal of Environmental Biology, 28*, 591–595.
Sonneman, J. A., Walsh, C. J., Breen, P. F., & Sharpe, A. K. (2001). Effects of urbanization on streams of the Melbourne region, Victoria, Australia. *Freshwater Biology, 46*, 553–565.
Sooknah, R. D., & Wilkie, A. C. (2004). Nutrient removal by floating aquatic macrophytes cultured in anaerobically digested flushed dairy manure wastewater. *Ecological Engineering, 22*, 27–42.
Srivastava, N., Agarwal, M., & Tyagi, A. (2003). Study of physico-chemical characteristics of water bodies around Jaipur. *Journal of Environmental Biology, 24*, 177–180.
Sriyaraj, K., & Shutes, R. B. E. (2001). An assessment of the impact of motorway runoff on a pond, wetland and stream. *Environment International, 26*, 433–439.
Storelli, M. M., Storelli, A., D'ddabbo, R., Marano, C., Bruno, R., & Marcotrigiano, G. O. (2005). Trace elements in loggerhead turtles (*Caretta caretta*) from the eastern Mediterranean Sea: Overview and evaluation. *Environmental Pollution, 135*, 163–170.
Suelee, A. L., HasanSNMS, K. F. M., Yusuff, F. M., & Ibrahim, Z. Z. (2017). Phytoremediation potential of Vetiver Grass (*Vetiveria zizanioides*) for treatment of metal-contaminated water. *Water, Air, and Soil Pollution, 228*, 158. https://doi.org/10.1007/s11270-017-3349-x.
Thorat, S. R., & Sultana, M. (2000). Pollution status of Salim Ali Lake, Aurangabad (M.S.). *Journal of Pollution Research, 19*, 307–309.

Tilak, K. S., Veeraiah, K., & Milton Prema Raju, J. (2007). Effects of ammonia, nitrite and nitrate on haemoglobin content and oxygen consumption of freshwater fish, *Cyprinus carpio* (Linnaeus). *Journal of Environmental Biology, 28*, 45–47.

Tremp, H., & Kohler, A. (1995). The usefulness of macrophyte monitoring systems, exemplified on eutrophication and acidification of running waters. *Acta Botanica Gallica, 142*, 541–550.

Truong P (1999) *Vetiver grass technology for mine tailings rehabilitation*. Proceedings of the first Asia-Pacific conference on ground and water bioengineering for erosion control and slope stabilization, Manila.

Truong P, Stone R (1996) *Vetiver grass for landfill rehabilitation: Erosion and leachate control*. Report to DNR and Redland Shire Council, Queensland, Australia.

Tuzen, M. (2003). Determination of heavy metals in fish samples of the Mid Dam Lake Black Sea (Turkey) by graphite furnace atomic absorption spectrometry. *Food Chemistry, 80*, 119–123.

US Environmental Protection Agency (USEPA), Office of Water. (1998). *National water quality inventory: 1996 report to congress* (p. 305). Washington, DC: US Environmental Protection Agency (USEPA).

USEPA. (1999). *Phytoremediation resource guide*. EPA/542/B-99/003, available online at http://www.epa.gov/tio

USEPA. (2000). *Introduction to phytoremediation*. EPA/600/R-99/107.

Van Aardt, W. J., & Erdmann, R. (2004). Heavy metals (Cd, Pb, Cu, Zn) in mudfish and sediments from three hard water dams of the Mooi River catchment area, South Africa. *Water SA, 30*, 211–218.

Van, A., & Clijsters, H. (1990). Effect of metals on enzyme activity in plants. *Plant, Cell and Environment, 13*, 195–206.

Van Steveninck, R., van Steveninck, M., & Fernando, D. (1992). Heavy metal (Zn, Cd) tolerance in selected clones of duckweed (Lemna minor). *Plant and Soil, 146*, 271–280.

Vardanyan, L., & Ingole, B. S. (2006). Studies on heavy metal accumulation in aquatic macrophytes from Sevan (Armenia) and Carambolim (India) lake systems. *Environment International, 32*, 208–218.

Vidali, M. (2001). Bioremediation: An overview. *Pure and Applied Chemistry, 73*, 1163–1172.

Vorreiter L, Hickey C (1994) *Incidence of the first flush phenomenon in catchments of the Sydney region*, vol. 3, 359–364. National Conference Publication- Institute of Engineers, Australia.

Wagh, V. M., Ghole, V. S., Wavde, P. N., Todkar, V. V., Kokate, K. K. (2008). *Assessment of water quality of Mutha River in Pune City*. GCE: Indo-Italian International Conference on Green and Clean Environment March 20–21, MAEER's MIT College of Engineering, Pune.

Walsh, C. J. (2000). Urban impact on the ecology of receiving waters: A framework for assessment, conservation and restoration. *Hydrobiologia, 431*, 107–114.

Wang, W., Zhen, W., & Lin, P. (1997). Content and dynamics of five heavy metal elements in the leaves of five mangrove species in Jiulong Estuary. *Journal of Oceanography in Taiwan Strait, 16*, 233–238.

Weerasinghe, A., Ariyawnasa, S., & Weerasooriya, R. (2008). Phytoremediation potential of Ipomoea aquatica for Cr (VI) mitigation. *Chemosphere, 70*, 521–524.

Wetzel, R. G. (1975). Primary production. In B. A. Whitton (Ed.), *River ecology* (pp. 81–105). Berkeley: University of California Press.

Wickramasinghe, S., & Jayawardana, C. K. (2018). Potential of aquatic macrophytes eichhornia crassipes, pistia stratiotes and salvinia molesta in phytoremediation of textile wastewater. *Journal of water security, 4*. https://doi.org/10.15544/jws.2018.001.

Wu, Q. T., Wei, Z. B., & Quyang, Y. (2007). Phytoextraction of metal contaminated soil by Sedum alfredi H: Effects of chelator and coplanting. *Water, Air, and Soil Pollution, 180*, 131–139.

Wu, Q., Hu, Y., Li, S. Q., Peng, S., & Zhao, H. B. (2016). Microbial mechanisms of using enhanced ecological floating beds for eutrophic water improvement. *Bioresource Technology, 211*, 451–456.

WWDR. (2003). *Water for people, water for life* (The UN world water development report) (p. 9). New York: UNESCO.

Yamane, Y., Nagashima, I., Izumikawa, S., & Murakami, A. (1993). Stormwater runoff of petroleum hydrocarbons in the Nogawa river basin in Tokyo. *Journal of Japan Society on Water Environment, 16*, 251–260.

Yang, H., Wang, G., Yang, Y., & XueB, W. B. (2014). Assessment of the impacts of land use changes on nonpoint source pollution inputs upstream of the three gorges reservoir. *The Scientific World Journal, 2014*, 1–15.

Yeh, N., Yeh, P., & Chang, Y. H. (2015). Artificial floating islands for environmental improvement. *Renewable and Sustainable Energy Reviews, 47*, 616–622.

Yin, C. Q., & Mao, Z. P. (2002). Nonpoint pollution control for rural areas of China with ecological engineering technologies. *Chinese Journal of Applied Ecology, 13*(2), 229–232.

Zayed, A., Gowthaman, S., & Terry, N. (1998). Phytoremediation of trace elements by wetland plants. *Journal of Environmental Quality, 27*, 715–721.

Zhang, M., Cui, L., Sheng, L., & Wang, Y. (2009). Distribution and enrichment of heavy metals among sediments, waterbody and plants in Hengshihu wetlands of northern China. *Ecological Engineering, 35*, 563–569.

Zhang, R., Jiang, D., Zhang, L., Cui, Y., Li, M., & Xiao, L. (2014). Distribution of nutrients, heavy metals and PAH's affected by sediment dredging in the Wujin'gang River basin flowing into Meiliang bay of Lake Taihu. *Environmental Science and Pollution Research, 21*, 2141–2153.

Zhu, Y. L., Zayed, A. M., Qian, J. H., Souza, M., & Terry, N. (1999). Phytoaccumulation of trace elements by wetland plants. *Journal of Environmental Quality, 28*, 339–344.

Zolotukhina, E. Y., & Gavrilenko, E. E. (1989). Heavy metals in aquatic plants: Accumulation and toxicity. *Biology Sciences, 9*, 93–106.

Aquatic Pollution Stress and Role of Biofilms as Environment Cleanup Technology

Shakeel Ahmad Dar and Rouf Ahmad Bhat

Abstract The burden of pollutants in water is growing at an alarming rate. The Pollution load is the chief element responsible for ecological stress in aquatic ecosystems. Causative agents for stress in the aquatic ecosystem are usually heavy metals, limiting or excessive nutrient availability, pesticides, pharmaceuticals and changing water properties. Biofilms are collective populations of microbes embodied in the extracellular polymeric substance and are mostly found on different surfaces. These are good indicators of pollution as well as the best candidates for the treatment of pollution load in water bodies. Microbial populations from biofilm have been successfully characterized and used for bioremediation and removal of nutrients from polluted waters due to their unique mechanism of binding with pollutants and high tolerance limit. Biofilm based bioreactors are in use today for cleaning polluted water and have been proved to be more efficient than conventional pollution treatment plants. In this section an effort has been made to evaluate biofilm as the best available option for environmental cleanup of pollution in aquatic ecosystems.

Keywords Biofilm · Aquatic ecosystem · Pollution · Bioremediation · Wastewater · Microbes

1 Introduction

Maximum portion of the earth is covered by water. Approximately 98% of Earth's water is present in oceans and other saline water bodies, whereas the bulk of the left over freshwater is ice-covered in the form of ice sheets and glaciers. Easily reached freshwater sources such as river water, lakes, wetlands and aquifers contribute less than 1% to the total water supply. But this valuable source supports a huge diversity of life, and is critical for human survival. Demand for uses of water rises with the

S. A. Dar (✉) · R. A. Bhat
Division of Environmental Sciences, Sher-e-Kashmir University of Agricultural Sciences and Technology of Kashmir, Srinagar, India

H. Qadri et al. (eds.), *Fresh Water Pollution Dynamics and Remediation*,
https://doi.org/10.1007/978-981-13-8277-2_16

rise in global population. In the meantime, anthropogenic activity, pollution, stress and changing climate are disturbing natural water cycle, putting freshwater ecosystems under tremendous stress. The effect of harmful pollutants on aquatic life is one of the crucial environmental problems. Study of aquatic ecosystems plays a significant part in screening the degree of environmental pressure and the efficacy of restoring actions. Methods that involve living communities are thought to be perfect for pollution stress assessment and rapid investigation at community level with numerous endpoints is a challengeable but meaningful assignment (Xuemei et al. 2010; DeForest et al. 2016). There is growing attention at global level in the stress monitoring and restoration of aquatic ecosystems. Fruitful monitoring needs the skill to label ecological change precisely using measurable indicators (Ryder and Miller 2005; Lear et al. 2009). In streams running through agricultural areas, mixtures of chemicals derived from agricultural activity may directly or indirectly affect biofilm community structure and function (Boivin et al. 2006). Considering the buildup of contaminants in the biofilms and their distinguishing reaction to foremost fluctuations in water quality, these living communities are extensively used in monitoring studies (Gold et al. 2002; Mages et al. 2004; Kropfl et al. 2006). "Biofilms are complex communities composed mainly of photoautotrophic (algae) and heterotrophic microorganism (bacteria, fungi, protozoa) which accumulate at surfaces of artificial or natural substrata and are typically surrounded by their secretory products such as the matrix of extracellular polymeric substances (EPS)" (Sekar et al. 2002; Kropfl et al. 2006; Denkhaus et al. 2007). EPS regulate the structural and functional integrity of microbial biofilms and contribute significantly to the organization of the biofilm community (Branda et al. 2005). Biofilms inhabit the base trophic level of the food chain of streams and help in fueling energy to higher trophic levels driving carbon and nutrient cycles (Battin et al. 2008). Microbial biofilms actively take part in the degradation of plant and animal remains, cycling of nutrients and elimination of suspended sediments in the aquatic environment. Biofilms hold many of the traits required for community level monitoring studies: (1) they are extensively disseminated; (2) they are sessile, thus can imitate the actual circumstances of habitat; (3) they react more quickly to environmental variations because of their short life cycle than higher level organisms; (4) these communities are composed of diverse taxonomic populations with varying environmental tolerance; (5) biofilm samples can be pretty easily collected (Kropfl et al. 2006; Nocker et al. 2007; Porsbring et al. 2007; Xuemei et al. 2010). Biofilms are good bioindicators and biomarkers, offering a suitable tool to screen metal pollution in water bodies that fallouts from mine tailing spots and the expansion of biofilm based substitutions for metal bioavailability is convenient and that combination of the effects of hardness and pH in this metal pollution monitoring tool is important. Benthic biofilms possess manifold functions in the stream and riverine ecosystems and their development and production are frequently restricted by dissolved inorganic nutrient availability, chiefly N or P (Reisinger et al. 2016). Autotrophic and heterotrophic organisms' living inside a given biofilm are often restricted by various nutrients despite experiencing analogous physical and chemical circumstances provided by superimposing river water (Johnson et al. 2009; Hoellein et al. 2010; Reisinger et al.

2016). The nutrients and organic contaminants impact the configuration and function of biofilm and the variation of biofilm indicators can be used to reveal the impact of pollutants on the health of aquatic ecosystem (Xuemei et al. 2010). The buildup of metal pollutants by biofilm is sensitive to the defensive effect of major cations and protons (as for many aquatic living organisms) (Leguay et al. 2016).The use of the structure of diatom communities to measure impacts of metal pollution on freshwater sources has been debated by numerous authors and Periphytic diatom distribution pattern through the occurrence of explicit species highlight metal tolerant indicator diatom groups which will be significant for monitoring pollution in natural aquatic systems (Duong et al. 2008). The quantity of pigments such as chlorophyll shows the dominance of green algae in the biofilms. Pigment composition changes after brief exposure to pollution load can be used as a biochemical marker of toxic effects (Sabater et al. 2007; Xuemei et al. 2010). The enzymatic activity can be deliberated as an indicator of the potential of microbes to degrade polymers and their extent of metabolism in the aquatic environment (Denkhaus et al. 2007). Stream ecosystems primarily obtain nutrients and organic carbon from terrestrial ecosystems and this process is reliant on the land use of the adjacent landscape. The aquatic impacts of anthropogenic land use are often first witnessed by benthic biofilms (Qu et al. 2017). Previous studies have demonstrated the variations in benthic algal distribution pattern in relation to land cover change from the forest to agriculture and urban areas (Teittinen et al. 2015; Smucker et al. 2013), close relationships between longitudinal patterns of stream biofilm biomass and pasture degradation (Ren et al. 2013) and an influence of land cover conditions on biofilm stoichiometry (O'Brien and Wehr 2010; Qu et al. 2017).

Water quality governs and balances the life in aquatic systems. The tolerance limit of aquatic organisms against pollution load in the water bodies depends on their adoptability and acclimatization potential and different organisms respond nonlinearly to the pollution load. Keeping in mind the increased pollution stress in aquatic life and potential of aquatic biofilms to be used as indicators of pollution and purification agents, the role of biofilms in avoiding pollution stress and cleaning water is discussed.

2 Aquatic Ecosystem Environmental Pollution Stressors

In the absence of anthropogenic activities, water bodies usually receive nutrients from different natural sources and nutrient loading in such cases remains below permissible limits. Naturally occurring nutrients are also easily consumed in ecosystem processing. Nutrients such as nitrates and phosphates in natural conditions are important for aquatic ecosystem management as the supply of such nutrients is limited and balanced. However, increased anthropogenic pollution stress on aquatic systems has disrupted this balance. The diverse pollutants are nowadays introduced in water bodies leading to detrimental effects on ecosystem health and functioning. Some of the leading pollutants stressors are discussed below.

2.1 *Heavy Metals as Pollution Stressors*

Metals exist in the environment naturally and some of them are essential for living organisms as well as they are produced from a variety of anthropogenic sources in such quantities that render most of them unfit for ecosystem health. Heavy metals are usually defined as "those metals which possess a specific density of more than 5 g/cm^3 and adversely affect the environment and living organisms" (Jarup 2003). Metals occur in nature with a series of oxidation states and coordination numbers and with this chemical property metals become toxic to living organisms (Pinto et al. 2003). The main metal sources are soil erosion, normal weathering process of the earth's crust, excavating, industrial effluents, urban runoff, sewage release, biocides applied to crops etc. (Morais et al. 2012). When rain falls, contaminants splashed from rooftops, roads, and other surfaces particularly in urban areas find their way into water bodies and may contain heavy metals such as (copper (Cu), lead (Pb) and zinc (Zn)) and polycyclic aromatic hydrocarbons (PAHs). Mine and mill waste water contain sources of contaminations (Adriano 2001; Lim et al. 2008; Brown and Peake 2006).

2.2 *The Toxicity Mechanism of Heavy Metals*

It is believed that toxicity of heavy metals is linked to creation of reactive oxygen species (ROS) and ROS is believed to be responsible for disruption of cellular redox potential balance and this leads to oxidative stress. A number of oxidative species such as superoxide anion ($O_2^{\cdot -}$), hydrogen peroxide (H_2O_2), singlet oxygen (O-) and the hydroxyl radical (^{-}OH) exist in aerobic conditions as regular derivatives of oxidative metabolism that can pose a constant threat to aerobic organisms (Pinto et al. 2003). Under such conditions, cells of such organisms usually produce defense mechanism by using certain phenolics, Glutathione, carotenoids, tocopherols and enzymatic catalysts etc. However, most of the researchers are of the view that oxidative stress in living cells is produced by the unevenness between the creation of ROS and the production of antioxidants to reclaim the reactive intermediates as shown in Fig. 1 (Jaishankar et al. 2014). The very high concentration of ROS owing to the presence of pollutants such as heavy metals may cause cell structural damage and damage to protein, nucleic acid and lipid membranes leading to the stressed condition in the living cells (Mathew et al. 2011). Heavy metals usually show bioaccumulation in the higher levels of the food chain. There are certain heavy metals that inhibit enzyme activation and also damage the nervous system of aquatic organisms.

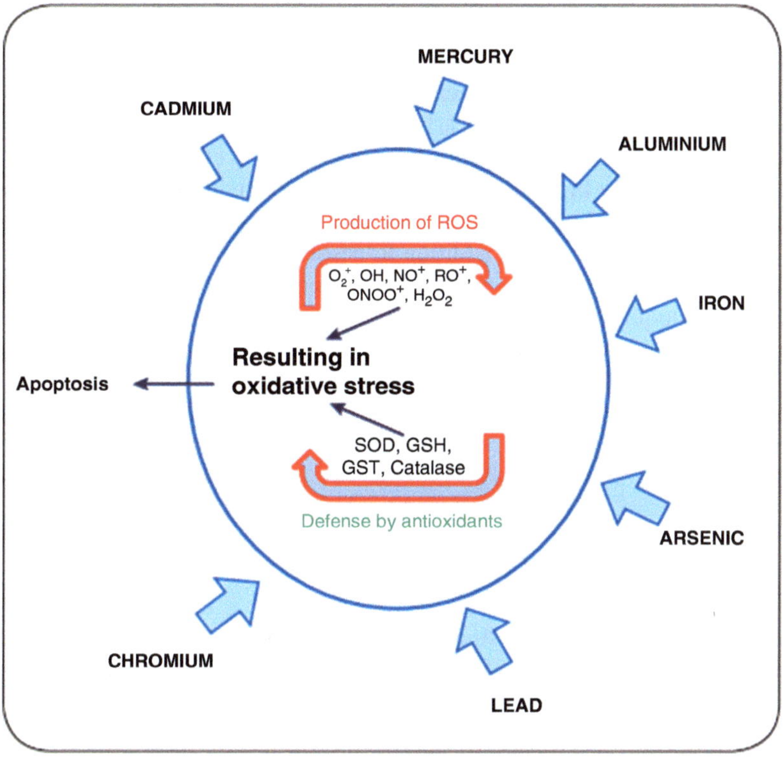

Fig. 1 Showing attack of heavy metals on the cell and the balance between ROS production and the subsequent defense presented by antioxidants. (Jaishankar et al. 2014)

3 Pesticides as Pollution Stressors

The extent of impact induced by pesticides depends on the kind of pesticide and these stressors can affect aquatic ecosystems by directly involving the organisms or indirectly through biotic interactions. Four main ways through which pesticides can enter the water bodies are: (1) It can drift outside of the projected area when it is gushed; (2). It can penetrate, or percolate, through the soil; (3) it can be carried to the water as runoff and (4). It may be spilled unintentionally or through carelessness. They may also be conceded to water by eroding soil. Some of the properties that may decide the potential of pesticide to pollute water are solubility in water, the distance of the application site from water body, nature of soil, crop type and application method. When pesticides enter water bodies, they cause harmful effects on aquatic food chains (Zacharia 2011).

3.1 The Direct Effect of Pesticides on Aquatic Life

The effect of same pesticide concentration on aquatic organisms at different life stages vary and organisms in their early life stages are more susceptible (Mann et al. 2009; Schafer et al. 2011). There may be timely delay in effects, for example, when organochlorines show bio-magnification in the food chain, organisms at the top of the food chain show relatively higher magnitude of pesticides per lipid (Borga et al. 2001). Presence of stressors in addition to pesticides such as UV-rays, parasitism, predation and food scarcity guarantee induction of synergetic effect (Lydy and Austin 2005; Duquesne and Liessm 2003; Coors and DeMeester 2008; Beketov and Liess 2006; Beketov and Liess 2005). Climate and disturbance region may also affect pesticide effect and disturbance recovery. In general ecological, physico-chemical, geographical and temporal factors affect pesticide stress on aquatic ecosystem (Schafer et al. 2011).

3.2 Indirect Effects of Pesticides on Aquatic Life

Researchers have extensively studied manifold indirect effects of pesticide use. Primary producers are at higher risk of pesticide pollution than higher organisms in aquatic system and reduction in the number and health status of primary producers could lead to secondary effects on herbivores and so on (Denoyelles et al. 1982). The population of *Daphnia pulex* and *Simocephalus serrulatus* declined under the influence of atrazine herbicide used in artificial pond due to reduction in the phytoplankton biomass (Denoyelles et al. 1982). Direct and Indirect effects of pesticide stressors are summed up in Fig. 2 (Schafer et al. 2011).

3.3 Pesticides as Oxidative Stressors in the Aquatic Ecosystem

As already discussed that under oxidative stress the critical balance between oxidants and antioxidants is troubled due to unwarranted addition of reactive oxygen species (Scandalios 2005). Freshwater characid fish, *Bryconcephalus*, exposed to organophosphorus insecticide *Folisuper*-600 (methyl parathion) proved that methyl parathion pesticide brings oxidative stress in *Bryconcephalus* (Monteiro et al. 2006). The general mechanism of oxidative stress induced by pesticides is depicted in Fig. 3.

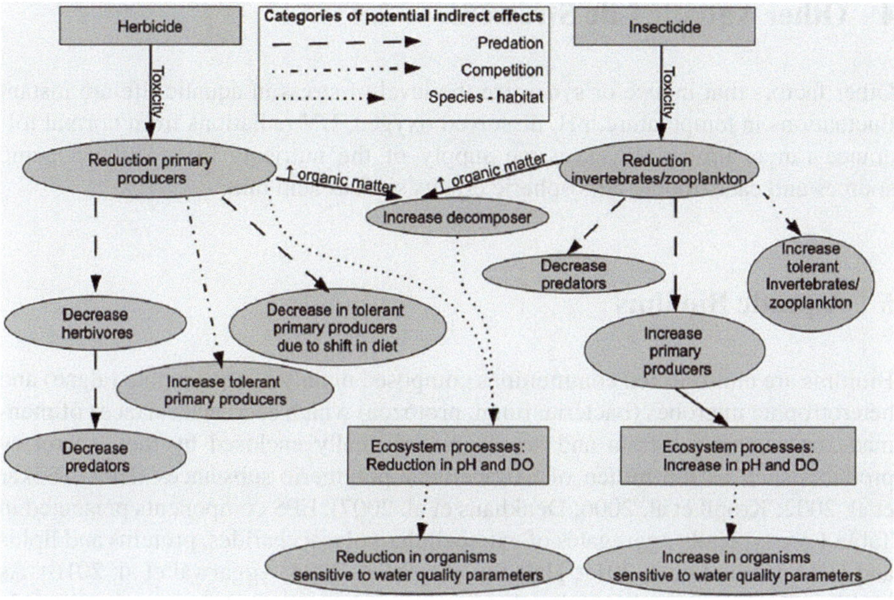

Fig. 2 Showing schematic representation of direct (solid line) and indirect (dashed and dotted lines) potential effects of pesticides in freshwater ecosystems. (Schafer et al. 2011)

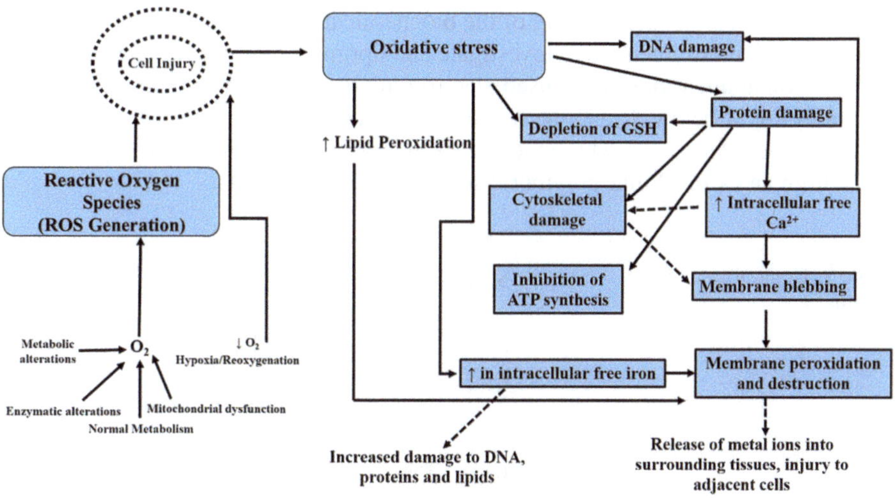

Fig. 3 Mechanism of Oxidative cellular harm tempted by pesticide. (Rehman et al. 2014)

4 Other Aquatic Life Stressors

Other factors that induce or synergize the level of stress in aquatic life are instant fluctuations in temperature, pH, dissolved oxygen, UV radiations from normal tolerance range, limited or excessive supply of the nutrients from anthropogenic sources and catastrophic atmospheric effects such as acid rain.

5 Aquatic Biofilms

Biofilms are multifaceted communities composed mainly of autotrophic (algae) and heterotrophic microbes (bacteria, fungi, protozoa) which accrue at surfaces of man-made or natural substrata and are characteristically enclosed by their secretory products such as the milieu of extracellular polymeric substances (EPS) (Sekar et al. 2002; Kropfl et al. 2006; Denkhaus et al. 2007). EPS components presented in Table 1 are typically aggregates of extracellular polysaccharides, proteins and lipids and DNA (Daniel et al. 2010; Hall-Stoodley et al. 2004; Aggarwal et al. 2016). As biofilms possess three dimensional structures and represent a community lifestyle for microorganisms, they are metaphorically considered as "Cities for Microbes" (Watnick and Kolter 2000).

Expolymeric matrix of microbial biofilms developed from river water and supplied with methanol has been extensively studied by (Lawrence et al. 2003) with an aim to improve the understanding of the biochemical basis for biofilm organization and to assist studies intended to investigate and optimize biofilms for environmental remediation applications. They used electron microscopes for mapping biofilms as shown in Fig. 4 below. It may be noted that structure and composition of biofilms vary with the substrate on which they develop; however, biofilms are always embedded within an extracellular polymeric substance (Photoplate 1).

Table 1 Composition of a biofilm

S. No	Component	Percentage of Matrix
1	Water	97%
2	Microbial cells	2–5%
3	Polysaccharides	1–2%
4	Proteins	<1–2% (includes enzymes)
5	DNA and RNA	<1–2%
6	Ions	Bound and free

Source: http://microbewiki.kenyon.edu/index.php/Stream_biofilm

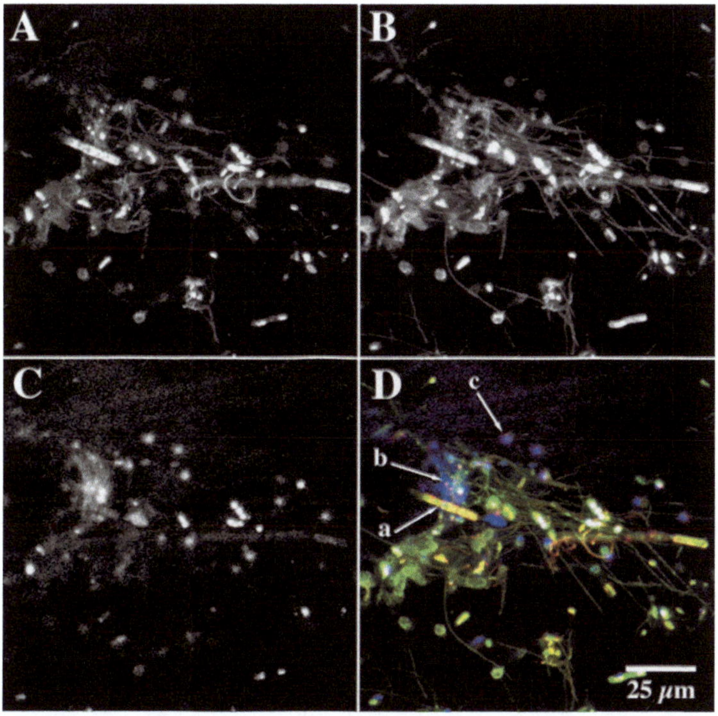

Fig. 4 Biofilm with Three-channel imaging stained with Sypro Orange (protein) (**a**), Syto9 (nucleic acids) (**b**), and Nile Red (hydrophobic-lipid rich) (**c**), and the three-color mixture of the networks displaying the localization and colocalization of protein (red), nucleic acids (green) and lipid-hydrophobic regions (blue) (**d**). Arrows specify protein plus DNA (**a**), protein plus lipid (**b**), and areas rich in lipid alone in the biofilm (**c**). (Source: Lawrence et al. 2003)

5.1 The Life of Biofilms

Biofilms are quite tolerant to external stress. For example, biofilms can bear antibiotic doses up to 1000 times greater than the doses that are enough to kill free living planktonic bacteria (MSU Center for Biofilm Engineering 2008). There are five stages of development of the biofilms viz., initial, irreversible, maturation-I, maturation-II and Dispersal (Cogen and Keener 2004) (Photoplate 2).

5.1.1 The Initial Attachment

When the surface is exposed to an aqueous medium, the initial stage begins and is characterized by the development of coating of polymers on the exposed surface. The layer formed is called as the conditioning layer. This film attaches quickly and colonization increases with the surface roughness (Donlan 2002) and organisms attach themselves to this layer (Cogen and Keener 2004).

Photoplate 1 A dense coating of biofilm on a cobblestone from a nutrient-enriched stream. (Source: Hatfield Consultants 2007)

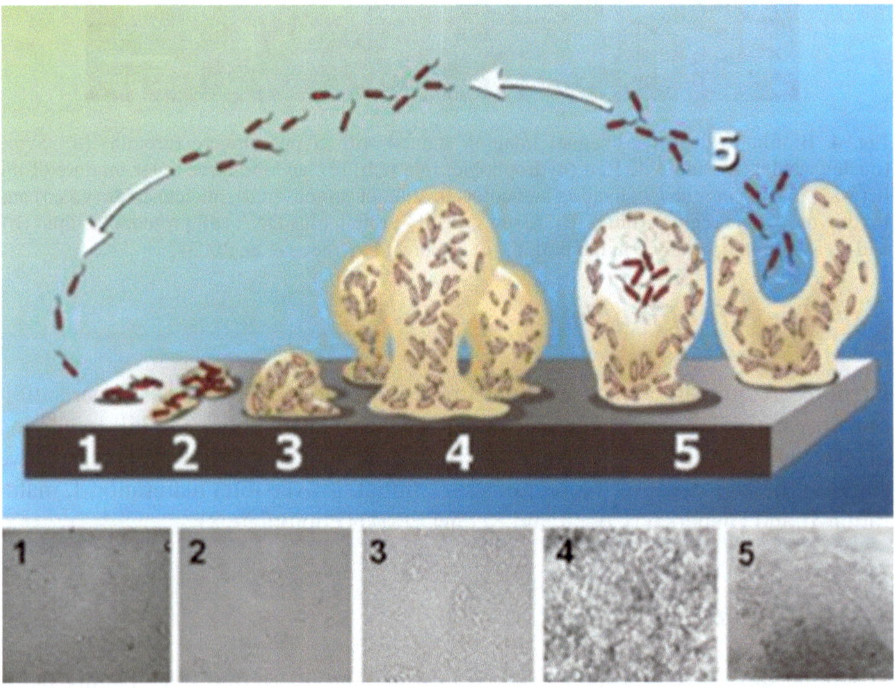

Photoplate 2 Stages of Biofilm Development. (Cogen and Keener 2004)

5.1.2 Irreversible Attachment

The microorganisms now start producing an extracellular polymeric substance (EPS) which has been said to act as the hour of the biofilms (Flemming et al. 2007). The poly-hydroxyl groups in EPS attach the bacterial population present in the biofilm to the surface through hydrogen bonding (Kjelleberg and Givskov 2007). The microbes now become immobile till the last stage of growth (Kolari et al. 2001).

5.1.3 Maturation-I and II

In the first stage of maturation, the main aim is to grow in the dimensions. The biofilm developing takes the debris from the surrounding environment along with new bacterial populations. In the second stage, the biofilm reproduces and may become several inches thick in some cases.

5.1.4 Dispersal

The last stage includes dispersion and detaching of cells and is also termed as "expansion". Bacterial biofilms can also involve active dispersion events in which sessile, matrix-encased biofilm cells change to free-swimming planktonic bacteria (Webb 2007). The cells in biofilm communicate via quorum sensing.

5.2 Biofilm Diversity and Distribution Patterns

Biofilms are the assemblage of adverse group of microorganisms, such as bacteria, archae, algae, fungi, protozoa and viruses and all of them form important part of the biofilm community and add to the diversity of aquatic ecosystems (Stoodley et al. 2002; Battin et al. 2007; Jackson and Jackson 2008). The diversity of biofilms often depends on the form of substrate and aquatic medium in which they are formed. The bacterial groups of prime importance in fresh water biofilms are *Proteobacteria*, *Bacteroidetes* and *cyanobacteria*. *Beta-Proteobacteria* typically dominates in streams, rivers and lakes (Battin et al. 2001; Besemer et al. 2012; Olapade and Leff 2005). However, *Alpha-protobacteria* is more copious in marine ecosystems (Besemer 2015). Other taxonomic groups include *Acidobacteria*, *Actinobacteria*, *Firmicutes*, *Gemmatimonadetes*, *Gamma and Delta-proteobacteria*, *Verrucomicrobia*, *Planctomycetes* and *Deinococcusthermus* (Besemer 2015). Among algae, most abundant species are diatoms and the most common algal groups are *Bacillariophyta* and *Chlorophyta*. These groups provide exudates and other products and are used as carbon source by heterotrophic biofilm microbes

(Battin et al. 2001; Romaní et al. 2004). Diverse fungal groups, such as *Ascomycota* is considered as the structural component of biofilms and help in decomposition of organic matter submerged in water (Das et al. 2004; Heino et al. 2014). Protists such as flagellates, ciliates, amoeba and viruses govern biofilm development and diversity. Environmental influences that regulate the conformation and diversity of microbes in aquatic biofilms are tabulated in Table 2.

Table 2 Environmental variables potentially driving biofilm community composition and biodiversity in different habitats (Besemer 2015)

Environmental variable	Community parameters affected	Habitat type	References
Temperature of Water	Bacterial community structure	Benthic (epilithic and epipsammic), hyporheic, experimental	Anderson-Glenna et al. (2008), Romaní et al. (2014), Wang et al. (2012), Hullar et al. (2006), Wilhelm et al. (2013), Rubin and Leff (2007) and Bouletreau et al. (2014)
	Bacterial and algal diversity	Benthic (epilithic and epipsammic)	Piggott et al. (2015)
pH	Bacterial community structure	Benthic (epilithic and epipsammic), hyporheic, fine benthic organic matter	Wilhelm et al. (2014), Anderson-Glenna et al. (2008), Lear et al. (2009), Fierer et al. (2007), Wilhelm et al. (2013), Freimann, et al. (2014) and Xia et al. (2014)
	Bacterial diversity	Benthic (epilithic), leaf litter	Lear et al. 2009 and Heino et al. (2014)
	Fungal community structure	Leaf litter	Heino et al. (2014)
Inorganic nutrients	Bacterial community structure	Benthic (epilithic), leaf litter,	Olapade and Leff (2005), Heino et al. (2014) and Rubin and Leff 2007
	Bacterial diversity	Experimental	Burgos-Caraballo et al. (2014)
	Fungal community structure	Benthic (epilithic)	Lear et al. (2009)
	Algal community structure	Leaf litter	Heino et al. (2014)
		Experimental	Olapade and Leff (2005)
Dissolved organic –carbon	Bacterial community structure	Benthic (epilithic and epipsammic), hyporheic, aquifer	Wilhelm et al. (2014), Hullar et al. (2006), Li et al. (2012) and Findlay et al. (2003)
	Microbial (bacterial and archaeal) diversity	Aquifer	Li et al. (2012)

5.3 Role of Aquatic Biofilms in Environmental Cleanup

Biofilms play a noteworthy part in industrial and ecological significance and decontamination of polluted water. Researchers have also revealed the potential of biofilms in the degradation of industrial chemical pollutants by using them as a carbon source (Sgountzos et al. 2006). Biofilms play an outstanding role in bioremediation of heavy metals and degradation of some harmful chemicals in the environment and can be used as bio-indicators of pollution. There are different groups of microbes in aquatic bodies that have been tested for decontamination of waste water (Srivastava et al. 2017). Figure 5 shows the role of commonly occurring microbes as biofilms in aquatic ecosystems.

5.4 Microbial Isolates Derived From Biofilms for Use in Water Decontamination

Biofilms have been efficiently used for natural treatment of water and much of microbial fauna has been isolated and characterized from the contaminated sites. Microbes from biofilms could be used to remove or reduce different types of nutrients and heavy metals including pharmaceuticals in an easy and ecofriendly way. Some of the microorganisms that have been evaluated for the elimination of different kinds of contaminants from polluted water along with their functional role in the environment are listed in Table 3. The application of microbes for wastewater treatment is effective and low cost technology and has tremendous scope in the future.

5.5 Mechanism of Elimination of Contaminants From Water by Microorganisms

The process of elimination of contaminants from contaminated water by the application of living organisms such as microorganisms from biofilms involves the basic mechanism of assimilation, adsorption and biodegradation.

5.5.1 Assimilation of Nutrients

Every organism requires some sort of nutrients for better survival and growth. Nutrients such as nitrogen, phosphorous are compulsory for microbial growth. Microbes use inorganic forms of nitrogen as a sole source of nitrogen for growth. Nitrate and phosphate assimilation occurs during the agitation period of the anaerobic-aerobic activated sludge process in the first reactor and during aeration and denitrification in the second reactor (Yariv 2001; Villaverde 2004; Wu et al.

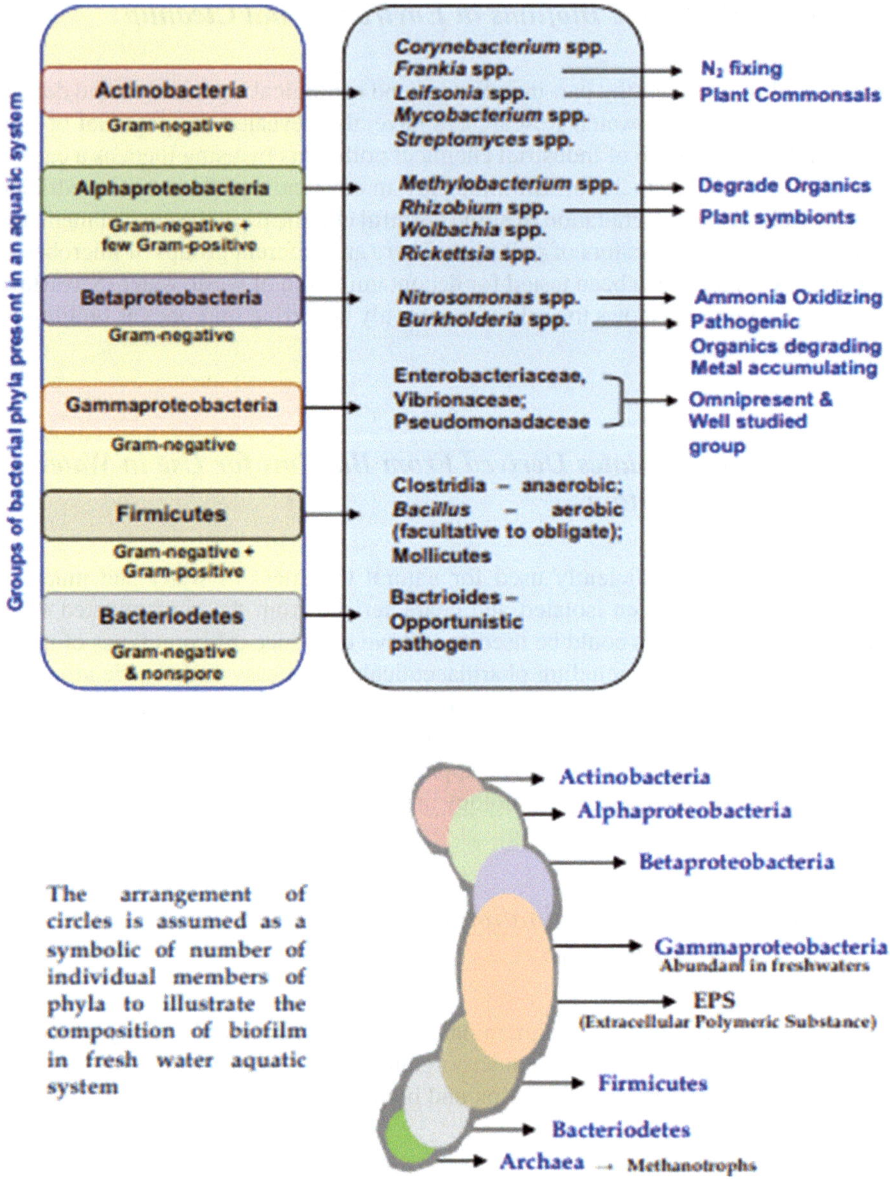

Fig. 5 Commonly present microbial groups with their role in aquatic ecosystem and microbial assemblages in biofilms. (Srivastava et al. 2017)

2012). Most of the organic materials absorbed by biofilms are converted into cellular components such as, cytoplasm which in term help in production of materials such as extracellular polymeric substances. However, all the material is not converted into cytoplasm or cellular material, but much of the organic material is stored

Table 3 The list of microorganisms as biofilms for the removal of different kinds of pollutants

Microbial isolates from biofilms	Role in environmental cleanup	References
Pseudomonas, Chryseomonas, Sphingomonas and *Burkholderiae* species	Biodegradation of phenol and pyridine and the highest biodegradation capacity (1700 mg/L and 3000 mg/L of phenol and pyridine respectively) is shown by *Pseudomonas MT1* isolate.	Rakaiby et al. (2012)
Pseudomonas stutzeri, Aeromonascaviae, Sphingobacteriumthalpophilum, Fusariumudum and *Rhodotorula mucilaginosa*	Reduction of BOD and COD values of wastewater.	Bestawy et al. (2014) and Rozitis and Strade (2014)
Bacillus amyloliquefaciens (S1), *E. coli* (Rz6) and their mixed culture. *Pseudomonas otitidis*	Removal of Total Suspended Solids, Fat, Oil, Grease and Total Coliform from waste water. *Pseudomonas otitidis* has been evaluated for crude oil degradation	Bestawy et al. (2014) and Dasgupta et al. (2013)
Pseudomonas stutzeri B. denitrificans B79 and *A. hydrophila L6, Rheinheimera pacifica, Thauera* sp.	Biological denitrification of wastewater and the river water	SriuNaik and PydiSetty (2011), Andersson (2009) and Jiang et al. (2008)
Comamonas, Thauera, Paracoccus, Paracoccus sp. and *Azoarcus*	Act as heterotrophic nitrifiers in aquatic systems and at a laboratory scale for nitrification	Wang et al. (2014), Cydzik-Kwiatkowska (2015) and Ma et al. (2015)
Nitrosomonas sp. and *Candidatus kuenenia*	Removal of Ammonium from highly concentrated streams	Park et al. (2014)
Phosphorous Accumulating Organisms (PAO) such as Accumulibacter sp., *Tetrasphaera* sp. and *Dechloromonas* sp.	Phosphorous removal from contaminated water	Oehmen et al. 2007, Nielsen et al. (2010), Nguyen et al. (2011) and Kong et al. (2007)
Marine strain P. mendocina NR802, *Pseudomonas paucimobilis, Sphingomonasbisphenolicum* and *Sphingomonas* sp. AO1	Biodegradation of PAHs from polluted water and other micro-pollutants such as PAH and Bisphenol A.	Mangwani et al. (2013) and Kwiatkowska and Zielinska 2016
Pseudomonas putida, Geobacter metallireducens	Bioremediation of heavy metals from metal polluted water.	Singh and Cameotra (2004)
Acinetobacter sp., *Graphium* sp., *Fusarium* sp., *Candida tropicalis*	Decontamination of phenol and m-cresol containing wastewater	Wang et al. (2007) and Santos and Linardi (2004)
P. aeruginosa ASU 6a (Gram-negative) and *Bacillus cereus* AUMC B52 (Gram-positive)	Low cost and effective bio-sorbants for removal of Zn (II) from wastewater	Joo et al. (2010)

(continued)

Table 3 (continued)

Microbial isolates from biofilms	Role in environmental cleanup	References
Acinetobacter calcoaceticus, Erwiniaherbicola, P. aeruginosa and *Pseudomonas maltophilia*	Affinity for bio-sorption of Gold (Au)	Tsuruta (2004)
Escherichia coli	Elimination of numerous heavy metals, such as lead (Pb), copper (Cu), cadmium (Cd), and zinc (Zn)	Kao et al. 2006)
Microcystins degrading bacteria such as *Sphingpoyxis* sp. and *Sphingomonas* sp.	Removal of some specific compounds such as microsytin-RR, aliphatic homopolyesters and aliphatic-aromatic copolyesters.	Wu et al. (2010) and Abou-Zeid et al. 2004)
Basidiomycetes, A. niger and *Trichoderma* sp.	Biosorption of dyes from contaminated water such as Congo red, Orange G. etc.	Tatarko and Bumpus (1998) and Sivasamy and Sundarabal (2011)

and used later during growth. These contain dissolved and suspended materials. The accumulation of polysaccharides and polyhydroxy butyrate inside the cells are used during denitrification (Badireddy et al. 2010; Laspidou and Rittmann 2002; Sheng et al. 2010).

5.5.2 Adsorption of Contaminants

Adsorption of microorganisms and their aggregates is usually called as bio-sorption. This process is essential for metal removal from the water. Bio-sorption in biofilms is shown by bacteria (e.g. *Pseudomonas aeruginosa*), fungi (*Aspergillusniger*) and algae (*Chaetomorphalinum*) (Joo et al. 2010; Fu and Wang 2011). The mechanism of bio-sorption of metals and other dies involves the use of special features of these microorganisms such as adhesion and flocculation properties. Bio-sorption does not produce toxic materials and biofilms maintain their heterogeneity by extracellular polymeric substance (Wu et al. 2012). The structure of biofilms is important in metal adsorption and porous structure of microbial aggregates in biofilms enables active bio-sorption (Wu et al. 2010). There are many functional groups such as carboxyl, phosphoric, sulfhydryl, phenolic and hydroxyl groups in extracellular polymeric substances of biofilms that bind metals (Sheng et al. 2010; Wu et al. 2012). Since the ion exchange is the basic mechanism by which divalent cations such as Ca^{2+} and Mg^{2+} interact with EPS (Extracellular Polymeric Substances) of the biofilms. When metals are bound to the biofilms, there is continuous release of these kinds of divalent cation in the solution which indicates the removal of metals.

5.5.3 Biodegradation of Contaminants

Biodegradation involves the chemical disintegration of organic materials by microorganisms. The contaminants may be biodegraded in an aerobic environment or anaerobic environment or may be mineralized into inorganic minerals. The biofilms are used in wastewater treatment for biodegradation and bioremediation of contaminants in bioreactors (Liong 2011). The removal of nitrogenous compounds from water involves simply nitrification and denitrification process. Biofilms remove phosphorous through the capacity of some microorganisms to store phosphates as orth-ophosphorous in intercellular spaces as polyphosphates. These microbes pile polyhydroxybutyrate (PHB) in absence of oxygen, which is oxidized in a phase with an electron acceptor such as oxygen or nitrate present (Villaverde 2004; Wu et al. 2012). Sulfurous compounds are microbial converted by different consortia of sulfur reducing and sulfur oxidizing bacteria and phototrophic sulfur bacteria. In aerobic biofilms sulfur reduction helps in the mineralization and sulfide oxidation (Zhang et al. 2009; Villaverde 2004). Phenol is the most reluctant contaminant in the wastewater and it can be removed by using biofilms. The basic mechanism is supported by lignin decomposing enzymes (*Manganese peroxidase* (MnP) and *Lacase*) mainly secreted by *basidiomycetes* (Kang et al. 2006). MnP oxidizes phenolic complexes in presence of Mn (II) and H_2O_2.Whereas, *Lacase* oxidizes phenolic mixtures by reducing oxygen to water. Quinoline which is a heterocyclic organic compound being used to produce pharmaceuticals and pesticides causes severe damage to the ecosystem. Some scientists have used mathematical models to describe the biological degradation of quinoline. For example, *B. pickettii* restrained on a hybrid carrier was used to decompose quinoline, with the successive deprivation process defined by a zero-order reaction rate equation when the initial quinoline concentration was in the range of 50–500 mg/L (Wang et al. 2002).The whole process of removal of contaminants from aquatic system is described in Fig. 6 (Wu et al. 2012)

6 Waste Water Treatment Techniques Involving Biofilms

Biofilms have become the important part of biological treatment of municipal waste water and there are various technologies used for treatment of waste water. A few of them are presented here.

6.1 Biofilm Reactors

Biofilm reactors are units which involve biological processes based on biofilm rich sludge used in wastewater treatment. Biofilm reactors are classification is based on (1). The number of phases involved (2) As per biofilm attached to a fixed or moving carrier within the reactor and (3) How electron donors or accepters are used

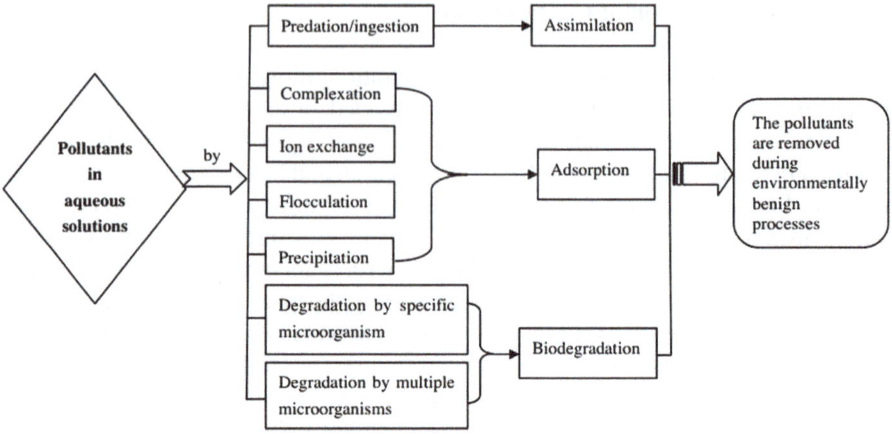

Fig. 6 The method of contaminant removal from aqueous solutions by the conjunct mechanisms of assimilation, adsorption and biodegradation. (Wu et al. 2012)

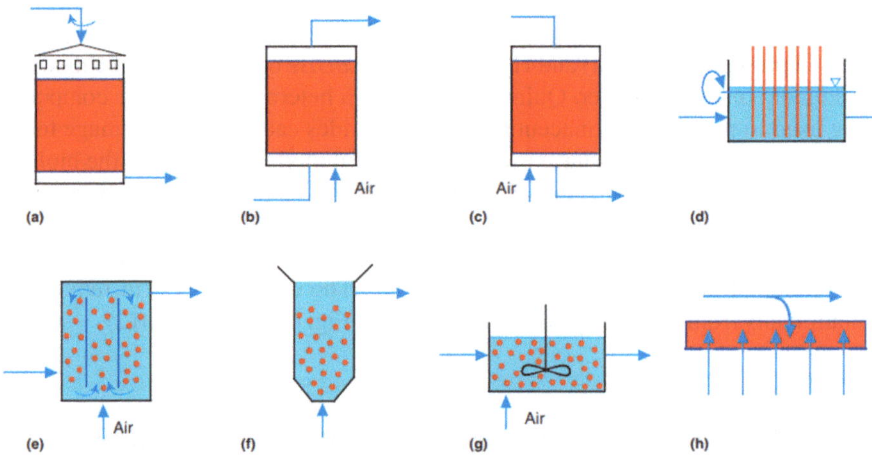

Fig. 7 Types of biofilm reactors: (**a**) trickling filter; (**b**) submerged fixed bed biofilm reactor operated as up flow or (**c**) down flow mode; (**d**) rotating biological contactor; (**e**) suspended biofilm reactor including airlift reactor; (**f**) fluidized bed reactor; (**g**) The moving bed biofilm reactor; and (**h**) The membrane attached biofilm reactors. (Morgenroth 2008; Lewandowski et al. 2011)

(Lewandowski et al. 2011). Lewandowski et al. 2011 has mentioned following types of biofilm reactors:

- **Trickling filter (Three-phase system – fixed biofilm-laden carrier, bulk water, and air.):** Here water drips over the surface of the biofilm and air is subjected to pass upward or downward in the third phase Fig. 7a.
- **Submerged fixed bed biofilm reactor operated as up flow or down flow (Three-phase system – fixed (or semi fixed) biofilm-laden carrier, bulk water,**

and air). Water and gas bubbles (Aerobic and bioactive filter) flow through the biofilm reactor. Here gravel is aimmovable media and polystyrene beads act as semi-fixed media Fig. 7b, c.

- **Aerobic moving bed biofilm reactor or MBBR (Three-phase system – moving biofilm-laden carrier, bulk water, and air. Water flows through the biofilm reactor):**Here water flows through the biofilm reactor and air is introduced with gas bubbles Fig. 7g. MBBRs can be used as single stage or multistage reactors. They are efficient enough to meet water quality standards for carbon oxidation, nitrification, and denitrification.
- **Denitrification fluidized bed biofilm reactor or FBBR (Two-phase system – moving biofilm-laden carrier and bulk water):**Water is allowed to flow through the biofilm reactor having electron donor and accepter Fig. 7g.
- **Denitrification filter (Two-phase system – fixed biofilm-laden carrier material and bulk water):** Water is allowed to flow through the electron donor and accepter Fig. 7b, c.
- **Membrane biofilm reactor or MBfR (Three-phase membrane system):** It is made of a microporous hollow membrane having water and biofilm on one side of the membrane and gas on the other side and gas is allowed to diffuse through the membrane to the biofilm Fig. 7h.
- **The biofilm-based microbial fuel cell or MFC (Two-phase membrane system):** It contains a proton exchange membrane which splits a classified biofilm-laden anode from a classified cathode with water on both sides. The electron donor is also separated from electron acceptor by this membrane.

Other methods of waste water treatment include constructed wetlands and lagoons where the application of biofilm is vital in nutrient removal along with certain macrophytes.

7 Comparison Between Conventional Membrane Bioreactor and Biofilm Membrane Based Bioreactor for Waste Water Treatment

Subtil et al. (2014) compared conventional membrane bioreactor with the biofilm based bioreactor by evaluating organic matter, nitrogen removal efficiency and impact on membrane fouling. The conformations studied included a Conventional Membrane Bioreactor (C-MBR) and a Biofilm Membrane Bioreactor (BF-MBR20) operated under similar conditions, both fed with domestic wastewater. It was noted that BF-MBR produced lower ammonia and total nitrogen concentrations with removal efficiency of 98% and 73% respectively. Thus BF-MBR showed 6% increase in total nitrogen removal. The fouling rate was also 35% higher in C-MBR than BF-MBR. As a result, the operational cycle length increased around 7 days in the BF-MBR compared to the C-MBR. From this study it is obvious that biofilm based bioreactors for wastewater treatment is ecofriendly, reliable and efficient.

Biofilm based treatment systems are advantageous because the microbial communities are impervious to changing environmental conditions, which makes them resilient to variation in toxicity concentrations. (Clark Ehlers and Susan 2012).

8 Limitations of Biofilm Based Treatment Plants

Some biofilm based bioreactors are costly. The layer of biofilm developed needs continuous watch and needs timely maintenance. However, this limitation could be tackled by using several physical methods like back washing, mechanical scrubbing etc. Another problem is biofouling of pipes. Despite of limitations, biofilm based bioreactors and bioremediation technologies are preferred and a naturally feasible way of combating pollution of our diminishing water bodies.

9 Conclusion

Biofilms are complex structured porous and most tolerant communities of the aquatic ecosystem that includes environmentally efficient strains of algae, fungi, bacteria, actinomycetes and viruses. These microbial isolates have been extensively studied for assimilation, bio-sorption and biodegradation of almost all sorts of organic and inorganic pollutants in aquatic ecosystems. Biofilms decontaminate water by the mechanism of adhesion, flocculation, ion exchange and complexation with the pollutants. Aerobic microorganisms and algal species in biofilms utilize and store excessive nutrients from water and thus free it from nutrient loadings. Biofilm based bioreactors are in use from decades and are ascertaining to be best available technologies for waste water treatment. Further research is needed to bio-stimulate the promising strains of biofilms for pollution treatment. There is also need to use bio-technological interventions in identification of resistant genes that can be used against inorganic and organic pollutants.

References

Abou-Zeid, D. M., Muller, R. J., & Deckwer, W. D. (2004). Biodegradation of aliphatic homopolyesters and aliphatic–aromatic copolyesters by anaerobic microorganisms. *Biomacromolecules, 5*, 1687–1697.

Adriano, D. C. (2001). *Trace elements in terrestrial environments* (2nd ed.). New York: Springer.

Aggarwal, S., Stewart, P., & Hozalski, R. (2016). Biofilm cohesive strength as a basis for biofilm recalcitrance: Are bacterial biofilms overdesigned. *Sage Journals, 8*(2), 29–32.

Andersson, S. (2009). Characterization of bacterial biofilms for wastewater treatment. *Royal Institute of Technology Stockholm*, 1–77.

Anderson-Glenna, M. J., Bakkestuen, V., & Clipson, N. J. W. (2008). Spatial and temporal variability in epilithic biofilm bacterial communities along an upland river gradient. *FEMS Microbiology Ecology, 64*, 407–418.

Badireddy, A. R., Chellam, S., Gassman, P. L., Engelhard, M. H., Lea, A. S., & Rosso, K. M. (2010). Role of extracellular polymeric substances in bioflocculation of activated sludge microorganisms under glucose-controlled conditions. *Water Research, 44*, 4505–4516.

Battin, T. J., Wille, A., Sattler, B., & Psenner, R. (2001). Phylogenetic and functional heterogeneity of sediment biofilms along environmental gradients in a glacial stream. *Applied and Environmental Microbiology, 67*, 799–807.

Battin, T. J., Sloan, W. T., Kjelleberg, S., Daims, H., Head, I. M., & Curtis, T. P. (2007). Microbial landscapes: New paths to biofilm research. *Nature Reviews. Microbiology, 5*, 76–81.

Battin, T. J., Kaplan, L. A., Findlay, S., Hopkinson, C. S., Marti, E., & Packman, A. I. (2008). Biophysical controls on organic carbon fluxes in fluvial networks. *Nature Geoscience, 1*, 95–100.

Beketov, M. A., & Liess, M. (2005). Acute contamination with esfenvalerate and food limitation: Chronic effects on the mayfly, *Cloeon dipterum. Environmental Toxicology and Chemistry, 24*, 1281–1286.

Beketov, M. A., & Liess, M. (2006). The influence of predation on the chronic response of *Artemia* sp. populations to a toxicant. *Journal of Applied Ecology, 43*, 1069–1074.

Besemer, K. (2015). Biodiversity, community structure and function of biofilms in stream ecosystems. *Research in Microbiology, 20*, 1–8.

Besemer, K., Peter, H., Logue, J. B., Langenheder, S., Lindström, E. S., & Tranvik, L. J. (2012). Unraveling assembly of stream biofilm communities. *The ISME Journal, 6*, 1459–1468.

Bestawy, E., AL-Hejin, A., Amer, R., & Kashmeri, R. A. (2014). Decontamination of domestic wastewater using suspended individual and mixed bacteria in batch system. *Journal of Bioremediation and Biodegradation, 5*, 231.

Biofilm Engineering. (2008). *Biofilm basics*. http://www.biofilm.montana.edu/biofilm-basics/what_are_biofilms.html

Boivin, M. E. Y., Massieux, B., Breure, A. M., Greve, G. D., Rutgers, M., & Admiraal, W. (2006). Functional recovery of biofilm bacterial communities after copper exposure. *Environmental Pollution, 140*, 239–246.

Borga, K., Gabrielsen, G. W., & Skaare, J. U. (2001). Biomagnification of organochlorines along a Barents Sea food chain. *Environmental Pollution, 113*, 187–198.

Bouletreau, S., Lyautey, E., Dubois, S., CompinA, D. C., & Touron-Bodilis, A. (2014). Warming-induced changes in denitrifier community structure modulate the ability of phototrophic river biofilms to denitrify. *Science of the Total Environment, 466*, 856–863.

Branda, S. S., Vik, F. L., & Kolter, R. (2005). Biofilms: The matrix revisited. *Trends in Microbiology, 13*(1), 20–26.

Brown, J. N., & Peake, B. M. (2006). Sources of heavy metals and polycyclic aromatic hydrocarbons in urban stormwater runoff. *Science of the Total Environment, 359*, 145–155.

Burgos-Caraballo, S., Cantrell, S. A., & Ramirez, A. (2014). Diversity of benthic biofilms along a land use gradient in tropical headwater streams, Puerto Rico. *Microbial Ecology, 68*, 47–59.

Clark, E. G. A., & Susan, J. T. (2012). *Microbial biofilms current research and applications*. Norfolk: Caister Academic Press.

Coors, A., & DeMeester, L. (2008). Synergistic, antagonistic and additive effects of multiple stressors: Predation threat, parasitism and pesticide exposure in *Daphnia magna. Journal of Applied Ecology, 45*, 1820–1828.

Cogan, N. G., & Keener, J. P. (2004). The role of the biofilm matrix in structural development. *Mathematical Medicine and Biology, 21*(2), 147–166.

Cydzik-Kwiatkowska, A. (2015). Bacterial structure of aerobic granules is determined by aeration mode and nitrogen load in the reactor cycle. *Bioresource Technology, 181*, 312–320.

Daniel, L., Hera, V., & Roberto, K. (2010). Biofilms. *Cold Spring Harbor Perspectives in Biology, 2*(7), 1943–0264.

Das, M., Royer, T. V., & Leff, L. G. (2004). Diversity of fungi, bacteria and actinomycetes on leaves decomposing in a stream. *Applied and Environmental Microbiology, 73*, 756–767.

Dasgupta, D., Ghosh, R., & Sengupta, T. K. (2013). Biofilm-mediated enhanced crude oil degradation by newly isolated pseudomonas species. *Hindawi Publishing Corporation ISRN Biotechnology, 250749*, 1–13.

DeForest, J. L., Drerup, S. A., & Vis, M. L. (2016). Using fatty acids to fingerprint biofilm communities: A means to quickly and accurately assess stream quality. *Environmental Monitoring and Assessment, 188*, 277.

Denkhaus, E., Meisen, S., Telgheder, U., & Wingender, J. (2007). Chemical and physical methods for characterization of biofilms. *Microchimica Acta, 158*(1), 1–27.

Denoyelles, F., Kettle, W. D., & Sinn, D. E. (1982). The responses of plankton communities in experimental ponds to atrazine, the most heavily used pesticide in the United States. *Ecology, 63*, 1285–1293.

Duong, T. T., Morinc, S., Herlory, O., Feurtet-Mazel, A., Coste, M., & Boudou, A. (2008). Seasonal effects of cadmium accumulation in periphytic diatom communities of freshwater biofilms. *Aquatic Toxicology, 90*, 19–28.

Donlan, R. M. (2002). Biofilms: Microbial life on surfaces. *Emerging Infectious Diseases, 8*(9), 881–890.

Duquesne, S., & Liessm, M. (2003). Increased sensitivity of the macroinvertebrate *Paramoreawalkeri* to heavy metal contamination in the presence of solar UV radiation in Antarctic shoreline waters. *Marine Ecology Progress Series, 255*, 183–191.

Fierer, N., Morse, J. L., Berthrong, S. T., Bernhardt, E. S., & Jackson, R. B. (2007). Environmental controls on the landscape-scale biogeography of stream bacterial communities. *Ecology, 88*, 2162–2173.

Findlay, S. E. G., Sinsabaugh, R. L., Sobczak, W. V., & Hoostal, M. (2003). Metabolic and structural response of hyporheic microbial communities to variations in supply of dissolved organic matter. *Limnology and Oceanography, 248*, 1608–1617.

Flemming, H. C., Neu, T. R., & Wozniak, D. J. (2007). The EPS matrix: The "house of biofilm cells". *Journal of Bacteriology, 189*(22), 7945–7947.

Freimann, R., Burgmann, H., Findlay, S. E. G., & Robinson, C. T. (2014). Spatio-temporal patterns of major bacterial groups in alpine waters. *PLOS One, 9*, 3511–3524.

Fu, F., & Wang, Q. (2011). Removal of heavy metal ions from wastewaters: A review. *Journal of Environmental Management, 92*, 407–418.

Gold, C., Feurtet-Mazel, A., Coste, M., & Boudou, A. (2002). Field transfer of periphytic diatom communities to assess short-term structural effects of metals (Cd, Zn) in rivers. *Water Research, 36*, 3654–3664.

Hall-Stoodley, L., Costerton, J. W., & Stoodley, P. (2004). Bacterial biofilms: From the natural environment to infectious diseases. *Nature Reviews Microbiology, 2*(2), 95–108.

Hatfield Consultants. (2007). *Regional aquatic monitoring programme*. http://www.ramp-alberta.org/river/gallery.aspx?galleryimage=1166

Heino, J., Tolkkinen, M., Pirttilea, A. M., Aisala, H., & Mykrea, H. (2014). Microbial diversity and community-environment relationships in boreal streams. *Journal of Biogeography, 41*, 2234–2244.

Hoellein, T. J., Tank, J. L., Kelly, J. J., & Rosi-Marshall, E. J. (2010). Seasonal variation in nutrient limitation of microbial biofilms colonizing organic and inorganic substrata in streams. *Hydrobiologia, 649*, 331–345.

Hullar, M. A. J., Kaplan, L. A., & Stahl, D. A. (2006). Recurring seasonal dynamics of microbial communities in stream habitats. *Applied and Environmental Microbiology, 72*, 713–722.

Jackson, E. F., & Jackson, C. R. (2008). Viruses in wetland ecosystems. *Freshwater Biology, 53*, 1214–1227.

Jaishankar, M., Tseten, T., Anbalagan, N., Mathew, B. B., & Beeregowda, K. N. (2014). Toxicity mechanism and health effects of some heavy metals. *Interdisciplinary Toxicology, 7*(2), 60–72.

Jarup, L. (2003). Hazards of heavy metal contamination. *British Medical Bulletin, 68*(1), 167–182.

Jiang, X., Ma, M., & Li, J. (2008). Bacterial diversity of active sludge in waste-water treatment plant. *Earth Science Frontiers, 15*, 163–168.

Johnson, L. T., Tank, J. L., & Dodds, W. K. (2009). The influence of land use on stream biofilm nutrient limitation across eight North American ecoregions. *Canadian Journal of Fisheries and Aquatic Sciences, 66*, 1081–1094.

Joo, J. H., Hassan, S. H. A., & Oh, S. E. (2010). Comparative study of biosorption of Zn^{2+} by *Pseudomonas aeruginosa* and *Bacillus cereus*. *International Biodeterioration and Biodegradation, 64*, 734–741.

Kang, J. H., Katayama, Y., & Kondo, F. (2006). Biodegradation or metabolism of bisphenol A: From microorganisms to mammals. *Toxicology, 217*, 81–90.

Kao, W. C., Chiu, Y. P., Chang, C. C., & Chang, J. S. (2006). Localization effect on the metal biosorption capability of recombinant mammalian and fish metallothioneins in *Escherichia coli*. *Biotechnology Progress, 22*, 1256–1264.

Kjelleberg, S., & Givskov, M. (2007). *The biofilm mode of life: Mechanisms and adaptations*. Wymondham: Horizon Bioscience. http://www.open-accessbiology.com/biofilms/biofilmsch2.pdf.

Kolari, M., Nuutinen, J., & Salkinoja-Salonen, M. S. (2001). Mechanisms of biofilm formation in paper machine by Bacillus species: The role of Deinococcus geothermalis. *Journal of Industrial Microbiology & Biotechnology, 27*, 343–351.

Kong, Y., Xia, Y., Nielsen, J. L., & Nielsen, P. H. (2007). Structure and function of the microbial community in a full-scale enhanced biological phosphorus removal plant. *Microbiology, 153*(12), 4061–4073.

Kropfl, K., Vlada, P., Szabo, K., Acs, E., Borsodi, A. K., Szikora, S., Caroli, S., & Zaray, G. (2006). Chemical and biological characterization of biofilms formed on different substrata in Tisza river (Hungary). *Environmental Pollution, 144*, 626–631.

Kwiatkowska, A. C., & Zielinska, M. (2016). Bacterial communities in full-scale wastewater treatment systems. *World Journal of Microbiology and Biotechnology, 32*, 66.

Laspidou, C. S., & Rittmann, B. E. (2002). A unified theory for extracellular polymeric substances, soluble microbial products, and active and inert biomass. *Water Research, 36*, 2711–2720.

Lawrence, J. R., Swerhone, G. D. W., Leppard, G. G., Araki, T., & Zhang, X. (2003). Scanning transmission X-Ray, laser scanning, and transmission electron microscopy mapping of the exopolymeric matrix of microbial biofilms. *Applied and Environmental Microbiology, 69*(9), 5543–5554.

Lear, G., Boothroyd, I. K. G., Turner, S. J., Roberts, K., & Lewis, G. D. (2009). A comparison of bacteria and benthic invertebrates as indicators of ecological health in streams. *Freshwater Biology, 54*, 1532–1543.

Leguay, S., Lavoie, I., Levy, J. I., & Fortin, C. (2016). Using biofilms for monitoring metal contamination in lotic ecosystems: The protective effects of hardness and pH on metal bioaccumulation. *Environmental Toxicology and Chemistry, 35*(6), 1489–1501.

Lewandowski, Z., Bozeman, M. T., & Boltz, J. P. (2011). *Biofilms in water and wastewater treatment* (pp. 529–567). Tampa: CH2M Hill Inc.

Li, D., Sharp, J. O., Saikaly, P. E., Ali, S., Alidina, M., & Alarawi, M. S. (2012). Dissolved organic carbon influences microbial community composition and diversity in managed aquifer recharge systems. *Applied and Environmental Microbiology, 78*, 6819–6828.

Lim, H. S., Lee, J. S., Chon, H. T., & Sager, M. (2008). Heavy metal contamination and health risk assessment in the vicinity of the abandoned Songcheon Au–Ag mine in Korea. *Journal of Geochemical Exploration, 96*, 223–230.

Liong, M. T. (2011). *Bioprocess sciences and technology*. Hauppauge: Nova Science Publishers.

Lydy, M. J., & Austin, K. R. (2005). Toxicity assessment of pesticide mixtures typical of the Sacramento-San Joaquin Delta using *Chironomu stentans*. *Archives of Environmental Contamination and Toxicology, 48*, 49–55.

Ma, Q., Qu, Y., Shen, W., Zhang, Z., Wang, J., Liu, Z., Li, D., Li, H., & Zhou, J. (2015). Bacterial community compositions of coking wastewater treatment plants in steel industry revealed by Illumina high-throughput sequencing. *Bioresource Technology, 179*, 436–443.

Mages, M., Ovari, M., VonTumpling, W., & Kropfl, K. (2004). Biofilms as bio-indicator for polluted waters: Total reflection X-ray fluorescence analysis of biofilms of the Tisza river (Hungary). *Analytical and Bioanalytical Chemistry, 378*, 1095–1101.

Mangwani, N., Shukla, S. K., Rao, T. S., & Das, S. (2013). Calcium-mediated modulation of *Pseudomonas mendocina* NR802 biofilm influences the phenanthrene degradation. *Colloids and Surfaces, B: Biointerfaces, 114*, 301–309.

Mann, R. M., Hyne, R. V., Choung, C. B., & Wilson, S. P. (2009). Amphibians and agricultural chemicals: Review of the risks in a complex environment. *Environmental Pollution, 157*, 2903–2927.

Mathew, B. B., Tiwari, A., & Jatawa, S. K. (2011). Free radicals and antioxidants: A review. *Journal of Pharma Research, 4*(12), 4340–4343.

Monteiro, D. A., Alves de Almeida, J., Rantin, F. T., & Al, K. (2006). Oxidative stress biomarkers in the freshwater characid fish, Brycon cephalus, exposed toorganophosphorus insecticide Folisuper 600 (methyl parathion). *Comparative Biochemistry and Physiology, 143*, 141–149.

Morais, S., Costa, F. G., & Pereira, M. L. (2012). Heavy metals and human health. In J. Oosthuizen (Ed.), *Environmental health – Emerging issues and practice* (pp. 227–246). Rijeka: InTech.

Morgenroth, E. (2008). Modelling biofilm systems. In M. Henze, M. C. M. van Loosdrecht, G. Ekama, & D. Brdjanovic (Eds.), *Biological wastewater treatment – Principles, modelling, and design* (pp. 457–492). London: IWA Publishing.

Nguyen, H. T., Le, V. Q., Hansen, A. A., Nielsen, J. L., & Nielsen, P. H. (2011). High diversity and abundance of putative polyphosphate-accumulating Tetrasphaera related bacteria in activated sludge systems. *FEMS Microbiology Ecology, 76*(2), 256–267.

Nielsen, P. H., Mielczarek, A. T., Kragelund, C., Nielsen, J. L., Saunders, A. M., Kong, Y., Hansen, A. A., & Vollertsen, J. (2010). A conceptual ecosystem model of microbial communities in enhanced biological phosphorus removal plants. *Water Research, 44*(17), 5070–5088.

Nocker, A., Lepo, J. E., Martin, L. L., & Snyder, R. A. (2007). Response of estuarine biofilm microbial community development to changes in dissolved oxygen and nutrient concentrations. *Microbial Ecology, 54*(3), 532–542.

O'Brien, P. J., & Wehr, J. D. (2010). Periphyton biomass and ecological stoichiometry in streams within an urban to rural land-use gradient. *Hydrobiologia, 657*, 89–105.

Oehmen, A., Lemos, P. C., Carvalho, G., Yuan, Z., Keller, J., Blackall, L. L., & Reis, M. A. (2007). Advances in enhanced biological phosphorus removal: From micro to macro scale. *Water Research, 41*(11), 2271–2300.

Olapade, O. A., & Leff, L. G. (2005). Seasonal response of stream biofilm communities to dissolved organic matter and nutrient enrichments. *Applied and Environmental Microbiology, 71*, 2278–2287.

Park, H., Murthy, S., & Bott, C. (2014). Nationwidemetagenome survey of anammox processes via high-throughput next generation sequencing (NGS): 2012–2013. *Proceedings of the Water Environment Federation, 2014*, 2366–2371.

Piggott, J. J., Salis, R. K., Lear, G., Townsend, C. R., & Matthaei, C. D. (2015). Climate warming and agricultural stressors interact to determine stream periphyton community composition. *Global Change Biology, 21*, 206–222.

Pinto, E., Morse, D., & Calepicolo, P. (2003). Heavy metal induced oxidative stress in algae. *Journal of Phycology, 39*, 1008–1018.

Porsbring, T., Arrhenius, S., Backhaus, T., Kuylenstierna, M., Scholze, M., & Blanck, H. (2007). The SWIFT periphyton test for high-capacity assessments of toxicant effects on microalgal community development. *Journal of Experimental Marine Biology and Ecology, 349*(2), 299–312.

Qu, X., Ren, Z., Zhang, H., Zhang, M., Zhang, Y., Liu, X., & Peng, W. (2017). Influences of anthropogenic land use on microbial community structure and functional potentials of stream benthic biofilms. *Scientific Reports, 7*, 15117.

Rakaiby, M. E., Essam, T., & Hashem, A. (2012). Isolation and characterization of relevant algal and bacterial strains from Egyptian environment for potential use in photosynthetically aerated wastewater treatment. *Journal of Bioremediation & Biodegradation, S8*(001), 1–5.

Rehman, H., Aziz, A. T., Saggu, S., Khurshid, Z. A., Mohan, A., & Ansari, A. A. (2014). Systematic review on pyrethroid toxicity with special reference to deltamethrin. *Indian Journal of Entomology and Zoology Studies, 2*(5), 01–06.

Reisinger, A. J., Tank, J. L., & Dee, M. M. (2016). Regional and seasonal variation in nutrient limitation of river biofilms. *Freshwater Science, 35*(2), 474–489.

Ren, Z., Jiang, Z. Y., & Cai, Q. H. (2013). Longitudinal patterns of periphyton biomass in Qinghai–Tibetan Plateau streams: An indicator of pasture degradation. *Quaternary International, 313*, 92–99.

Romaní, A. M., Guasch, H., Munoz, I., Ruana, J., Vilalta, E., & Schwartz, T. (2004). Biofilm structure and function and possible implications for riverine DOC dynamics. *Microbial Ecology, 47*, 316–328.

Romaní, A. M., Borrego, C. M., Díaz-Villanueva, V., Freixa, A., Gich, F., & Ylla, I. (2014). Shifts in microbial community structure and function in light- and dark-grown biofilms driven by warming. *Environmental Microbiology, 16*, 2550–2567.

Rozitis, D. Z., & Strade, E. (2014). COD reduction ability of microorganisms isolated from highly loaded pharmaceutical wastewater pre-treatment process. *Journal of Mater Environmental Sciences, 6*(2), 507–512.

Rubin, M. A., & Leff, L. G. (2007). Nutrients and other abiotic factors affecting bacterial communities in an Ohio River (USA). *Microbial Ecology, 54*, 374–383.

Ryder, D. S., & Miller, W. (2005). Setting goals and measuring success: Linking patterns and processes in stream restoration. *Hydrobiology, 552*, 147–158.

Sabater, S., Guasch, H., Ricart, M., Romani, A., Vidal, G., Klunder, C., & Schmitt-Jansen, M. (2007). Monitoring the effect of chemicals on biological communities. The biofilm as an interface. *Analytical and Bioanalytical Chemistry, 387*(4), 1425–1434.

Santos, V. L., & Linardi, V. R. (2004). Biodegradation of phenol by a filamentous fungi isolated from industrial effluents-identification and degradation potential. *Process Biochemistry, 39*, 1001–1006.

Scandalios, J. G. (2005). Oxidative stress: Molecular perception and transduction of signals triggering antioxidant gene defenses. *Brazilian Journal of Medical and Biological Research, 38*, 995–1014.

Schafer, R. B., Van Den Brink, P. J., & Liess, M. (2011). Impacts of pesticides on freshwater ecosystems. *Ecological Impacts of Toxic Chemicals, 2011*, 111–137.

Sekar, R., Nair, K. V. K., Rao, V. N. R., & Venugopalan, V. P. (2002). Nutrient dynamics and successional changes in a lentic freshwater biofilm. *Freshwater Biology, 47*, 1893–1907.

Sgountzos, I. N., Pavlou, S., Paraskeva, C. A., & Payatake, A. C. (2006). Growth kinetics of *Pseudomonas flourescens* in sand beds during biodegradation of phenol. *Biochemical Engineering Journal, 30*(2), 164–173.

Sheng, G. P., Yu, H. Q., & Li, X. Y. (2010). Extracellular polymeric substances (EPS) of microbial aggregates in biological wastewater treatment systems: A review. *Biotechnology Advances, 28*, 882–894.

Singh, P., & Cameotra, S. S. (2004). Enhancement of metal bioremediation by use of microbial surfactants. *Biochemical and Biophysical Research Communications, 319*, 291–297.

Sivasamy, A., & Sundarabal, N. (2011). Biosorption of an azo-dye by *Aspergillusniger* and *Trichoderma* sp. fungal biomasses. *Current Microbiology, 62*, 351–357.

Smucker, N. J., Detenbeck, N. E., & Morrison, A. C. (2013). Diatom responses to watershed development and potential moderating effects of near-stream forest and wetland cover. *Freshwater Science, 32*, 230–249.

SriuNaik, S., & PydiSetty, Y. (2011). Biological Denitrification of wastewater in aFBBRd by immobilization of *Pseudomonas stutzeri* using poly propylene granules. *International Journal of Biotechnology Applications, 3*, 106–109.

Srivastava, J. K., Chandra, H., Kalra, S. J. S., Mishra, P., Khan, H., & Yadav, P. (2017). Plant–microbe interaction in aquatic system and their role in the management of water quality: A review. *Applied Water Science, 7*, 1079–1090.

Stoodley, P., Sauer, K., Davies, D. G., & Costerton, J. W. (2002). Biofilms as complex differentiated communities. *Annual Review of Microbiology, 56*, 187–209.

Subtil, E. L., Mierzwa, J. C., & Hespanhol, I. (2014). Comparison between a conventional membrane bioreactor (C-MBR) and a biofilm membrane bioreactor (BF-MBR) for domestic wastewater treatment. *Brazilian Journal of Chemical Engineering, 31*(3), 683–691.
Tatarko, M., & Bumpus, J. A. (1998). Biodegradation of Congo Red by *Phanerochaetechrysosporium. Water Research, 32*, 1713–1717.
Teittinen, A., Taka, M., Ruth, O., & Soininen, J. (2015). Variation in stream diatom communities in relation to water quality and catchment variables in a boreal, urbanized region. *Science of the Total Environment, 530*, 279–289.
Tsuruta, T. (2004). Biosorption and recycling of gold using various microorganisms. *The Journal of General and Applied Microbiology, 50*, 221–228.
Villaverde, S. (2004). Recent development on biological nutrient removal processes for wastewater treatment. *Reviews in Environmental Science and Biotechnology, 3*, 171–183.
Wang, J., Quan, X., Han, L., Qian, Y., & Werner, H. (2002). Microbial degradation of quinoline by immobilized cells of *Burkholderiapickettii. Water Research, 36*, 2288–2296.
Wang, Y., Tian, Y., Han, B., Zhao, H. B., Bi, J. N., & Bl, C. (2007). Biodegradation of phenol by free and immobilized *Acinetobacter* sp. strain PD12. *Journal of Environmental Sciences, 19*, 222–225.
Wang, J., Soininen, J., He, J., & Shen, J. (2012). Phylogenetic clustering increases with elevation for microbes. *Environmental Microbiology Reports, 4*, 217–226.
Wang, Z., Zhang, X. X., Lu, X., Liu, B., Li, Y., Long, C., & Li, A. (2014). Abundance and diversity of bacterial nitrifiers and denitrifiers and their functional genes in tannery wastewater treatment plants revealed by high-throughput sequencing. *PLoS One, 9*(11), 113–603.
Watnick, P., & Kolter, R. (2000). Biofilm, city of microbes. *Journal of Bacteriology, 182*(10), 2675–2679.
Webb, J. S. (2007). Differentiation and dispersal in biofilms. *Journal of Microbiological Methods, 9*(3), 165–174.
Wilhelm, L., Singer, G. A., Fasching, C., Battin, T. J., & Besemer, K. (2013). Microbial biodiversity in glacier-fed streams. *The ISME Journal, 7*, 1651–1660.
Wilhelm, L., Besemer, K., Fasching, C., Urich, T., Singer, G. A., & Quince, C. (2014). Rare but active taxa contribute to community dynamics of benthic biofilms in glacier-fed streams. *Environmental Microbiology, 16*, 2514–2524.
Wu, Y., He, J., & Yang, L. (2010). Evaluating adsorption and biodegradation mechanisms during the removal of microcystin-RR by periphyton. *Environmental Science & Technology, 44*, 6319–6324.
Wu, Y., Li, T., & Yang, L. (2012). Mechanisms of removing pollutants from aqueous solutions by microorganisms and their aggregates: A review. *Bioresource Technology, 107*, 10–18.
Xia, N., Xia, X., Liu, T., Hu, L., Zhu, B., Zhang, X., et al. (2014). Characteristics of bacterial community in the water and surface sediment of the Yellow River, China, the largest turbid river in the world. *Journal of Soils and Sediments, 14*, 1894–1904.
Xuemei, W., Jingling, L., Muyuan, M., & Zhifeng, Y. (2010). Response of freshwater biofilm to pollution and ecosystem in Baiyangdian Lake of China. *Procedia Environmental Sciences, 2*, 1759–1769.
Yariv, C. (2001). Biofiltration-the treatment of fluids by microorganisms immobilized into the filter bedding material: A review. *Bioresource Technology, 77*, 257–274.
Zacharia, J. T. (2011). Ecological effects of pesticides. In D. M. Stoytcheva (Ed.),. ISBN: 978-953-307-458-0 *Pesticides in the modern world – Risks and benefits*. Rijeka: InTech.
Zhang, L., Keller, J., & Yuan, Z. (2009). Inhibition of sulfate-reducing and methanogenic activities of anaerobic sewer biofilms by ferric iron dosing. *Water Research, 43*, 4123–4132.

Wonders of Nanotechnology for Remediation of Polluted Aquatic Environs

Dig Vijay Singh, Rouf Ahmad Bhat, Moonisa Aslam Dervash, Humaira Qadri, Mohammad Aneesul Mehmood, Gowhar Hamid Dar, Mehvish Hameed, and Nowsheeba Rashid

Abstract On earth, all forms of life wholly and solely depend upon the clean water sources for their survival. The freshwater ecosystems are home for large number of organisms from microscopic to macroscopic species. However, water pollution has changed the history of freshwater ecosystems due to addition of variety of pollutants. The problem of water pollution is getting worsened year after year which ultimately affects the limited freshwater resources. The anthropogenic activities have created a situation that may, in the coming years, cause permanent damage to the balanced structure of freshwater ecosystems. There are numerous techniques available for wastewater treatment prior to its discharge into recipient water bodies. But, due to one or other reasons, these conventional techniques fail to meet the demands of treating the wastewaters. Besides, efficiency of these available conventional techniques is also a matter of concern. The literature cited in this chapter suggests that nanotechnology could be a valuable, efficient and clean technology to treat the wastewaters. It is not selective to cleanup only organic based pollutants but efficient

D. V. Singh
School of Environmental Science, DES, Babasaheb Bhimrao Ambedkar University, Lucknow, India

R. A. Bhat (✉) · M. A. Dervash
Division of Environmental Science, Sher-e-Kashmir University of Agricultural Sciences and Technology, Shalimar, India

H. Qadri · M. A. Mehmood · G. H. Dar
Department of Environmental Sciences, School of Sciences, Sri Pratap College Campus, Cluster University of Srinagar, Srinagar, Jammu and Kashmir, India

M. Hameed
College of Agricultural Engineering, Division of Soil and Water Engineering, Sher-e-Kashmir University of Agricultural Sciences and Technology, Shalimar, India

N. Rashid
Amity Institute of Food Technology, Amity University, Noida, Uttar Pradesh, India

H. Qadri et al. (eds.), *Fresh Water Pollution Dynamics and Remediation*,
https://doi.org/10.1007/978-981-13-8277-2_17

to remediate heavy metals (Cd^{2+}, Pb^{2+}, Zn^{2+}, Hg^{2+} and Cr^{3+}) and pesticides in wastewaters. Furthermore, due to an excellent adsorption and catalytic properties of nanomaterials, it has proven to have marvellous antimicrobial activity, pathogen detection and disinfectant quality for the treatment wastewaters.

Keywords Water pollution · Heavy metals · Nanotechnology · TiO_2 · Adsorption · Antimicrobial · Disinfectant

1 Introduction

Water one of the most precious liquid substance for all organisms living on the earth (Qu et al. 2013). Majority of the water (97%) on the surface of the ground is salt water while only 3% is available in the form of freshwater. Two-third of the 3% is in the frozen form and only 1% is in the assessable form (Radwan et al. 2011; Jose et al. 2018) and this dearth in reality a grim quandary for under developed cities (Smith 2009). The large share of world population (780 million) is still lacking the basic pure drinking water facility (WHO 2012; Vijayageetha et al. 2018). The uncontrolled rise in population is adding more problems on the water bodies as per one estimate the population of the world will reach upto 9 billion by 2050 and this rising population will increase water pollution by increasing amount of waste accumulation in water bodies (Sharma and Sharma 2012). The conventional methods like physico-chemical and microorganism based processes for removal of hazardous substances from wastewater (Bellona and Drewes 2007) are not economical and are also less efficient (Zekic et al. 2018). That is why new approaches like nanotechnology are continuously being examined to improve the available techniques for waste water treatment methods (Sharma and Sharma 2013). This technology relies on the relevance of materials on the nano level, so that new structures, components and materials can be built at this (atomic) level (Lens et al. 2013; Ayanda and Petrik 2014). The emergence of nanotechnology occurs billions of years ago in the nature when life began to start on the earth (El Saliby et al. 2009). Plants through the process of evolution, mutation and adaptation were capable of converting CO_2 into carbohydrate in the company solar energy (sunlight) (Roco 1999; El Saliby et al. 2009). Another example of a natural nanotechnology is "chemical catalysis" through "catalysts" or "enzymes" (Arora and Mathur 2017) as enzymes increase the rate of reaction (Smith 1997) and are also vital for the completion of the specific reactions. It is generally accepted that nanotechnology implies management of those materials and particles of at least one dimension ranging from 1–100 nm (Watlington 2005) and these particles and material are called nanomaterials or nanoparticles (El Saliby et al. 2009).

Nanotechnology got a major boost in 1980s by the coming of cluster science and scanning tunnelling microscope (STM). The current tools for the measurement of the nanotechnology include STM, scanning probe microscope, atomic force microscope and molecular beam epitaxy (Roco 1999; El Saliby et al. 2009). Nanotechnology is one of the broad research areas of today's era that involve the use of nano sized building blocks for the creation of new material and particle (Hu and Shaw 1998). Molecules which are made up of building blocks are arranged in nanostructures and nano materials with dimensions of 1–100 nm. The arranging and then assembling of building blocks in order to form large material is referred as "bottom up" approach (Wu et al. 1993). Materials at nano scale level are having different properties than normal size equivalents (Zekić et al. 2018). The main properties like stability, morphological characteristics, adsorption ability and degree of catalysis are maintained by the approaches utilised for the structuring of nanomaterials (Zekić et al. 2018). Nano-sized materials have unique properties completely different from the equivalent structures on the macro-level. The most important feature is the large surface to volume ratio, which is why they are suitable for different forms of water treatment *viz.,* adsorption, photocatalysis, membrane processes (Zekić et al. 2018). Thus nanostructures have completely different optical, electrical, magnetic properties, greater reactivity with neighbouring (polluting) atoms faster chemical processes (Hu and Shaw 1998; Lens et al. 2013; Chatuverdi et al. 2012; Zekić et al. 2018) and all these characteristics of nanomaterials make this technology attractive in terms of eliminating contaminants and making nanotechnology as a suitable technique for wastewater treatment (Kanchi 2014; Zekić et al. 2018). Nanotechnological wastewater treatment processes can be divided into three main groups including (a) treatment and remediation (b) sensing and detection (c) pollution prevention (Watlington 2005). Nanomaterials and nanoparticles are the recent and significant approaches for the remediation and treatment of wastewater (Hu et al. 2005; Mohan and Pittman 2007). The major environmental benefits of nanotechnology include stronger and lighter nonmaterial's, treatment and remediation of wastewater (Theron et al. 2008; Savage et al. 2008). The benefits of using nanotechnology in environment and industrial area has been observed but in future nanotechnology might be helpful in bringing innovations that can be helpful in solving the specific problems (Theron et al. 2008; Savage et al. 2008; El Saliby et al. 2009). Hence, nanotechnology is going to play an important role in addressing fundamental issues of health, energy and water (Binks 2007; El Saliby et al. 2009) but greatest emphasis has currently been placed on the treatment and remediation of wastewater (Zekić et al. 2018). Some nano particles destroy contaminants (oxidation in the presence of nanocatalysts), while other separate and isolate these contaminants (nano membrane filtration). Carbon nanotubes have been recognized for playing beneficial role in the adsorption of dioxins and are much more efficient than the conventional methods (Ren et al. 2011). The multiple benefits like high efficiency, economical waste water treatment and less utilization of large infrastructure of nanotechnology has made this technology more reliable for solving of the various emerging problems of the world (Qu et al. 2013; Vijayageetha et al. 2018).

2 Various Allied Techniques of Treating Contaminated Freshwater Based on Nanoproducts

2.1 *Adsorption*

Adsorption being economically viable (Onundi et al. 2010) is commonly employed for removal of organic and inorganic contaminants from wastewater (Qu et al. 2013). Adsorption involves the use of different materials like zeolites, alumina silicates, clay for removal of metals from solutions (Issabayeva et al. 2007; Nouri et al. 2009; Gupta et al. 2012). The high surface area and aspect ratio of the nanomaterial makes them one of the new adsorbent with superior performance (Yano et al. 2018). The use of multiwall carbon nanotubes is one of the best options as these have higher metal-ion sorption capacity (3–4 times) than the powder and granular activated carbon (Li et al. 2003; Yano et al. 2018). The available conventional adsorbent are less efficient as compared to nano-adsorbents as nano-adsorbent have high surface area, short intraparticle diffusion distance and tunable pore size which make them efficient for adsorption (Qu et al. 2013).

2.2 *Nano-adsorbents Based on Carbon Substances*

2.2.1 Removal of Organic Susbstances

Carbon nanotubes (CNT) are an aromatic surface when the carbon atoms are in a sp^2-hybridization wrapped up in a tubular structure (Terronesa et al. 2010). The carbon nanotubes have unique electronic and mechanical properties which are dependent on chirality and diameter. The exceptional properties of carbon nanotubes make them one of the attractive option for various application like devices for energy storage, adsorption for many practical applications, such as, the development of devices for energy storage (Agnihotri et al. 2006) and adsorption with high sensitivity, selectivity and efficiency (Gupta and Saleh 2013; Ren et al. 2011). CNTs are efficient in the process of adsorption of various organic chemicals compared to activated carbon (Pan and Xing 2008) and high adsorption is because of large surface area and varied contaminant-CNT interactions. The available surface area for adsorption on individual CNTs is their external surfaces (Yang and Xing 2010). The adsorption capacity of the carbon nanostructure can be easily determined by its textural properties (Gupta and Saleh 2013). The surface of the carbonaceous nanomaterials surface can be photeric and other functional groups along with oxygen are added which provide new sites for chemical adsorption (Ren et al. 2011). CNTs forming loose bundles in the aqueous solution are due to the hydrophobicity of graphitic surface thereby reducing the efficiency of adsorption, while aggregates formation contain interstitial spaces which increases the adsorption sites thus increasing the efficiency of adsorption of various organic molecule (Pan et al. 2008). The activated carbon and CNT bundles have almost same specific surface area but activated carbon have micropores which are not assessable to bulky organic molecules (Ji et al. 2009). CNTs have large pores which increase the adsorption of bulky molecules thus have higher adsorption capacity because of more accessible sorption sites. CNTs strongly adsorb many of these polar organic compounds due to the diverse contaminant eCNT interactions including hydrophobic effect, pep interactions, hydrogen bonding, covalent bonding, and electrostatic interactions (Yang and Xing 2010). The p electron rich CNT surface allows pep interactions with organic molecules (Chen et al. 2007; Lin and Xing 2008). Organic compounds which have -COOH, -OH, $-NH_2$ functional groups could also form hydrogen bond with the graphitic CNT surface which donates electrons (Yang et al. 2008). The strong interactions between CNT and organic molecules occur through non-covalent forces such as hydrogen bonding, π–π stacking, electrostatic forces, van der Waals forces, and hydrophobic interactions (Rengaraj et al. 2007). Electrostatic attraction facilitates the adsorption of positively charged organic chemicals such as some antibiotics at suitable pH (Ji et al. 2009). Furthermore, the selectivity and stability of the system can also be increased as CNT allows the addition of one or more functional group that helps in increasing the efficiency of the system (Tarun et al. 2009; Sze et al. 2008). The chemical functional groups may be anchored to the CNT surface through functionalization or purification processes. The carbon nanotubes can be single-walled or multi-walled carbon nanotubes (MWCNT), depending on preparation condition (Li et al. 2003) and because of high stability and surface area, the CNT have attained great attention for the remediation of waste water. CRTs by the process of adsorption helps in the removal of heavy metals (Cd^{2+}, Pb^{2+}, Zn^{2+} and Cu^{2+}) from the waste water. CNT can be used not only as adsorbents but also as a support for adsorption materials (Wang et al. 2007).

2.2.2 Heavy Metal Removal

Oxidized CNTs contain various functional groups that increase the adsorption capacity for metal ions from the waste water (Rao et al. 2007). CNTs are better adsorbents than activated carbon for heavy metals (Li et al. 2003; Lu et al. 2006) as due to the unique properties; the adsorption is fast on CNTs. In order to fulfill these application, small quantity of materials is required and are independent of the material cost. The recent reports have shown that the removal of Hg and a bulky dye molecule was efficiently done by sand granules coated with graphite oxide (Rhodamine B) and was almost comparable to the efficiency of the activated carbon (Gao et al. 2011). The various metallic pollutants present in the waste water and industrial effluent can be easily removed by the nanoparticles of zero-valent form (Ponder and Darab 2000; Kanel et al. 2005; Ponder et al. 2001). Nanoparticles have also been used for the removal of many halogenated hydrocarbons, radionuclides and organic compounds. The removal rate of nanoscale zero-valent iron is 30 times higher for Cr(VI) and Pb(II) as compared to the iron powder. Iron oxide nano-adsorbent is helpful in removing of both the forms of arsenic (As (V) and As (III)) and removal rate is 5–10 times higher than the micron size counterpart. The major advantage is that by the application of magnetic field, iron oxide adsorbent can be easily separated. Nano alumina particles because of low cost and high stability is being used for the removal of metals like Pb, Cr, Cu, Cd and Hg (Pacheco and Rodriguez 2001) from the wastewater (Valente et al. 2004; Baumgarten and Dusing 1994). Nanoparticles known for its unique properties can effectively remove metals and other pollutants from the waste water (Sharma et al. 2009) Because of large number of active site in nano particles, the removal of the different chemical species from wastewater become easy (Hristovski et al. 2007). The difficulty in removing the heavy metals from waste water as no technology is available which can efficiently remove the metals from the wastewater but the use of nanotechnology is providing good alternative for the management of wastewater (Li et al. 2006a, b; Vaseashta et al. 2007).

2.2.3 Regeneration and Reuse

The adsorbent used is cost effective as it can be regenerated. The reduction of the pH is helpful in reversing the metal adsorption on CNTs and at pH less than 2, the recovery rate is almost about 100% (Li et al. 2006a, b; Lu et al. 2006). It has been observed that nano-adsorbent can he regenerated as well as reused for numerous times without effecting the adsorption capacity of the adsorbent (Lu et al. 2007).

2.3 Nano-adsorbents Based on Metals

Oxides of metals (Fe, Ti and alumina) are cheap and effective approach for the adsorption of toxic metals. The dissolved significantly adsorbs the concerned metal ions (Koeppenkastrop and Decarlo 1993; Trivedi and Axe 2000). Nanoparticles cleans the contaminates sites due to presence of large number of active sites show higher adsorption and with the decrease in the size the capacity of adsorption increases (Yean et al. 2005). Surface area of nanoadsorbents plays crucial role and enhances the capacity to remove the pollutants from wastewaters (Auffan et al. 2008, 2009). The "nanoscale effect" helps in the creation of the new adsorption site by changing the magnetite surface structure (Lucas et al. 2001; Auffan et al. 2009). The magnet based nanoparticles as the core material adsorbents provides the required role as presented in Fig. 1.

The volume and size of the pore can be controlled by adjusting the consolidation pressure (Sharma et al. 2009). Activated carbon acts as adsorbent for many contaminants but has limited capacity to absorb arsenic As (V) from the contaminated environment. The various metal oxide nanomaterials have good adsorption capacity as compared to the activated carbon (Deliyanni et al. 2003; Mayo et al. 2007). Metal oxide nanoparticles also can be impregnated onto the skeleton of acti-

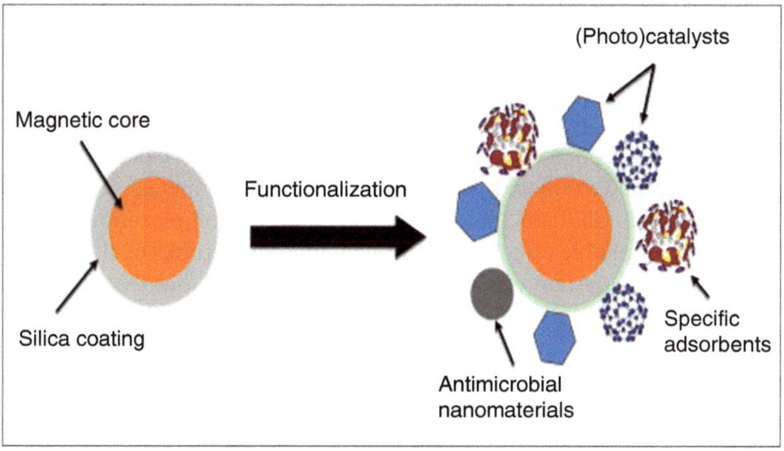

Fig. 1 Magnetic based nanoparticle coated with silica for wastewater treatment

vated carbon or other porous materials to achieve simultaneously removal of arsenic and organic co-contaminants, which favours point-of use (POU) applications (Hristovski et al. 2009a, b).

Photocatalytic degradation has attracted great attention since 1972 when Fujishima and Honda (Fujishima and Honda 1972) observed electrochemical photolysis of water on TiO_2 semiconductor electrode. This technique has been successfully applied in the degradation of contaminants from the wastewater. The contaminants can be easily degraded in the presence of light and catalyst into low molecular weight products which ultimately is converted into CO_2, H_2O, and anions (NO_3^-, PO_4^{3-} and Cl^-). The various kinds of photocatalysts are used but among them TiO_2 is most commonly studied because of its cheapness, photostability and biological stability (Guesh, et al. 2016; Rawal et al. 2013). Because of energy gap, ultraviolet radiation is required to separate charge and TiO_2 on irradiation with UV light generate reactive oxygen species (ROS) which can completely degrade contaminants in very short reaction time. Besides, TiO_2 NPs show little selectivity and thus are suitable for the degradation of all kinds of contaminants, such as chlorinated organic compounds (Ohsaka et al. 2008), polycyclic aromatic hydrocarbons (Guo et al. 2015), dyes (Lee et al. 2008), phenols (Nguyen et al. 2016), pesticides (Alalm et al. 2015), arsenic (Moon et al. 2014), cyanide (Kim et al. 2016), and heavy metals (Chen et al. 2016). The removal of heavy metals from the contaminated environment by using the nanoparticles (iron oxides nanoparticles and nonmagnetic hematite) has increased as these nanoparticles are simple and easily available. Thus these nanoparticles can be easily used for removal of heavy metals from the wastewater system (Lei et al. 2014; Ngomsik et al. 2012).

2.4 Polymeric Nano-adsorbents

Dendrimers are the adsorbents that have the capacity of removing both organics as well as heavy metals from the water. Their interior shells can be hydrophobic for sorption of organic compounds while the exterior branches can be tailored (e.g., hydroxyl- or amine-terminated) for adsorption of heavy metals. The sorption of metals and organic pollutants on adsorbent is based on complexation, electrostatic interactions, hydrophobic effect and hydrogen bonding (Crooks et al. 2001). A dendrimer ultrafiltration system was designed to recover metal ions from aqueous solutions (Diallo et al. 2005). The dendrimers laden with metals is recovered by the process of ultrafiltration and by decreasing the pH to 4, the dendrimers can be regenerated.

2.5 Potential Application in Water Treatment

Nano-adsorbents because of numerous benefits can be easily integrated into existing treatment processes for the batter efficiency of the system (Vijayageetha et al. 2018). Nano-adsorbent applied in the powder form in slurry reactors can be highly efficient (Vijayageetha et al. 2018) but for the recovery of the nanoparticles, the additional system is required to be attached (Aragon et al. 2007; Sylvester et al. 2007; Vijayageetha et al. 2018; Li et al. 2018). Nano-adsorbents can also be utilised by loading beads or granules with the nano adsorbents in fixed or fluidized adsorbers (Qu et al. 2013). Fixed-bed reactors are usually associated with mass transfer limitations and head loss but it does not need future separation process (Qu et al. 2013; Miklos et al. 2018). Nano-adsorbents have been widely used for the removal of arsenic as nano-adsorbent show good performance and are also economical as compared to the other adsorbents (Aragon et al. 2007; Qu et al. 2013; Li et al. 2018; Vijayageetha et al. 2018). ArsenXnp an iron oxide nanoparticles and polymer (Vijayageetha et al. 2018) while ADSORBSIA is a nanocrystalline titanium dioxide medium as both are used for the removal of arsenic from the water (Aragon et al. 2007; Sylvester et al. 2007; Qu et al. 2013; Vijayageetha et al. 2018). ArsenXnp and ADSORBSIA have been employed in small to medium scale drinking water treatment systems and were proven to be cost-competitive (Qu et al. 2013; Miklos et al. 2018 Vijayageetha et al. 2018).

3 Nanofiltration

Nanofiltration (NF) is recent membrane filtration process for treatment of wastewater and is based on charge-based repulsion property and high rate of permeation (Kolpin et al. 2002; Sharma and Sharma 2012). Nano filtration is separation process whose cut off lies between that of reverse osmosis and ultrafiltration. Nanofilteration is also cost effective in comparison to reverse osmosis because of its lower operating pressure and high flow rate (Sharma and Sharma 2012). Nanofilteration membranes helps to remove hardness from the water to great extent as monovalent ions are partially permeable and shoe total impermeability towards bivalent salts (Sharma and Sharma 2012). NF can lower total dissolved solids (TDS) and hardness, reduce color and odor, and remove heavy metal ions from ground water (Kolpin et al. 2002; Koyuncu et al. 2001). The high efficiency of membrane filtration helps in removing of the different types of pollutants from the water and can also achieve desired water purification standards (Zekic et al. 2018). Nanofiltration is a high-pressure membrane treatment process but due to low drive pressure (Liu et al. 2008), less energy is consumed as compared to reverse osmosis (Zekic et al. 2018). Centrifugal pumps are most often used for the pressure and circulation of wastewater within the nanomembrane (Liu et al. 2008). The plant consists of a large number of modules, with different membrane configurations within each module (Kolpin et al. 2002; Sharma and Sharma 2012). In nanofiltration, the usual length of the module varies from 0.9 to 5.5 m, and the diameter ranges from 100 to 300 mm (Tchobanoglous et al. 2014). Nanofiltration produces water that meets highly stringent requirements in terms of water reuse (Zekic et al. 2018). Since this process remove all the types of pollutants present in the water and the further disinfection of water is less required (Liu et al. 2008; Sharma and Sharma 2012; Zekic et al. 2018).

3.1 Nanomaterials for Water Disinfection

Some nanomaterial like chitosan, silver nanoparticles, titanium dioxide, fullerene nanoparticles, carbon nanotubes (Duhan et al. 2017; Zekic et al. 2018) have great adsorption capacity but are also known for their antimicrobial activity (Feng et al. 2000; Inoue et al. 2002; Kumar et al. 2004; Yamanaka et al. 2005; Amin et al. 2014). Being mild oxidant and inert, these nanoparticles does not create any harmful by-product in the water (Buzea et al. 2007; Zekic et al. 2018). Nanomaterials can be applied in various manners like by direct action on bacterial cell, oxidation of cellular component for the treatment of the water (Li et al. 2008, 2011; Zekic et al. 2018). The use of nanotechnology for the disinfection of wastewater has certain drawbacks (Chorawala and Mehta 2015; Zekic et al. 2018) as for the removal of micro-organism from the wastewater, the nanoparticle must come in direct contact with their cell membrane (Feng et al. 2000; Inoue et al. 2002; Kumar et al. 2004; Yamanaka et al. 2005; Chorawala and Mehta 2015; Zekic et al. 2018). Therefore, some nanomaterials (carbon nanotubes) need to be strongly connected to the reactive surface (Khan et al. 2017; Zekic et al. 2018). Also, the deficiency of nanotechnology in disinfection processes is that there is no residual, i.e., subsequent antimicrobial activity in wastewater (Lens et al. 2013).

3.2 Membranes Processes

The basic goal of water treatment is to remove undesired constituents from water (Qu et al. 2013). The use of membrane for wastewater treatment helps to remove the pollutants thus allowing the use of water from the unconventional source (Qu et al. 2013). A major challenge of the membrane technology is the inherent trade off between membrane selectivity and permeability (Khan et al. 2017; Zekic et al. 2018). The major drawback in the utilization of the pressure driven membrane processes is the higher energy consumption which ultimately affects the life of the membrane (Qu et al. 2013). The performance of membrane systems is largely decided by the membrane material (Gehrke et al. 2012; Feng et al. 2013; Qu et al. 2013). Various function (permeability, fouling resistance) can be improved by the inclusion of functional nanomaterials into membranes and this also help in degradation of the different pollutants (Fathizadeh et al. 2011; Kim and Deng 2011; Gehrke et al. 2012; Feng et al. 2013).

3.3 Nanofiber Membranes

The use of polymers, ceramics or metals in order to produce ultra fine fibre is known as electrospinning. The process is very simple and economical to make fibre and the produced fibre have high porosity as well as surface area (Cloete et al. 2010; Jayavarthanan et al. 2017). The diameter, morphology, composition, secondary structure, and spatial alignment of electrospun nanofibers can be easily manipulated for specific applications (Jayavarthanan et al. 2017). Although nanofiber membranes have not been largely used for treatment of water but have been widely exploited for air filtration application (Nowack et al. 2011; Kim et al. 2012). Nanofibre membrane with high rejection rate can easily remove small size particles without affecting the life of the membrane (Ramakrishna et al. 2006). Thus they have been proposed to be used as pre-treatment prior to ultrafiltration or reverse osmosis (Nora and Mamadou 2005; Quang et al. 2013). For example, by incorporating ceramic nanomaterials or specific capture agents on the nanofiber scaffold, affinity nanofiber membranes can be designed to remove heavy metals and organic pollutants during filtration (Ursino et al. 2018).

3.4 Nanocomposite Membranes

The large number of studies on nanotechnology has aimed to create synergism by adding hydrophilic metal oxide nanoparticle (Al_2O_3, TiO_2, and zeolite) into inorganic membrane. The main goal of adding hydrophilic metal oxide nanoparticles is to reduce fouling by increasing the hydrophilicity of the membrane. The addition of alumina (Maximous et al. 2010), zeolite (Pendergast et al. 2010) and TiO_2 increase hydrophilicity as well as permeability of the membrane. The addition of metal oxide nanoparticles including alumina, silica (Bottino et al. 2001), and zeolite to polymeric ultrafiltration membranes has been shown to increase membrane surface hydrophilicity, water permeability, or fouling resistance. By addition of inorganic nanoparticle, the stability of the membrane increases thereby reducing the impact of heat on the membrane (Ebert et al. 2004; Pendergast et al. 2010). Antimicrobial nanomaterials (nano-Ag and CNTs) doped on the membrane surface can be helpful in reducing the biofilm formation and inhibit the attachment of bacteria on the membrane thus prevent the biofoiling of the membrane (Mauter et al. 2011; Zodrow et al. 2009) on the membrane surface as well as inactivate viruses (De Gusseme et al. 2011). The membrane doped with photocatalytic nanoparticle results in combining of the unique properties for the degradation of the contaminants and efforts are also running for the development of photocatalytic inorganic membranes consisting of nanophotocatalysts (Choi et al. 2006).

4 Thin Film Nanocomposite (TFN) Membranes

The main focus in the development of thin film nanocomposite is the doping of nanomaterial (zeolites, nano-Ag, nano-TiO_2 and CNTs) on the thin film nanocomposite. The properties of the membrane is totally dependent on the nanoparticle which is added to the membrane. The most widely used dopant which help to increase the permeability and create negative charge (Lind et al. 2009a) of the membrane is nano-zeolites (Fathizadeh et al. 2011; Kim and Deng 2011; Gehrke et al. 2012; Feng et al. 2013). One study reported water permeability increased up to 80% over the TFC membrane, with the salt rejection largely maintained (Jeong et al. 2007). TFN membranes doped with 250 nm nano-zeolites at 0.2 wt% achieved moderately higher permeability and better salt rejection (>99.4%) than commercial RO membranes (Lind et al. 2010). It was hypothesized that the small, hydrophilic pores of nano-zeolites create preferential paths for water. The permeability of water increased with pores filled with zeolite can be due to the problem at zeolite polymer interface. Nano-zeolites were also used as carriers for antimicrobial agents such as Agþ, which imparts anti-fouling property to the membrane (Lind et al. 2009b). Incorporation of nano-TiO_2 (up to 5 wt%) into the TFC active layer slightly increased the membrane rejection while maintaining the permeability (Lee et al. 2008). TiO_2 under UV irradiation can degrade contaminants and also inhibit the activity of microorganisms (Qu et al. 2013) and also reduces the biological fouling (Fathizadeh et al. 2011; Kim and Deng 2011; Gehrke et al. 2012; Feng et al. 2013; Qu et al. 2013). However, the close adjacency between the photocatalyst and the membrane may also lead to detrimental effects on polymeric membrane materials, which needs to be addressed for long-term efficacy (Chin et al. 2006). CNTs (unaligned) also found their application in TFN membranes due to their antimicrobial activities (Fathizadeh et al. 2011; Kim and Deng 2011; Gehrke et al. 2012; Feng et al. 2013;

Qu et al. 2013). Covalently bonded SWNTs to a TFC membrane surface (Tiraferri et al. 2011). This approach is advantageous as it uses relatively small amount of the nanomaterial and minimizes perturbation of the active layer.

5 Photocatalysis

Nanocatalysts can effectively be used for chemical oxidation of organic and inorganic pollutants in water in advanced oxidation processes (Lens et al. 2013). These processes are based on formation of highly reactive radicals that react easily with pollutant molecules. The application of this process is often limited because of the extremely high costs of providing required energy (UV lamps, ozonators, ultrasonicators, etc.) (Lens et al. 2013). Photocatalysis is the most significant oxidation process. This is a chemical reaction change that is induced by adsorption of a photon whose energy is greater than the energy needed to overcome the interstitial of two electron shells (valentine and conductive) of a semiconductor. When the photon illuminates the catalytic surface, the electron (negatively charged particle) is transferred from the valentine shell to the empty conduction shell and leaves a "hole" behind it with a positive charge. This "e-h" pair ("electron-hole") creates highly reactive radicals that bind the molecules of pollution and thus break them down (Bora and Dutta 2014). However, there are several technical challenges that have to be met to enable broader practical application of this process, including – optimization of catalysts in the exploitation of available light energy – more efficient separation of nanocatalysts after treatment and re-application improvement of selective properties during chemical reactions.

6 Trace Contaminant Detection

In trace organic or inorganic contaminant detection, nanomaterials can be used in both concentration and detection. CNTs have great potential for environmental analysis of trace metal or organic pollutants as they offer high adsorption capacity and recovery rate as well as fast kinetics. The pre-concentration factors for metal ions were found to be between 20 and 300 with fast adsorption kinetics (Duran et al. 2009). CNTs have also been extensively studied for pre-concentrating a variety of organic compounds, many of which were done in real water samples (Cai et al. 2003; Chin et al. 2006). Adsorption of charged species to CNTs results in changes of conductance, providing the basis for the correlation between analyte concentration and current fluctuation (Mauter and Elimelech 2008). Other nanomaterials such as nano-Au and QDs have alsobeen used. Nano-Au was used to detect pesticides at ppb levels in a colorimetric assay (Lisha and Anshup Pradeep 2009), modified nano-Au was shown to detect Hg^{2+} and CH3Hgþ rapidly with high sensitivity and selectivity (Lin and Tseng, 2010). QD modified TiO_2 nanotubes lowered the detection limits of PAHs to the level of pica-mole per liter based on fluorescence resonance energy transfer (Yang et al. 2010a; da Silva et al. 2011). A nanosensor based on CoTe QDs immobilized on a glassy carbon electrode surface was reported to detect Bisphenol A in water at concentrations as low as 10nM within 5 s.

7 Nanomaterials for Adsorption of Pollutants

Nanoparticles possess two important characteristics that make them very good adsorbents. These are the large specific surface of nanomaterials and surface multifunctionality or the ability to easily chemically react and bind to different adjacent atoms and molecules (Fig. 1). These characteristics make nanoparticles not only effective adsorbents for various contaminants in wastewater but also allow for long-term stability, as this also results in adsorbent degradation (with the addition of catalytic properties of nanoparticles) and improves the adsorption efficiency. With the discovery of carbon nanotubes (Iijima 1991), a new carbon-based adsorption material was introduced to the world. Compared to the best known such material - activated carbon - carbon nanotubes possess approximately the same large specific surface, but their great advantage lies in the structure of nanomaterials and a much better arrangement of carbon atoms. In addition, nanomaterials possess unique mechanical, electrical, chemical, optical and many other characteristics that allow them to have much better adsorption properties for some contaminants (heavy metals and organic pollutants). This is why they are called the "material of the 21st century" (Ren et al. 2011). Besides carbon nanotubes, metal based nanoparticles also have adsorption characteristics. The most common metal oxides used as adsorbents are iron oxides (Fe_xO_y), silicon (Si), titanium (Ti) and tungsten (W). They are mainly used for adsorption of heavy metals and radionuclides (unstable nuclides). The adsorption process is based on the electrostatic interaction of dissolved metals in wastewater and the nanoadsorbent surface. Changing the pH of the solution can significantly affect the strength of this interaction. Thus, the surface of the nano adsorbent may be: acidic, with positive charge attracting anions basic, with a negative charge attracting cations from waste water (Crane and Scott 2012; Qu et al. 2013)

8 Conclusion

Water pollution has created widespread problems and becomes a concern to all. There are limited resources of freshwater available for consumption and sustaining the life. But, these sources nowadays has badly damaged due to growing anthropogenic activities. Convention treatment technology has plenty to offer in the form of remediating the pollutants from wastewaters, but fail to deal with treating huge quantity of polluted waters ecofriendly. Nano technology has proven to be a valuable technology, which has least negative impacts on environment. It is not selective to cleanup only organic based pollutants but efficient to remediate heavy metals and pesticides in wastewaters. Furthermore, due to an excellent adsorption and catalytic properties of nanomaterials, it has proven to have marvellous antimicrobial activity, pathogen detection and disinfectant quality for the treatment wastewaters.

References

Agnihotri, S., Mota, J. P. B., Rostam-Abadi, M., & Rood, M. J. (2006). Structural characterization of single-walled carbon nanotube bundles by experiment and molecular simulation. *Langmuir, 21*, 896–904.

Alalm, M. G., Tawfik, A., & Ookawara, S. (2015). Comparison of solar TiO_2 photocatalysis and solar photo-Fenton for treatment of pesticides industry wastewater: Operational conditions, kinetics, and costs. *Journal of Water Process Engineering, 8*, 55–63.

Amin, M. T., Alazba, A. A., & Manzoor, U. (2014). A review of removal of pollutants from water/wastewater using different types of nanomaterials. *Advances in Materials Science and Engineering, 2014*. https://doi.org/10.1155/2014/825910.

Aragon, M., Kottenstette, R., Dwyer, B., Aragon, A., Everett, R., Holub, W., Siegel, M., & Wright, J. (2007). *Arsenic pilot plant operation and results*. Anthony: Sandia National Laboratories.

Arora, J., & Mathur, A. (2017). Role of Nano-technology in water and waste-water management. *Internation Journal of Advance Research in Science and Engineering, 6*(10), 161–168.

Auffan, M., Rose, J., Proux, O., Borschneck, D., Masion, A., Chaurand, P., Hazemann, J. L., Chaneac, C., Jolivet, J. P., Wiesner, M. R., Van Geen, A., & Bottero, J. Y. (2008). Enhanced adsorption of arsenic onto maghemites nanoparticles: As (III) as a probe of the surface structure and heterogeneity. *Langmuir, 24*(7), 3215–3222.

Auffan, M., Rose, J., Bottero, J. Y., Lowry, G. V., Jolivet, J. P., & Wiesner, M. R. (2009). Towards a definition of inorganic nanoparticles from an environmental, health and safety perspective. *Nature Nanotechnology, 4*(10), 634–641.

Ayanda, O. S., & Petrik, L. F. (2014). Nanotechnology: The breakthrough in water and wastewater treatment. *Internatioanl Journal of Chemical, Materijal and Enviromnmental Research, 1*, 1–2.

Baumgarten, E., & Dusing, U. K. (1994). Sorption of metal ions on alumina. *Journal of Colloid and Interface Science, 194*, 1–9.

Bellona, C., & Drewes, J. E. (2007). Viability of a low-pressure nanofilter in treating recycled water for water reuse applications: A pilot-scale study. *Water Research, 41*, 3948–3958.

Binks P (2007) *Nanotechnology & water: Opportunities and challenges*. Victorian water sustainability seminar.

Bora, T., & Dutta, J. (2014). Applications of nanotechnology in wastewater treatment – A review. *Journal of Nanoscience and Nanotechnology*, 613–626. https://doi.org/10.1166/jnn.2014.8898.

Bottino, A., Capannelli, G., D'Asti, V., & Piaggio, P. (2001). Preparation and properties of novel organic-inorganic porous membranes. *Separation and Purification Technology, 22*(23), 269–275.

Buzea, C., Blandino, I. I. P., & Robbie, K. (2007). Nanomaterials and nanoparticles: Sources and toxicity. *Biointerphases, 2*(4), MR17–MR172.

Cai, Y. Q., Jiang, G. B., Liu, J. F., & Zhou, Q. X. (2003). Multiwalled carbon nanotubes as a solid-phase extraction adsorbent for the determination of bisphenol a, 4-n-nonylphenol, and 4-tertoctylphenol. *Analytical Chemistry, 75*(10), 2517–2521.

Chatuverdi, S., Dave, P. N., & Shah, N. K. (2012). Applications of nanocatalyst in new era. *Journal of Saudi Chemical Society, 16*, 307–325.

Chen, W., Duan, L., & Zhu, D. Q. (2007). Adsorption of polar and nonpolar organic chemicals to carbon nanotubes. *Environmental Science and Technology, 41*(24), 8295–8300.

Chen, Z. P., Li, Y., Guo, M., et al. (2016). One-pot synthesis of Mn-doped TiO2 grown on graphene and the mechanism for removal of Cr(VI) and Cr(III). *Journal of Hazardous Materials, 310*, 188–198.

Chin, S. S., Chiang, K., & Fane, A. G. (2006). The stability of polymeric membranes in a TiO2 photocatalysis process. *Journal of Membrane Science, 275*(1–2), 202–211.

Choi, H., Stathatos, E., & Dionysiou, D. D. (2006). Sol-gel preparation of mesoporous photocatalytic TiO2 films and TiO2/Al2O3 composite membranes for environmental applications. *Applied Catalysis B-Environmental, 63*(1–2), 60–67.

Chorawala, K. K., & Mehta, M. J. (2015). Applications of nanotechnology in wastewater treatment. *International Journal of Innovative and Emerging Research in Engineering, 2*(1), 21–26.

Cloete, T. E., de Kwaadsteniet, M., Botes, M., & Lopez-Romero, J. M. (2010). *Nanotechnology in water treatment applications*. Wymondham: Caister Academic Press.

Crane, R. A., & Scott, T. B. (2012). Nanoscale zero-valent iron: Future prospects for an emerging water treatment technology. *Journal of Hazardous Materials, 211–212*, 112–125.

Crooks, R. M., Zhao, M. Q., Sun, L., Chechik, V., & Yeung, L. K. (2001). Dendrimer-encapsulated metal nanoparticles: Synthesis, characterization, and applications to catalysis. *Accounts of Chemical Research, 34*(3), 181–190.

da Silva, B. F., Perez, S., Gardinalli, P., Singhal, R. K., Mozeto, A. A., & Barcelo, D. (2011). Analytical chemistry of metallic nanoparticles in natural environments. *TrAC Trends in Analytical Chemistry, 30*(3), 528–540.

De Gusseme, B., Hennebel, T., Christiaens, E., Saveyn, H., Verbeken, K., Fitts, J. P., Boon, N., & Verstraete, W. (2011). Virus disinfection in water by biogenic silver immobilized in polyvinylidene fluoride membranes. *Water Research, 45*(4), 1856–1864.

Deliyanni, E. A., Bakoyannakis, D. N., Zouboulis, A. I., & Matis, K. A. (2003). Sorption of As(V) ions by akaganeite-type nanocrystals. *Chemosphere, 50*(1), 155–163.

Diallo, M. S., Christie, S., Swaminathan, P., Johnson, J. H., & Goddard, W. A. (2005). Dendrimer enhanced ultrafiltration. 1. Recovery of Cu (II) from aqueous solutions using PAMAM dendrimers with ethylenediamine core and terminal NH2 groups. *Environmental Science and Technology, 39*(5), 1366–1377.

Duhan, J. S., Kumar, R., Kumar, N., Kaur, P., Nehra, K., & Duhan, S. (2017). Nanotechnology: The new perspective in precision agriculture. *Biotechnology Report, 15*, 11–23.

Duran, A., Tuzen, M., & Soylak, M. (2009). Preconcentration of some trace elements via using multiwalled carbon nanotubes as solid phase extraction adsorbent. *Journal of Hazardous Materials, 169*(1e3), 466–471.

Ebert, K., Fritsch, D., Koll, J., & Tjahjawiguna, C. (2004). Influence of inorganic fillers on the compaction behaviour of porous polymer based membranes. *Journal of Membrane Science, 233*(1e2), 71–78.

El Saliby, I. J., Shon, H. K., Kandasamy, J., & Vigneswaran, S. (2009). Nanotechnology for wastewater treatment: In brief. In Vigneswaran S. (Ed.), *Water and wastewater treatment technologies* (1 pp).

Fathizadeh, M., Aroujalian, A., & Raisi, A. (2011). Effect of added NaXnano-zeolite into polyamide as a top thin layer of membrane on water flux and salt rejection in a reverse osmosis process. *Journal of Membrane Science, 375*, 88–95.

Feng, Q. L., Wu, J., Chen, G. Q., Cui, F. Z., Kim, T. N., & Kim, J. O. (2000). A mechanistic study of the antibacterial effect of silver ions on *Escherichia coli* and *Staphylococcus aureus*. *Journal of Biomedical Materials Research, 52*(4), 662–668.

Feng, C., Khulbe, K. C., Matsuura, T., Tabe, S., & Ismail, A. F. (2013). Preparation and characterization of electro-spun nanofiber membranes and their possible applications in water treatment. *Separation and Purification Technology, 102*, 118–135.

Fujishima, A., & Honda, K. (1972). Electrochemical photolysis of water at a semiconductor electrode. *Nature, 238*, 37–38.

Gao, W., Majumder, M., Alemany, L. B., Narayanan, T. N., Ibarra, M. A., Pradhan, B. K., & Jayan, P. M. (2011). Engineered graphite oxide materials for application in water purification. *ACS Applied Materials & Interfaces, 3*(6), 1821–1826.

Gehrke, I., Keuter, V., & Groß, F. (2012). Development of nanocomposite membranes with photocatalytic surfaces. *Journal of Nanoscience and Nanotechnology, 12*, 9163–9168.

Guesh, K., Mayoral, A., Alvarez, C. M. C. Y., & Diaz, I. (2016). Enhanced photocatalytic activity of TiO2 supported on zeolites tested in real wastewaters from the textile industry of Ethiopia. *Microporous and Mesoporous Materials, 225*, 88–97.

Guo, M., Song, W., Wang, T., Li, Y., Wang, X., & Du, X. (2015). Phenyl-functionalization of titanium dioxide-nanosheets coating fabricated on a titanium wire for selective solid-phase microextraction of polycyclic aromatic hydrocarbons from environment water samples. *Talanta, 144*, 998–1006.

Gupta, V. K., & Saleh, T. A. (2013). Sorption of pollutants by porous carbon, carbon nanotubes and fullerene – An overview. *Environmental Science and Pollution Research, 20*, 2828–2843.

Gupta, V. K., Ali, I., Saleh, T. A., Nayak, A., & Agarwal, S. (2012). Chemical treatment technologies for waste-water recycling – An overview. *RSC Advances, 2*, 6380–6388.

Hristovski, K., Baumgardener, A., & Westerhoff, P. (2007). Selecting metal oxide nanomaterials for arsenic removal in fixed bed columns: From nanoparticles to aggregated nanoparticles media. *Journal of Hazardous Materials, 147*, 265–274.

Hristovski, K. D., Nguyen, H., & Westerhoff, P. K. (2009a). Removal of arsenate and 17-ethinyl estradiol (EE2) by iron oxide modified activated carbon fibers. *Journal of Environmental Science and Health Part A-Toxic/Hazardous Substances & Environmental Engineering, 44*(4), 354–361.

Hristovski, K. D., Westerhoff, P. K., Moller, T., & Sylvester, P. (2009b). Effect of synthesis conditions on nano-iron (hydr)oxide impregnated granulated activated carbon. *Chemical Engineering Journal, 146*(2), 237–243.

Hu, E. L., & Shaw, D. T. (1998). Synthesis and assembly. In R. W. Siegel, E. Hu, & M. C. Roco (Eds.), *Nanostructure science and technology*. Dordrecht: Kluwer academic publishers.

Hu, J., Chen, G., & Lo, I. M. C. (2005). Removal and recovery of Cr(VI) from wastewater by maghemite nanoparticles. *Water Research, 39*, 4528–4536.

Iijima, S. (1991). Helical microtubules of graphitic carbon. *Nature, 354*, 56–58.

Inoue, Y., Hoshino, M., Takahashi, H., et al. (2002). Bactericidal activity of Ag-zeolite mediated by reactive oxygen species under aerated conditions. *Journal of Inorganic Biochemistry, 92*(1), 37–42, 2002.

Issabayeva, G., Aroua, M. K., & Sulaiman, N. M. (2007). Continuous adsorption of lead ions in a column packed with palm shell activated carbon. *Journal of Hazardous Materials, 155*(1–2), 109–113.

Jayavarthanan, R., Nanda, A., & Bhat, M. A. (2017). The impact of nanotechnology on environment. In B. K. Nayak, A. Nanda, & M. Bhat (Eds.), *Integrating biologically-inspired nanotechnology into medical practice* (167p). Hershey: IGI Global.

Jeong, B. H., Hoek, E. M. V., Yan, Y. S., Subramani, A., Huang, X. F., Hurwitz, G., Ghosh, A. K., & Jawor, A. (2007). Interfacial polymerization of thin film nanocomposites: A new concept for reverse osmosis membranes. *Journal of Membrane Science, 294*(1–2), 1–7.

Ji, L. L., Chen, W., Duan, L., & Zhu, D. Q. (2009). Mechanisms for strong adsorption of tetracycline to carbon nanotubes: A comparative study using activated carbon and graphite as adsorbents. *Environmental Science and Technology, 43*(7), 2322–2327.

Jose, A. J., Jacob, A. M., Manjush, K. C., & Kappen, J. (2018). Chitosan in water purification technology. In S. Ahmad & C. M. Hussain (Eds.), *Green and sustainable advanced materials: Applications*.

Kanchi, S. (2014). Nanotechnology for water treatment. *International Journal of Environmental Analytical Chemistry, 1*(2). https://doi.org/10.4172/jreac.1000e102.

Kanel, S. R., Charlet, B., & Choi, L. (2005). Removal of As(III) from groundwater by nanoscale zerovalent iron. *Environmental Science & Technology, 39*, 1291–1298.

Khan, I., Saeed, K., & Khan, I. (2017). Nanoparticles: Properties, applications and toxicities. *Arabian Journal of Chemistry*. https://doi.org/10.1016/j.arabjc.2017.05.011.

Kim, E. S., & Deng, B. (2011). Fabrication of polyamide thin-film nano-composite (PA-TFN) membrane with hydrophilized ordered mesoporous carbon (H-OMC) for water purifications. *Journal of Membrane Science, 375*, 46–54.

Kim, E. S., Hwang, G., El-Din, M. G., & Liu, Y. (2012). Development of nanosilver and multi-walled carbon nanotubes thin-film nanocomposite membrane for enhanced water treatment. *Journal of Membrane Science, 394*, 37–48.

Kim, S. H., Lee SW Lee, G. M., Lee, B. T., Yun, S. T., & Kim, S. O. (2016). Monitoring of TiO2-catalytic UV-LED photo-oxidation of cyanide contained in mine wastewater and leachate. *Chemosphere, 143*, 106–114.

Koeppenkastrop, D., & Decarlo, E. H. (1993). Uptake of rare-earth elements from solution by metal-oxides. *Environmental Science and Technology, 27*(9), 1796–1802.

Kolpin, D. W., Furlong, E. T., Meyer, M. T., Thurman, E. M., Zaugg, S. D., Barber, L. B., & Buxton, H. T. (2002). Pharmaceuticals, hormones, and other organic wastewater contaminants in U.S. Streams, 1999–2000: A National Reconnaissance. *Environmental Science & Technology, 36*(6), 1202–1211.

Koyuncu, I., Kural, E., & Topacik, D. (2001). Pilot scale nanofiltration membrane separation for waste management in textile industry. *Water Science and Technology, 43*(10), 233–240.

Kumar, V. S., Nagaraja, B. M., Shashikala, V., et al. (2004). Highly efficient Ag/C catalyst prepared by electro-chemical deposition method in controlling microorganisms in water. *Journal of Molecular Catalysis A: Chemical, 223*(1–2), 313–319, 2004.

Lee, Y., Kim, S., Venkateswaran, P., Jang, J., Kim, H., & Kim, J. (2008). Anion co-doped Titania for solar photocatalytic degradation of dyes. *Carbon letters, 9*, 131–136.

Lei, Y., Chen, F., Luo, Y., & Zhang, L. (2014). Three-dimensional magnetic graphene oxide foam/ Fe3O4 nanocomposite as an efficient absorbent for Cr(VI) removal. *Journal of Materials Science, 49*(12), 4236–4245.

Lens, P. N. L., Virkutye, J., Jegatheesan, V., Kim, S. H., & Al-Abed, S. (2013). *Nanotechnology for water and wastewater treatment*. IWA Publishing.

Li, Y. H., Ding, J., Luan, Z. K., Di, Z. C., Zhu, Y. F., Xu, C. L., Wu, D. H., & Wei, B. Q. (2003). Competitive adsorption of Pb2+, Cu2+ and Cd2+ ions from aqueous solutions by multiwalled carbon nanotubes. *Carbon, 41*, 2787–2792.

Li, L., Fan, M., Brown, R. C., Leeuwen, J. V., Wang, J., Wang, W., Wang, W., Song, Y., & Zhang, P. (2006a). Synthesis, properties and environmental application of nanoscale iron-based materials: A review. *Critical Reviews in Environmental Science and Technology, 36*, 405–431.

Li, X. Q., Elliot, D. W., & Zhang, W. X. (2006b). Zerovalent iron nanoparticles for abatement of environmental pollutants: Materials and engineering aspects. *Critical Reviews in Solid State and Materials Sciences, 31*, 111–122.

Li, Q., Mahendra, S., Lyon, D. Y., Brunet, L., Liga, M. V., Li, D., & Alvarez, P. J. J. (2008). Antimicrobial nanomaterials for water disinfection and microbial control: Potential applications and implications. *Water Research, 42*, 4591–4602. https://doi.org/10.1016/j.watres.2008.08.015.

Li, X., Xu, H., Chen, Z., & Chen, G. (2011). Biosynthesis of nanoparticles by microorganisms and their applications. *Journal of Nanomaterials, 2011*. https://doi.org/10.1155/2011/270974.

Li, J., Liu, H., & Chen, J. P. (2018). Microplastics in freshwater systems: A review on occurrence, environmental effects, and methods for microplastics detection. *Water Research, 137*(15), 362–374.

Lin, Y. H., & Tseng, W. L. (2010). Ultrasensitive sensing of Hg(2þ) and CH(3)Hg(þ) based on the fluorescence quenching of lysozyme type VI-stabilized gold nanoclusters. *Analytical Chemistry, 82*(22), 9194–9200.

Lin, D. H., & Xing, B. S. (2008). Adsorption of phenolic compounds by carbon nanotubes: Role of aromaticity and substitution of hydroxyl groups. *Environmental Science and Technology, 42*(19), 7254–7259.

Lind, M. L., Ghosh, A. K., Jawor, A., Huang, X. F., Hou, W., Yang, Y., & Hoek, E. M. V. (2009a). Influence of zeolite crystal size on zeolitepolyamide thin film nanocomposite membranes. *Langmuir, 25*(17), 10139–10145.

Lind, M. L., Jeong, B. H., Subramani, A., Huang, X. F., & Hoek, E. M. V. (2009b). Effect of mobile cation on zeolite-polyamide thin film nanocomposite membranes. *Journal of Materials Research, 24*(5), 1624–1631.

Lind, M. L., Suk, D. E., Nguyen, T. V., & Hoek, E. M. V. (2010). Tailoring the structure of thin film nanocomposite membranes to achieve seawater RD membrane performance. *Environmental Science and Technology, 44*(21), 8230–8235.

Lisha, K. P., & Anshup Pradeep, T. (2009). Enhanced visual detection of pesticides using gold nanoparticles. *Journal of Environmental Science and Health Part B-Pesticides Food Contaminants and Agricultural Wastes, 44*(7), 697–705.

Liu, F., Zhang, G., Meng, Q., & Zhang, H. (2008). Performance of nanofiltration and reverse osmosis membranes in metal effluent treatment. *Chinese Journal of Chemical Engineering, 16*(3), 441–445.

Liu, S. B., Zeng, T. H., Hofmann, M., Burcombe, E., Wei, J., Jiang, R. R., Kong, J., & Chen, Y. (2011a). Antibacterial activity of graphite, graphite oxide, graphene oxide, and reduced graphene oxide: Membrane and oxidative stress. *ACS Nano, 5*(9), 6971–6980.

Liu, S. W., Yu, J. G., & Jaroniec, M. (2011b). Anatase TiO(2) with dominant high-energy {001} facets: Synthesis, properties, and applications. *Chemistry of Materials, 23*(18), 4085–4093.

Liu, Z. Y., Bai, H. W., Lee, J., & Sun, D. D. (2011c). A low-energy forward osmosis process to produce drinking water. *Energy & Environmental Science, 4*(7), 258–2585.

Lu, C., & Su, F. (2007). Adsorption of natural organic matter by carbon nanotubes. *Separation and Purification Technology, 58*, 113–121.

Lu, C. S., Chiu, H., & Liu, C. T. (2006). Removal of zinc (II) from aqueous solution by purified carbon nanotubes: Kinetics and equilibrium studies. *Industrial & Engineering Chemistry Research, 45*(8), 2850–2855.

Lucas, E., Decker, S., Khaleel, A., Seitz, A., Fultz, S., Ponce, A., Li, W. F., Carnes, C., & Klabunde, K. J. (2001). Nanocrystalline metal oxides as unique chemical reagents/sorbents. *Chemistry-A European Journal, 7*(12), 2505–2510.

Mauter, M. S., & Elimelech, M. (2008). Environmental applications of carbon-based nanomaterials. *Environmental Science and Technology, 42*(16), 5843–5859.

Mauter, M. S., Wang, Y., Okemgbo, K. C., Osuji, C. O., Giannelis, E. P., & Elimelech, M. (2011). Antifouling ultrafiltration membranes via post-fabrication grafting of biocidal nanomaterials. *ACS Applied Materials & Interfaces, 3*(8), 2861–2868.

Maximous, N., Nakhla, G., Wong, K., & Wan, W. (2010). Optimization of Al(2)O(3)/PES membranes for wastewater filtration. *Separation and Purification Technology, 73*(2), 294–301.

Mayo, J. T., Yavuz, C., Yean, S., Cong, L., Shipley, H., Yu, W., Falkner, J., Kan, A., Tomson, M., & Colvin, V. L. (2007). The effect of nanocrystalline magnetite size on arsenic removal. *Science and Technology of Advanced Materials, 8*(1e2), 71–75.

Miklos, D. B., Zemy, C., Jekel, M., Linden, K. G., Drewes, J. E., & Hübner, U. (2018). Evaluation of advanced oxidation processes for water and wastewater treatment – A critical review. *Water Research, 139*, 118–131.

Mohan, D., & Pittman, C. U. (2007). Arsenic removal from water/wastewater using adsorbents – A critical review. *Journal of Hazardous Materials, 142*, 1–53.

Moon, G., Kim, D., Kim, H., Bokare, A. D., & Choi, W. (2014). Platinum-like behavior of reduced graphene oxide as a cocatalyst on TiO2 for the efficient photocatalytic oxidation of arsenite. *Environmental Science & Technology Letters, 1*(2), 185–190.

Ngomsik, A. F., Bee, A., Talbot, D., & Cote, G. (2012). Magnetic solid-liquid extraction of Eu(III), La(III), Ni(II) and Co(II) with maghemite nanoparticles. *Separation and Purification Technology, 86*, 1–8.

Nguyen, A. T., Hsieh, C. T., & Juang, R. S. (2016). Substituent effectsm on photodegradation of phenols in binary mixtures by hybrid H2O2 and TiO2 suspensions under UV irradiation. *Journal of the Taiwan Institute of Chemical Engineers, 62*, 68–75.
Nora, S., & Mamadou, S. D. (2005). Nanomaterials and water purification: Opportunities and challenges. *Journal of Nanoparticle Research, 7*, 331–342.
Nouri, J., Khorasani, N., Lorestani, B., Karami, M., Hassani, A. H., & Yousefi, N. (2009). Accumulation of heavy metals in soil and uptake by plant species with phytoremediation potential. *Environment and Earth Science, 59*(2), 315–323.
Nowack, B., Krug, H. F., & Height, M. (2011). 120 years of nanosilver history: Implications for policy makers. *Environmental Science & Technology, 45*, 1177–1183.
Ohsaka, T., Shinozaki, K., Tsuruta, K., & Hirano, K. (2008). Photoelectrochemical degradation of some chlorinated organic compounds on n-TiO2 electrode. *Chemosphere, 73*(8), 1279–1283.
Onundi, Y. B., Mamun, A. A., Al Khatib, M. F., & Ahmed, Y. M. (2010). Adsorption of copper, nickel and lead ions from synthetic semiconductor industrial wastewater by palm Shell activated carbon. *International Journal of Environmental Science and Technology, 7*(4), 751–758.
Pacheco, S., & Rodriguez, R. (2001). Adsorption properties of metal ions using alumina nano particles in aqueous and alcoholic solution. *Journal of Sol-Gel Science and Technology, 20*, 263–273.
Pan, B., Lin, D. H., Mashayekhi, H., & Xing, B. S. (2008). Adsorption and hysteresis of bisphenol A and 17 alpha-ethinyl estradiol on carbon nanomaterials. *Environmental Science and Technology, 42*(15), 5480–5485.
Pan, B., & Xing, B. S. (2008). Adsorption mechanisms of organic chemicals on carbon nanotubes. *Environmental Science and Technology, 42*(24), 9005–9013.
Pendergast, M. T. M., Nygaard, J. M., Ghosh, A. K., & Hoek, E. M. V. (2010). Using nanocomposite materials technology to understand and control reverse osmosis membrane compaction. *Desalination, 261*(3), 255–263.
Ponder, S. M., & Darab, J. G. (2000). Remediation of Cr(VI) and Pb(II) aqueous solutions using nanoscale zerovalent iron. *Environmental Science & Technology, 34*, 2564–2569.
Ponder, S. M., Darab, J. G., Bucher, J. D., Craig, C. I., Davis, L., Stein, N. E., Lukens, W., Nitsche, H., Rao, L. F., Shuh, D. K., & Mallouk, T. E. (2001). Surface chemistry and electrochemistry of supported zerovalent iron nanoparticles in the remediation of aqueous metal contaminants. *Chemistry of Materials, 13*, 479.
Qu, X., Alvarez, P. J. J., & Li, Q. (2013). Application of nanotechnology in wastewater treatment. *Wastewater Research, 47*, 3931–3946.
Quang, D. V., Pradi, B., Sarawade, S. J., et al. (2013). Effective water disinfection using silver nanoparticle containing silica beads. *Applied Surface Science, 287*, 84–90.
Radwan, H., Elattar, S., & Khmes, R. (2011). Global water resources. In M. Aufleger & W. Rauch (Eds.), *Handshake across the Jordan: Water and understanding international conference 26.9. – 28.9.2010, Pella, Jordanien* (pp. 7–26).
Ramakrishna, S., Fujihara, K., Teo, W. E., Yong, T., Ma, Z. W., & Ramaseshan, R. (2006). Electrospun nanofibers: Solving global issues. *Materials Today, 9*(3), 40–50.
Rao, G. P., Lu, C., & Su, F. (2007). Sorption of divalent metal ions from aqueous solution by carbon nanotubes: A review. *Separation and Purification Technology, 58*(1), 224–231.
Rawal, S. B., Bera, S., Lee, D., Jang, D. J., & Lee, W. I. (2013). Design of visible-light photocatalysts by coupling of narrow bandgap semiconductors and TiO2: Effect of their relative energy band positions on the photocatalytic efficiency. *Catalysis Science and Technology, 3*(7), 1822–1830.
Ren, X., Chen, C., Nagatsu, M., & Wang, X. (2011). Carbon nanotubes as adsorbents in environmental pollution management: A review. *Chemical Engineering Journal, 170*, 395–410.
Rengaraj, S., Jei-Won, Y., Younghun, K., & Won-Ho, K. (2007). Application of Mg-mesoporous alumina prepared by using magnesium stearate as a template for the removal of nickel: Kinetics, isotherm and error analysis. *Industrial and Engineering Chemistry Research, 46*, 2834–2842.

Roco, M. C. (1999). *Nanotechnology, shaping the world atom by atom*. National Science and Technology Council, Committee on Technology, The Interagency Working Group on Nanoscience, Engineering and Technology, Washington, DC, USA.

Savage, N., Wentsel, R., et al. (2008). *Draft nanomaterial research strategy (NRS)* (pp. 1–2). Washington, DC: Environmental Protection Agency.

Sharma, V., & Sharma, A. (2012). Nanotechnology: An emerging future trend in wastewater treatment with its innovative products and processes. *International Journal of Enhanced Research in Science Technology and Engineering, 1*, 121–128.

Sharma, V., & Sharma, A. (2013). Nanotechnology: An emerging future trend in wastewater treatment with its innovative products and processes. *International Journal of Enhanced Research in Science Technology & Engineering, 1*, 2.

Sharma, V. K., Yngard, R. A., & Lin, Y. (2009). Silver nanoparticles: Green synthesis and their antimicrobial activities. *Advances in Colloid and Interface Science, 145*, 83–96.

Smith, A. (2009). Nanotechnology: An answer to the World's water crisis? *Chemistry International, 31*(4), 137–139.

Smith, A. D. (1997). *Oxford dictionary of biochemistry and molecular biology*. Oxford: Oxford University Press.

Sylvester, P., Westerhoff, P., Mooller, T., Badruzzaman, M., & Boyd, O. (2007). A hybrid sorbent utilizing nanoparticles of hydrous iron oxide for arsenic removal from drinking water. *Environmental Engineering Science, 24*(1), 104–112.

Sze, M. F. F., Lee, V. K. C., & McKay, G. (2008). Simplified fixed bed column model for adsorption of organic pollutants using tapered activated carbon columns. *Desalination, 218*, 323–333.

Tarun, K. N., Ashim, K. B., & Sudip, K. D. (2009). Adsorption of Cd(II) and Pb(II) from aqueous solutions on activated alumina. *Journal of Colloid and Interface Science, 333*, 14–26.

Tchobanoglous, G., Stensel, H. D., Tsuchihashi, R., & Burton, F. (2014). *Wastewater engineering: Treatment and resource recovery* (5th ed.). New York: McGraw-Hill.

Terronesa, M., Botello-Méndezb, A. R., Campos-Delgadoc, J., et al. (2010). Graphene and graphite nanoribbons: Morphology, properties, synthesis, defects and applications. *Nano Today, 5*(4), 351–372.

Theron, J., Walker, J. A., & Cloete, T. E. (2008). Nanotechnology and water treatment: Applications and emerging opportunities. *Critical Reviews in Microbiology, 34*, 43–69.

Tiraferri, A., Vecitis, C. D., & Elimelech, M. (2011). Covalent binding of single-walled carbon nanotubes to polyamide membranes for antimicrobial surface properties. *ACS Applied Materials & Interfaces, 3*(8), 2869–2877.

Trivedi, P., & Axe, L. (2000). Modeling Cd and Zn sorption to hydrous metal oxides. *Environmental Science and Technology, 34*(11), 2215–2223.

Ursino, C., Castro-Muñoz, R., Drioli, E., Gzara, L., Albeirutty, M. H., & Figoli, A. (2018). Progress of nanocomposite membranes for water treatment. *Membranes (Basel), 8*(2). https://doi.org/10.3390/membranes8020018.

Valente, S., Bokhimi, X., & Toledo, J. A. (2004). Synthesis and catalytic properties of nanostructured aluminas obtained by sol-gel method. *Appl Catal A, 264*, 175–181.

Vaseashta, V., Vaclavikova, M., Vaseashta, V., Gallios, G., Roy, P., & Pummakarnchana, O. (2007). Nanostructures in environmental pollution detection, monitoring and remediation. *Science and Technology of Advanced Materials, 8*, 47–59.

Vijayageetha, V. A., Annamalai, V., & Pandiarajan, A. (2018). A study on the nanotechnology in water and waste water treatment. *IOSR Journal of Applied Physics (IOSR-JAP), 10*(4), 28–31.

Wang, S. G., Gong, W. X., Liu, X. W., Yao, Y. W., Gao, B. Y., & Yue, Q. Y. (2007). Removal of lead (II) from aqueous solution by adsorption onto manganese oxide-coated carbon nanotubes. *Separation and Purification Technology, 58*, 17–23.

Watlington, K. (2005). *Emerging nanotechnologies for site remediation and wastewater treatment*. National network for environmental management studies fellow, North Carolina State University.

WHO. (2012). *Progress on drinking water and sanitation*. 2012 Update.

Wu, M. K., Windeler, R. S., Steiner, C. K., Bros, T., & Friedlander, S. K. (1993). Controlled synthesis of nanosized particles by aerosol processes. *Aerosol Science and Technology, 19*, 527–548.

Yamanaka, M., Hara, K., & Kudo, J. (2005). Bactericidal actions of a silver ion solution on Escherichia coli, studied by energy-filtering transmission electron microscopy and proteomic analysis. *Applied and Environmental Microbiology, 71*(11), 7589–7593.

Yang, K., & Xing, B. S. (2010). Adsorption of organic compounds by carbon nanomaterials in aqueous phase: Polanyi theory and its application. *Chemical Reviews, 110*(10), 5989–6008.

Yang, K., Wu, W. H., Jing, Q. F., & Zhu, L. Z. (2008). Aqueous adsorption of aniline, phenol, and their substitutes by multi-walled carbon nanotubes. *Environmental Science and Technology, 42*(21), 7931–7936.

Yang, L. X., Chen, B. B., Luo, S. L., Li, J. X., Liu, R. H., & Cai, Q. Y. (2010a). Sensitive detection of polycyclic aromatic hydrocarbons using CdTe quantum dot-modified TiO(2) nanotube array through fluorescence resonance energy transfer. *Environmental Science and Technology, 44*(20), 7884–7889.

Yano, H., Omura, H., Honma, Y., Okumura, H., Sano, H., & Nakatsubo, F. (2018). Designing cellulose nanofiber surface for high density polyethylene reinforcement. *Cellulose, 25*(6), 3351–3362.

Yean, S., Cong, L., Yavuz, C. T., Mayo, J. T., Yu, W. W., Kan, A. T., Colvin, V. L., & Tomson, M. B. (2005). Effect of magnetite particle size on adsorption and desorption of arsenite and arsenate. *Journal of Materials Research, 20*(12), 3255–3264.

Zekić, E., Vuković, Z., & Halkijević, I. (2018). Application of nanotechnology in wastewater treatment. *Građevinar, 70*(4), 315–323.

Zodrow, K., Brunet, L., Mahendra, S., Li, D., Zhang, A., Li, Q., & Alvarez, P. J. J. (2009). Polysulfone ultrafiltration membranes impregnated with silver nanoparticles show improved biofouling resistance and virus removal. *Water Research, 43*(3), 715–723.

MIX
Papier aus verantwortungsvollen Quellen
Paper from responsible sources
FSC® C105338

If you have any concerns about our products,
you can contact us on
ProductSafety@springernature.com

In case Publisher is established outside the EU,
the EU authorized representative is:
Springer Nature Customer Service Center GmbH
Europaplatz 3, 69115 Heidelberg, Germany

Printed by Libri Plureos GmbH
in Hamburg, Germany